CONTAINMENT STRUCTURES OF U.S. NUCLEAR POWER PLANTS

BACKGROUND, REGULATIONS, CODES AND STANDARDS, AND OTHER CONSIDERATIONS

Contributors

Hansraj Ashar

Nilesh Chokshi
Dan Naus
Dana Powers
Javeed Munshi
Jason Petti
Jim Xu

Co-Contributors

Goutam Bagchi
Abdul Sheikh
Shawn Burns
Shen Wang

ASME PRESS

ASME Press books are available at special quantity discounts to use as premiums or for use in corporate training programs. For more information, contact Special Sales at customercare@asme.org

Library of Congress Cataloging-in-Publication Data

Ashar, Hansraj.
Containment structures of U.S. nuclear power plants / technical editor, Hansraj Ashar ; contributors, Nilesh Chokshi [and nine others].
p. cm.
Includes bibliographical references and index.
ISBN 978-0-7918-6017-5
1. Nuclear reactors—United States—Containment. 2. Nuclear reactors—United States—Containment—History. 3. Nuclear energy—History. I. Ashar, Hansraj, editor of compilation. II. Chokshi, N. C. (Nilesh C.) editor of compilation. III. Title: U.S. nuclear power plants.

TK9211.C66 2013
621.48′35—dc23

2013022514

ACKNOWLEDGMENTS

I thank the principle contributors, Nilesh Chokshi, Jaweed Munshi, Dan Naus, Jason Petti, Dana Powers, and Jim Xu, for their cooperation and technical input during the development of this book. It required a number of back and forth communications and support of the co-authors: Goutam Bagchi, Shawn Burns, Abdul Sheikh, and Shen Wang in the development of the book. I sincerely appreciate their contributions.

I take this opportunity to thank retired NRC historian, Dr. Samuel Walker for his help in reconstructing the historical background related to nuclear reactors and containments.

After the draft chapter manuscripts were provided, ASME Press sought help from Mr. Bryan Erler for a technical review. I sincerely thank Mr. Erler for providing constructive technical comments on the book chapters.

I appreciate the constant support, quick response, and guidance provided by Mary Grace Stefanchik, and Tara Collins Smith of ASME Press.

Last, but not the least, I appreciate my computer savvy wife Kusum Ashar for supporting the book writing by optimizing my work place for light and sitting arrangement, and keeping track of the book progress.

CONTRIBUTOR BIOGRAPHIES

ASHAR, HANSRAJ, PE, FACI, FASCE

Mr. Ashar has a Bachelor of Civil Engineering degree from Gujarat University, India, and a Master of Science degree in Civil Engineering from the University of Michigan. He has worked with the Nuclear Regulatory Commission for 36 years as a Sr. Structural Engineer. Prior to that, Mr. Ashar has worked with a number of consultants in the U.S. and Germany designing Bridges and Buildings. Mr. Ashar has authored and presented over 30 papers related to the structures in nuclear power plants.

Mr. Ashar's participation in National and International Standards Organization includes membership of the NSO and INSO Committees, such as American Institute of Steel Construction (AISC), Chairman of Nuclear Specification Committee (January 1996 to March 2008), (AISC/ANSI N690); Member of Building Specification Committee, and Corresponding member of Seismic Provisions Committee.

Mr. Ashar's professional activities with the American Concrete Institute (ACI) 349 Committees include Member of the Main committee, Subcommittee 1 on General Requirements, Materials and QA, and Subcommittee 2 on Design. His professional activities also include American Society of Mechanical Engineers (ASME), Corresponding Member, Working Group on inservice Inspection of Concrete and Steel Containments (Subsections IWE and IWL of ASME Section XI Code), Member, ASME/ACI Joint Committee on Design, Construction, Testing and Inspection of Concrete Containments and Pressure Vessels (till December 2010); Member, RILEM Task Committee 160MLN: Methodology for Life Prediction of Concrete Structures in Nuclear Power Plants; Member, Federation Internationale du Beton (FIB) Task Group 1.3: Containment Structures, and Consultant to IAEA on Concrete Containment Database (2001 to 2005).

Mr. Ashar is a professional engineer in the States of Ohio and Maryland; He is a Fellow member of the American Concrete Institute; and American Society of Civil Engineers. He is a Professional Member of Post-tensioning Institute. Mr. Ashar has been a Peer Reviewer of the Papers to be published in ASCE Material Journal, Nuclear Engineering and Design Periodicals and in ACI Material Journal.

NILESH CHOKSHI, PH.D.

Currently, Dr. Nilesh Chokshi is Deputy Director, Division of Site and Environmental Reviews in the Office of New Reactors. During his 33 years at the NRC, Dr. Chokshi has managed several research and regulatory areas including seismic and structural engineering, materials engineering, risk analysis, and radiation protection. He has also been extensively involved in the area of probabilistic risk assessment, particularly in the development of external event methodology and the standard.

Dr. Chokshi has been Vice Chairman of Board of Directors of ASME Codes and Standards Technology Institute; past Chairman of CSNI Working Group on Integrity and Aging of Components and Structures; Dr. Chokshi is a member of the Advisory Board of International Association of Structural Mechanics in Reactor Technology (IASMiRT) and was Chairman of International Scientific Committee of 16th SMiRT Conference. Prior to joining the NRC, Dr. Chokshi worked at an architectural/engineering firm involved in designs of nuclear power plants. Dr. Chokshi obtained his Ph.D. in the field of civil engineering (with specialization in structural engineering) from Rice University and Masters degree from the University of Michigan.

JAVEED MUNSHI, PH.D., PE, SE, FACI

Javeed Munshi has over 20 years of experience in design, evaluation, and construction of concrete structures including heavy industrial (fossil, nuclear, and renewable) power structures, bridges, buildings, underground structures (tunnels), and environmental concrete structures. He is has contributed to seven books/design aids for concrete and published over 60 papers. He has received a Meritorious Publication Award from Structural Engineers Association of Illinois (SEAOI). Dr. Munshi is the Chair of Subcommittee A, ASME Section III, Div 2 (ACI 359), Code for Concrete Containments. He is also a member of ACI 349, Code Requirements for Nuclear Concrete Structures, ACI 350, Code Requirements for Environmental Engineering Concrete Structures, ACI 314 — Simplified Concrete Design, ACI 307 — Concrete Chimneys, ACI 437 — Strength Evaluation of Existing Concrete Structures and ACI 374 — Performance-Based Seismic Design of Concrete. He is also a member of the American Society of Civil Engineers (ASCE) and The Masonry Society (TMS). He has conducted concrete design seminars and training for the American Concrete Institute (ACI), the Portland Cement Association (PCA), and the Concrete Reinforcing Steel Institute (CRSI). He was named as Bechtel Distinguished Engineer (BDE) and Fellow of the American Concrete Institute (ACI). Dr. Munshi received his B.S. in civil engineering from National Institute of Technology and M.S. in earthquake engineering from the Indian Institute of Technology (IIT) Roorkee, India. He received his Ph.D. from Illinois Institute of Technology, Chicago. He is a licensed Professional Engineer in the states of New York and Wisconsin, and a licensed Structural Engineer in Illinois.

DAN NAUS, FACI, FAIC, FRILEM

Mr. Naus graduated from the University of Illinois with degrees in Civil Engineering and Theoretical and Applied Mechanics. He is currently a Distinguished Research Staff Member at the Oak Ridge National Laboratory where he has been employed for the last 37 years. Initially, during his employment at the Lab, he was involved in research addressing the development of prestressed concrete pressure vessels for application to high-temperature gas-cooled and gas-cooled fast breeder reactor concepts, and development of elevated temperature concrete properties for the Clinch River Breeder Reactor.

For the last 25 years, Mr. Naus has led a research program addressing aging management of nuclear power plant safety-related concrete and steel structures. He has also participated in activities sponsored by the International Atomic Energy Agency, RILEM, and the Nuclear Energy Agency of the Organization for Economic Cooperation and Development addressing aging of nuclear power plant containments and other structures. He is a fellow in ACI, ASCE, and RILEM, and a member of several technical committees, and has authored over 300 publications, technical reports, and papers.

DR. JASON PETTI

Jason Petti is a Principal Member of the Technical Staff (PMTS) and Test Director within the Mechanical Environments Department of the Validation and Qualification Sciences Experimental Complex at Sandia National Laboratories. Dr. Petti joined the staff at Sandia National Laboratories in January of 2004, spending his first 8 years in the Structural and Thermal Analysis Department of the Nuclear Energy Safety Technologies Group. The initial focus of his work centered on the application of finite element methods to perform assessments of structures and components related to the nuclear power and fossil fuel industries. His previous work at Sandia included the study of age-related degradation in nuclear power plant containment vessels for use in severe accident risk assessments, seismic assessments of spent nuclear fuel dry storage casks, studying fire-induced thermal effects on reinforced concrete structures, assessing aircraft impact effects on existing and new nuclear plants, supporting NRC in license renewal procedures for the Oyster Creek Nuclear Power Plant through containment stress and stability analyses, providing support for Taiwanese regulator for assessing the stability of spent nuclear fuel casks during seismic events and assessing the response of Liquefied Natural Gas vessels subjected to specific threats. Dr. Petti's current position focuses on directing large-scale experiments at Sandia's Rocket Sled Track and Aerial Cable Facility.

Prior to joining Sandia, Dr. Petti obtained his BS, MS, and Ph.D. in Civil Engineering (Structural Engineering) from University of Illinois at Urbana-Champaign (1998, 2000, and 2004, respectively). At Illinois, he worked under Professor Robert H. Dodds, Jr. and the Computational Fracture Mechanics Research Group. As a Research Assistant from 1999 to 2004, Dr. Petti developed methodologies for characterizing the brittle fracture behavior of ferritic steels in the low-to-mid region of the ductile-to-brittle transition. During his studies, Dr. Petti spent time working at the U.S. Naval Academy, Annapolis, MD, and the Naval Surface Warfare Center, West Bethesda, MD.

DR. DANA POWERS

D. A. Powers received his Bachelor of Science degree in chemistry from the California Institute of Technology in 1970. He received a Ph.D. degree in Chemistry, Chemical Engineering and Economics in 1975 from the California Institute of Technology. His research for this degree program included magnetic properties of basic iron compounds, catalyst characterization, and the rational pricing of innovative products. In 1974, Powers joined Sandia National Laboratories where he worked in the Chemical Metallurgy Division. His principal research interests were in high temperature and aggressive chemical processes.

In 1981, Dr. Powers became the supervisor of the Reactor Safety Research Division and conducted analytic and experimental studies of severe reactor accident phenomena in fast reactor and light water reactors. These studies included examinations of core debris interactions with concrete, sodium interactions with structural materials, fission product chemistry under reactor accident conditions, aerosol physics, and high temperature melt interactions with coolants. In 1991, Powers became the acting Manager of the Nuclear Safety Department at Sandia that was involved in the study of fission reactor accident risks and the development of plasma-facing components for fusion reactors. Powers has also worked on the Systems Engineering for recovery and processing of defense nuclear wastes and has developed computer models for predicting worker risks in Department of Energy nuclear facilities. Dr. Powers was promoted to Senior Scientist at Sandia in 1997. Dr. Powers is the author of 103 technical publications. He is a Fellow of the American Nuclear Society.

From 1988 to 1991, Dr. Powers served as a member of the Department of Energy's Advisory Committee on Nuclear Facility Safety (ACNFS). In 1994, he was appointed to the Advisory Committee on Reactor Safeguards (ACRS) for the U.S. Nuclear Regulatory Commission. He was Vice Chairman of the ACRS in 1997 and 1998. He was elected Chairman in 1999 and 2000. In 2001, Dr. Powers received the Distinguished Service Award from the US Nuclear Regulatory Commission. Dr. Powers has served on committees for the National Research Council involved with the safety of Department of Energy facilities and the nuclear safety of reactors in the former Soviet Union. He has been an instructor for courses on reactor safety and accident management held by the International Atomic Energy Agency in several countries. Dr. Powers was voted the Theos J. (Tommy) Thompson Award for Nuclear Safety in 2007 by the American Nuclear Society "in recognition of outstanding contributions to the field of nuclear reactor safety."

DR. JIM XU, PH.D.

Dr. Jim Xu currently is Chief of Structural Engineering Branch in the Office of New Reactors of United States Nuclear Regulatory Commission (USNRC), responsible for safety evaluations of applications for design certifications and combined licenses for new and advanced reactor installations. Dr. Xu joined the USNRC as senior structural engineer in 2007 and had led safety reviews and evaluations of seismic analysis and containment designs, as well as the design of Category I structures and foundations. He also led assessments of the seismic margin assessments and containment fragility associated with severe accident containment phenomena. Dr. Xu led NRC guidance developments for the PRA-based seismic margin assessment for generation III/III+ reactors and for enhancing Standard Review Plan to address seismic analyses and design of containment and foundations. Dr. Xu is the author of 85 technical publications. Dr. Xu is currently a member of American Society of Civil Engineers (ASCE) National Nuclear Standard Committee, and ASCE 1 and ASCE 4 Standard Development Committees.

Jim Xu received his Bachelor of Science degree in Engineering Mechanics from the Dalian University of Technology in 1982. He received a Ph.D. degree in Civil Engineering in 1988 from the City University of New York. His research for this degree included applications of Fredholm integral equation of the second kind to dynamic foundation impedance functions associated with two-phase foundation media. In 1985, Dr. Xu joined TransNuclears, Inc. (now as AREVA TN) as structural engineer. In 1987, Dr. Xu joined Brookhaven National Laboratory (BNL) where he spent the next 20 years as scientist conducting research and technical reviews for the USNRC and the Department of Energy (DOE) to address nuclear safety, structural mechanics, earthquake engineering and soil structure interactions, geotechnical engineering, as well as the seismic probabilistic risk assessments (SPRA) and seismic margin assessments associated with nuclear reactors and nuclear waste facilities in the U.S.

GOUTAM BAGCHI, PE, FASCE

Goutam Bagchi has over 45 years of experience in the nuclear power industry. He had hands on experience in the design of several containment structures at currently operating nuclear power plants. He retired from the Nuclear Regulatory Commission in September 2007 from the senior level service after 32 years of service in progressively senior positions, including his service in the senior executive position. He has extensive regulatory experience and proposed and participated in significant initiatives in containment structures inspection regulation, maintenance rule 10 CFR 50.65, the Individual Plant Examination for External Events program, seismic margins criteria, new reactors siting criteria and many other regulatory guide development. While serving as a branch chief in the Office of Research, Mr. Bagchi proposed and initiated a research program for testing scaled containment structures at the Sandia National Laboratories. This program produced significant insights in the behavior of reinforced and pre-stressed concrete and steel containment structures. The research results were published in several NUREG/CR documents. He participated in writing guidance documents for the International Atomic Energy Agency.

Mr. Bagchi served in several national code/standard writing committees. He is currently in the Nuclear Standards Committee of the American Society of Civil Engineers (ASCE), ASCE 1, and ASCE 4 standards.

Mr. Bagchi received his Bachelor of Civil Engineering from the Bengal Engineering College, Calcutta University in 1959, a Master of Structural Engineering from the Imperial College, London University in 1965 and a Master of Science in Mechanical Engineering from the Northeastern University, Boston in 1974. He is a fellow of the ASCE and a member of the American Society of Mechanical Engineers, and a member of the Earthquake Research Institute. He is a registered professional engineer.

DR. SHAWN BURNS

Dr. Burns is the manager of Sandia National Laboratories Risk and Reliability Analysis department. He received his Ph.D. in Mechanical Engineering from the University of Texas at Austin and joined the staff at Sandia in 1995. Dr. Burns has a background in high-performance scientific computing and finite element methods in thermal radiation transport. Since 2004, Dr. Burns has supported the U.S. Nuclear Regulatory Commission rule making as the principal investigator of funded research projects in severe accident progression and risk analysis. Dr. Burns has been the manager of the Risk and Reliability Analysis Department since 2010 and is responsible for the coordination of multidisciplinary research teams including probabilistic risk analysis, fire risk analysis, human reliability analysis, and infrastructure security.

ABDUL H. SHEIKH, PE

Abdul Sheikh received his Master of Science degree in structural engineering from University of Southampton in United Kingdom. He is a registered professional engineer in the State of California. Until 1975, he was involved with the design of bridges and Thames Barrier Project in England. He moved to Montreal, Canada in 1979 and was involved in the design of design of pressurized heavy water reactor CANDU containments for Gentilly Unit 2 plant in Quebec, and Wolsung 1 plant in Korea. In 1979, he joined Bechtel Power Corporation where he spent the next 25 years in the design, procurement, construction, startup, testing, and modifications of the nuclear power plants.

During this time, he worked on containment designs for Palo Verde. Vogtle, and four nuclear units in South Korea. He was involved in all facets of civil/structural design and construction from initial design in Los Angles to startup in Korea for two nuclear units. During the later part of his employment in Bechtel, Mr. Sheikh worked on the design and construction of steam generators replacement and reactor head replacement projects for different containments.

In 2004, Mr. Sheikh joined the USNRC and performed detailed structural analysis to determine the impact of aircrafts on existing and new reactor containments and spent fuel pools designs. In addition, he performed analysis to determine the structural behavior of containments during severe accidents. For the last 4 years, he has been responsible for review of license renewal applications for different nuclear power plants. This includes aging management of containment structures and materials. Mr. Sheikh represents USNRC at the International Atomic Energy Agency as co-chairman of the International Generic Aging Lessons Learned (IGALL) for civil/structural issues.

SHEN WANG, PH.D., PE

Shen Wang received his Bachelor and Master of Science degrees in Civil Engineering from Tongji University. He received his Ph.D. in Civil Engineering from Clarkson University in 2007. His research for this degree focuses on the cohesive fracture of quasi-brittle material, including in-situ field tests in McMurdo Sound, Antarctica.

Dr. Wang currently works as a Senior Civil Engineer and Technical Specialist with Bechtel Power Corporation. In past 6 years, Dr. Wang has designed various nuclear safety-related structures in eleven (11) nuclear power plants, and other types of critical structures in eleven (11) fossil power plants. Dr. Wang has published over 20 papers in fracture mechanics, nuclear structure design and missile impact analysis. He has served as a peer reviewer of over 10 papers for Engineering Fracture Mechanics and Mechanical Engineering Research. Dr. Wang is currently registered as Professional Engineer in Texas. Dr. Wang is a committee member in ASME BPV III, Div. 2/ACI 359 Joint Committee of Concrete Components for Nuclear Reactors and in ASME BPV III Code Committee Working Group Design Methodology. He is also an associate member in ACI 349 Committee of Concrete Nuclear Structures.

TABLE OF CONTENTS

PREFACE

We (the contributing authors) collectively recognized the renewed relevance of nuclear power in the US, after decades of stagnation. We felt that it was imperative to develop an up-to-date, scholarly work on containment structures, incorporating the underlying regulations, safety significance, history, design philosophy, design experience, operating experience, and application to new design. It would benefit the nuclear industry as it transitions to a new generation of designers, constructors, and regulators. We think the book will be a valuable asset to the nuclear utilities, nuclear regulators, A/Es, and international organizations involved in the design and construction of nuclear power plants (NPPs).

With this basic purpose, coupled with my extensive experience in various aspects of NPP containment design, construction, inspection, and testing, principally, with the Nuclear Regulatory Commission (NRC), where I worked for the last 36 years before retiring in December 2010, I thought about developing this type of book in early 2010. As I wanted to explore historical background, as well as how and when of the nuclear reactors and containments, I first contacted Dr. Samuel J. Walker, the NRC historian, and requested him, if he could write this preface for the book, or help me construct the historical background related to nuclear reactors and containments. At the time (~June 2010), when I contacted him, he was preparing to retire from the NRC and told me that he could not find time to write such a preface. However, he assured me that I would find the required historical information in two NRC published books: (1) Controlling the Atom, and (2) Containing the Atom, the books, he was involved in authoring. Most of the historical background that I have compiled in the Preface and in Chapters 1 and 2 are based on the contents of these books.

Among a number of potential technical publishers, such as American Society of Civil Engineers (ASCE), American Society of Mechanical Engineers (ASME), American Nuclear Society (ANS), and Elsevier, I finally corresponded with ASME, because of my long-term association with ASME in the development of the standards related to the NPP containment structures. I had also realized that I am not an expert in various specialized subjects (e.g. severe accident considerations), I started looking for appropriate experts in these areas. Fortunately, I found these experts in National Laboratories, NRC, and in Nuclear Industry.

This preface provides an overview of the historical developments relevant to the content of the book. It briefly provides historical background related to the development of commercial use of nuclear energy, as well as a brief description of the physical processes involved during the operation of nuclear reactors.

In the late 1930s, scientists had discovered that when an atom of uranium was bombarded by neutrons, the uranium atom would sometimes split or fission. Later, the scientists found that when the atom of uranium fissioned, additional neutrons were emitted and became available for further reaction with other uranium atoms. These facts demonstrated that it was possible to device perpetual chain reactions. In December 1942, underneath the West Stands of Stagg Field at the University of Chicago, a team of scientists led by Enrico Fermi created man's first controlled nuclear chain reaction. A crude reactor, remembered, as the first pile (CP-1), consisted of uranium embedded in a matrix of graphite. With sufficient uranium in the pile, the few neutrons emitted in a single fission may accidentally strike neighboring atoms, which in turn undergo fission and produce more neutrons. The atomic pile was controlled and prevented from burning itself by cadmium-plated rods which absorbed neutrons and stopped the process. The pile was square at the bottom and flattened sphere on the top. Around the pile, there was a tent of cloth fabric balloon provided so that the reactor could be sealed to minimize unproductive loss of neutrons.

Following the success of the CP-1 experiment, in February 1943, the U.S. Army moved the CP-1 pile to the south of Chicago, where it was reassembled as CP-2. CP-2 was considerably larger than CP-1, and had a 5-ft concrete shield building around it to protect the personnel working around the pile. The shield building can be termed as a containment structure that protected the general public and the staff against radiation hazard. However, readers should recognize that the early use of the technology was in developing atomic weapons during World War II. The neutrons that are produced in a fission reaction are fast neutrons and are less likely to cause fission than slower neutrons. As a consequence, in the most common type of power reactors, the kinetic energy of the fissionable neutrons is reduced to a value, where it is more likely to cause fission. This is accomplished by introducing a medium between the fuel rods that would slow down the fission neutrons. The medium used is called the moderator. Light water, graphite, heavy water, and other materials have been used as moderators in commercial and research reactors. Approximately, 90% of the energy released in a nuclear reactor manifests itself as heat energy near the point of fission in the core of the reactor.

Two major considerations associated with the products of fission process were (1) the products that include radio isotopes could damage the fuel elements and thus limit the time the fuel can be allowed to remain in the reactor, and (2) the fission products are the sources of most of the radioactivity in irradiated fuel. It is the second consideration that the reactor designers and operators have to control and provide containment for the fission products, under both the normal and abnormal conditions. Thus, maintaining adequate protection of the health and safety of the general public was a major requirement in exploring power reactor design for commercial use of nuclear energy.

The Atomic Energy Act of 1946, signed by President Truman, paved the way for transferring the function of civilian use of the Atom in the jurisdiction of Atomic Energy Commission (AEC). It was in December 1953, when President Dwight D. Eisenhower's "Atoms for Peace" speech to the United Nations General Assembly, envisaged peaceful nuclear technology which would be made available to all nations under appropriate international controls. Subsequently, the 1954 Atomic Energy Act made it possible to encourage the commercial use of atomic energy in the U.S. for producing power.

The earlier reactors built as research and demonstration reactors, e.g. Hanford, Savannah River, and the Idaho National Laboratory Reactor Testing Station were located at remote locations away from the population centers. These reactors normally had concrete shield buildings enclosing the reactors for protecting the working personnel against ionizing radiation. However, some other AEC facilities, constructed in early 1950s, such as, Argonne Research Reactor, near Chicago, and the Submarine Intermediate Reactor at West Milton, NY, indicated the need for reliance on engineered safety features that would compensate for their proximity to population centers. The General Electric (GE), designer of the West Milford reactor, sets a major safety precedent by enclosing the reactor in a large steel containment structure. Later, the Argonne Research Rector was enclosed in a leak tight concrete building. Containment was also a major design feature of the Westinghouse designed Shippingport reactor. Except for a few experimental reactors, constructed at remote sites, and some gas-cooled reactors, all power reactor facilities in the United States after that time included provisions for containment structures, as the major safety features of the reactor facilities.

This book is devoted to the subject of containment structures in the United States. The following is a brief description of the content of this book.

Readers should note that "containment structure" is a part of the containment or containment system. Sometimes, these phrases (containment, containment systems) are used interchangeably with containment structure, as the final physical barrier that would prevent release of the ionizing radiation. Containments are also described as "containment vessels," i.e., reinforced concrete containment vessels (RCCVs) and prestressed concrete containment vessels (PCCVs).

The book is divided into nine chapters. Each chapter describes specific aspect of containment structures. I am one of the seven prime contributors of the book. The four co-authors provided specialized inputs. The table below provides information regarding the chapter titles and the authors. It should be noted that Chapters 1 to 5 principally discuss History, Design, Construction, Inspection, and License Renewal aspects of operating reactors. Chapters 6 to 8 discuss more generic aspects of containment analysis under various loadings, and Chapter 9 discusses containment systems of Advanced Reactors.

Chapter Title	Author	Affiliation	Co-Authors
1. Evolution of Reactors and Containments	Hansraj Ashar	HKPrefessional, LLC	
2. Regulatory Requirements and Containments	Hansraj Ashar	HKPrefessional, LLC	
3. Design, Construction, Inspection and Testing of Containment Structures	Javeed Munshi	Bechtel Power Corporation	Shen Wang and Abdul Sheikh
4. Inservice Inspection and Leak Rate Testing of Containments	Hansraj Ashar	HKPrefessional, LLC	
5. License Renewal and Aging Management for Continued Service	Dan Naus	Oak Ridge National Laboratory	Hansraj Ashar
6. Containment Structure Testing, Modeling and Degradation	Jason Petti	Sandia National Laboratories	
7. Containment System Challenges Under Severe Accidents	Dana Powers	Sandia National Laboratories	Shawn Burns and Hansraj Ashar
8. Design Basis and Beyond Design Basis Considerations of Natural Phenomena	Nilesh Chokshi	Nuclear Regulatory Commission	Goutam Bagchi
9. Evolution of Containment Systems for Gen III Reactors	Jim Xu	Nuclear Regulatory Commission	

As the chapters are authored by 11 individuals, it is possible that there may be some duplication of information. However, extreme care has been taken to avoid excessive duplication. Each chapter has its own figures, tables, and references. Some chapters may have appendices to address some special information relevant to the chapter. The following is a summary of each chapter.

In Chapter 1, the author describes the development of commercial nuclear power plants, concept of containment, and historical debates that took place in constructing NPPs. The chapter describes a short historical background as to how light water reactors (LWRs) became the accepted technology for commercial nuclear power plants in the U.S. The chapter also describes various types of reactors and containment structures and how various designs of leaktight containments were developed.

In Chapter 2, the author describes the historical concept of regulations developed to ensure public health and safety. The chapter discusses regulations that dictate various factors to be considered in designing, constructing, and maintaining the containment structures, so they can perform their intended function under various natural phenomena and design basis accidents. The chapter describes the regulatory framework and some of the regulations that are important for the integrity of the containment system as a whole. The chapter also describes how the containment related regulations are implemented.

In Chapter 3, the authors describe the design and construction requirements together with quality assurance requirements, preoperational inspections and tests used to ensure the adequacy of the constructed containment structures. Specifically, the authors describe in detail, how the containments are analyzed, designed, and constructed using the requirements of national codes and standards. The chapter also describes how the containment structures are tested for structural adequacy, and leak rate tested prior to the start of the plant operation. The authors briefly describe how the impactive and impulsive loads are considered in the containments of operating reactors and in the containments of the standardized Advanced Reactors.

In Chapter 4, the author describes in-service inspection (ISI) requirements for steel and steel-lined concrete containments, and periodic leak rate testing (LRT) of the containment structure, as well as that of the system piping penetrating the containment structure. In combination, the ISI and the LRT provide containment availability and reliability. The chapter describes how the deterministic ASME Code requirements, in combination with the regulatory requirements, have been successfully used to monitor the condition of the containment structure. As the regulatory requirements for performing the leak rate testing have been transitioned from the deterministic to the performance based requirements, the chapter provides detailed discussion of the performance based criteria. The use of risk-informed approach for LRT requirements has been discussed.

In Chapter 5, the author describes the regulation and guidance developed to review the applications for extending the operational life of containments of NPPs. The chapter describes how the regulatory requirements in 10 CFR Part 54 are implemented to ensure the integrity of passive components, such as the safety related structures and components. The author also points out the use of the guidance documents prepared by the NRC and the nuclear industry. For certain structural components, subjected to time limited behavior, such as metal fatigue, and prestressing tendon forces, the author discusses the logical process. For trending analysis of prestressing tendons (required for Time-Limited Aging Analysis — TLAA), Appendix 5A provides historical discussion of the process. Appendix 5B provides a detailed discussion of the major degradation found in the U.S. containment structures.

In Chapter 6, the author provides information on the small-scale and large-scale containment models tested to understand the containment behavior under severe (beyond design basis) accident loadings. The results of a number of tests have been scrutinized through consensus process to predict the potential behavior under beyond design basis loadings. The model tests have been conducted on small scale steel containments, as well as on large scale reinforced concrete and prestressed concrete containments. Based on the test results, attempts have been made to generate probabilistic model to arrive at the containment response to severe accidents. In the same chapter, the author has discussed the models with degraded containments to assess the response of various containment designs to severe accident loading. Finally, the author provides comparison of responses (in terms of fragility curves) for degraded and non-degraded containment structures.

In Chapter 7, the authors describe severe accident processes that threaten the containment are examined. Also, in this chapter, processes within the containment that affect the inventory of radioactive material suspended in the containment atmosphere and available for release, should there be a loss of structural integrity, are examined. This chapter principally describes (1) severe accident phenomena that load the containment, (2) source terms developed during and after the core melt, and (3) the regulatory requirements and guidelines provided to address severe accident threats.

In Chapter 8, the authors discuss how the NPP systems including containments are affected by the design basis, as well as low probability extreme natural phenomena, such as the low probability earthquakes and tsunamis. The chapter describes the historical perspective of the beyond design basis Natural phenomena, evolution of understanding the hazards, and methods for evaluating beyond design basis loadings. The chapter describes the use of containment capacity and fragility analysis in evaluating containment response to various beyond design basis loadings. The authors also discuss current initiatives following the Fukushima event.

In Chapter 9, the author defines the Advanced Reactors as Generation III/III+ reactors and provides their descriptions compared to the Generation II reactors, that is, the existing operating reactors. The author discusses: (1) design and analysis considerations for Standard designs, (2) technical considerations and challenges in structural designs for standardized design and analysis, and (3) containment features of Generation III/III+ reactors, that includes BWRs as ABWR, and ESBWR, and PWRs, as AP 1000, US EPR, and US APWR.

Hansraj Ashar

CHAPTER

1

EVOLUTION OF POWER REACTORS AND CONTAINMENTS

Hansraj Ashar

This chapter describes the historical description of the evolution of nuclear reactors in the United States. A detailed historical narration is provided in Reference 1. The chapter also provides a brief description of the U.S. Atomic Energy Commission's demonstration program, in which a number of reactor designs, other than the light water reactor designs, had been investigated, and for one reason or the other, eliminated from consideration.

The chapter provides conceptual description of the basic functional characteristic of pressurized water reactors (PWRs) and boiling water reactors (BWRs) and how the concept of containment structure was brought into the design of commercial nuclear reactors. Additional information related to the containment functionality, and the regulations developed to assure that the containment functionality is maintained, is provided in Chapter 2 of this book. Furthermore, the chapter describes various designs of PWR containment structures and BWR containment structures.

In describing various containment concepts, the major emphasis is on the structural characteristics of containment design. The auxiliary systems, such as the containment ventilation system, have been briefly described to enable the readers to understand that the auxiliary systems are essential for the containment structure to perform its function as an essentially leak tight barrier under the postulated loadings.

Readers should note that the PWRs were designed and constructed by three nuclear steam supply system (NSSS) vendors, and the containment structures associated with these systems were designed and constructed by a number of architect/engineer companies. While General Electric Company was the only NSSS supplier for BWRs and design concepts evolved with time for better reliability and cost effectiveness. In spite of the variability in design concepts, the PWR designs also evolved for simplicity and cost-effectiveness.

Appendix 1A of this chapter provides containment-related information for all operating nuclear power plants in the United States.

Keywords: Nuclear reactors, containments, boiling water reactors, pressurized water reactors, large dry PWR containments, subatmospheric containments, ice-condenser containments, BWR Mk I containments, Mk II containments, Mk III containments

1.1 INTRODUCTION

On December 1953, President Dwight D. Eisenhower delivered an "Atoms for Peace" speech to the United Nations General Assembly in which he envisaged peaceful nuclear technology, which would be made available to all nations under appropriate international controls. The "Atoms for Peace" speech was the impetus for the formation of the U.S. commercial nuclear industry. Today, nuclear energy supplies more than 16%

of the world's electricity and 20% of total U.S. electric output. The following paragraphs briefly review the history of development of commercial nuclear reactors in the United States. Some of the information has been extracted from the book authored by Mazutan and Walker [1].

In 1954, Congress passed a legislation that for the first time permitted the wide use of atomic energy for peaceful purposes. The 1954 Atomic Energy Act redefined the atomic energy program by ending the government monopoly on technical data and making the growth of a private commercial nuclear industry an urgent national goal. The measure directed the Atomic Energy Commission (AEC, also called the Agency) "to encourage widespread participation in the development and utilization of atomic energy for peaceful purposes." At the same time, the Congress instructed the agency to prepare regulations that would protect public health and safety from radiation hazards. Thus, the 1954 act assigned the AEC three major roles: (1) to continue its weapons program, (2) to promote the private use of atomic energy for peaceful applications, and (3) to protect public health and safety from the hazards of commercial nuclear power. Those functions were in many respects inseparable and incompatible, especially when combined in a single agency.

The AEC's regulatory program was most directly affected by the agency's commitment to encourage the rapid growth of civilian nuclear power. The initial impetus for peaceful development of nuclear power came mostly from considerations of other than meeting America's energy demands. In the early 1950s, projections of future energy requirements predicted that atomic power would eventually play an important role in the nation's energy supplies, but they did not suggest an immediate need to construct atomic power reactors. The prevailing sense of urgency among government leaders led to the development of 1954 Atomic Energy Act. The Act provided an incentive to the growth of commercial nuclear power. The urgency for the implementation of the Act was mostly derived from the fear of falling behind other nations in fostering the peaceful use of atomic power. The strides that Great Britain and the Soviet Union (USSR) were making in the field of civilian power development provided the required ammunition. AEC Commissioner Thomas E. Murray described "nuclear power race" in a 1953 speech and warned that the "stakes are high," once we become fully conscious of the possibility that power hungry countries will gravitate toward the USSR, if it wins the nuclear power race. Many government officials emphasized that surrendering America's lead in expanding the peaceful applications of atomic energy would be a severe blow to its international prestige and world scientific dominance.

A broad political consensus viewed the development of nuclear energy for civilian purposes as a vital goal. The Atomic Energy Act of that year (1954) resulted partly from perceptions of the long-range need for new energy sources, but mostly from the immediate commitment to maintain: (1) America's world leadership in nuclear technology, (2) enhance its international prestige, and (3) demonstrate the benefits of peaceful atomic energy. In particular, the act mandated the atomic power program with a sense of urgency, and in that atmosphere, the AEC established its developmental and regulatory policies. The 1954 act gave the AEC wide discretion on how to proceed. Despite the general agreement on ultimate objectives, the means by which they should be accomplished were not quite apparent.

1.2 STEPS TOWARD DEVELOPMENT OF COMMERCIAL NUCLEAR REACTORS

The AEC favored a partnership between government and industry in which private firms would play an integral role in demonstrating and expanding the use of atomic power. To accomplish its objectives, in January 1955, the AEC announced a "power demonstration reactor program." The agency offered to perform research and development on power reactors in its national laboratories to supplement research undertaken by the industry under fixed-sum contracts. The AEC also waived the established fuel use charges and the loan of fissionable materials for 7 years. However, the government would continue to own fissionable material. For their part, private utilities and vendors would supply the capital for construction of nuclear plants and pay operating expenses other than fuel charges. The purpose of the demonstration program was to stimulate private participation and investment in exploring the technical and economic feasibility of different reactor designs.

At that time, no single reactor design had clearly emerged as the most promising of the several that had been considered. The pressurized light water reactor, being constructed at Shippingport in Pennsylvania, seemed to have a lead over other alternatives. However, the AEC's demonstration program was intended to show which one of several designs being considered was practical and reliable. Consequently, the program encouraged research on many reactor designs between 1955, when the program began, and 1963, when it ended. This was called the "first round" of demonstration program.

Four industry proposals came in "round one." One of the four proposals came from the industry's Nuclear Power Group (NPG). However, later on, in 1955, NPG decided to waive its request for research and development aid from the AEC, and proceeded privately. Eventually, the project culminated in the construction of a nuclear power reactor in Dresden, Illinois. It was the first large-scale boiling light-water reactor and effectively demonstrated the concept's feasibility for electric power generation. The project was designed and built by General Electric (GE). Readers should note that in addition to its own research facilities in New York State and California, GE had drawn heavily from AEC research on BWR at Argonne National Laboratory.

The second proposal was from Yankee Atomic Electric Company to build a 600 megawatt-thermal (MWt) pressurized water reactor at Rowe, Massachusetts. Design and construction went relatively smoothly, and Yankee Atomic began operation in 1960. The third proposal was from the Consumers Power District of Nebraska to build a small sodium graphite reactor at Hallam, Nebraska. The reactor was designed by Atomic International. The project went through certain financial problems. However, under "round two" of the demonstration program, AEC took ownership of the reactor. The reactor started operating in 1962. The fourth proposal was from the Power Reactor Development Company led by the Detroit Edison Company of Michigan for building a 200 MWt fast breeder reactor at Newport, Michigan in mid-1956. The project got involved in legal and political dispute. However, the reactor went into operation in mid-1963 as Fermi 1.

Concurrently, with the evaluation of "round one" proposals, in May 1956, the AEC granted construction permit to Consolidated Edison Company (Con-Ed) of New York to build a PWR on the Hudson River, at Indian Point, NY, 24 miles north of center of New York City. Con-Ed did not participate in AEC's Power Demonstration Reactor Program. The Indian Point 1 (IP-1) began operating in 1962.

In "round two" of the AEC's demonstration program, at the desire of publicly and privately owned utilities to construct small (less than 100 MWt) size reactors, the AEC approved (1) a closed-cycle BWR (58 MWt) manufactured by Allis-Chalmers near Elk River, Minnesota, and (2) organic-cooled and moderated reactor (46 MWt) owned by the city of Piqua, Ohio.

In "round three," the AEC gave operating license to (1) a 240-MWt BWR at Big Rock Point, Michigan, (2) a 190 MWt BWR at Pathfinder, Sioux falls, SD, and (3) a 115-MWt Peach Bottom, in Delta, PA.

Finally, in 1962, the AEC announced a modified third round of the demonstration program calling for proposals for large (> 250 MWt) electric generating facilities that would demonstrate reactors as reliable sources of electric power. Table 1.1 provides information related to the plants licensed in the three rounds of the AEC's demonstration program. The following paragraph explains column 2 of Table 1.1.

In a nuclear fission process, scientists have known that the high speed neutrons (> 5 million electron volts — Mev) have to be slowed down to the speed of thermal neutrons (< 0.3 ev) for the chain reaction to continue effectively. What is required is a material that has the ability to slow down the high-speed neutrons quickly and has little tendency to absorb neutrons. Such a material is called a "moderator." Effective moderators are: ordinary water, graphite, beryllium, and certain organic materials. For example, in column 2, for Hallam reactor, sodium was used as a coolant and graphite was used as a moderator.

The following paragraphs provide a brief description of the non-light water reactors that did not succeed in the United States.

Hallam reactor, experimental graphite-moderated sodium cooled reactor operated from January 19, 1962, to September 1964. It stopped operating after facing a number of issues with the cladding, corrosion, and stress corrosion cracking, which led to the failure of the cladding cans. Correcting the failure would have been expensive so the owner decommissioned the plant and sealed the nuclear components in concrete.

TABLE 1.1
PLANTS LICENSED DURING AEC's DEMONSTRATION PROGRAM

Plant Name (US Location)	Reactor Type (MWt)	NSSS Vendor	OL-Issued Shutdown	Notes
Dresden 1 (Morris, IL)	BWR 700	GE	09/1959 10/1978	Round 1
Yankee-Rowe (Rowe, MA)	PWR 600	Westinghouse	12/1963 10/1991	Round 1
Hallam (Hallam, NE)	Sodium-Graphite 256	Atomic-Interna (AI)	01/1962 09/1964	Round 1
Firmi 1 (Newport, MI)	Sod cooled Breeder 200	Combustion-Eng-CE	05/1963 09/1972	Round 1
Indian Pt 1 (Buchanan, NY)	PWR 615	B & W	03/1962 10/1974	Independent of the Program
Elk River (Elk River, MI)	BWR 58	Allis Chalmers (AC)	11/1962 02/1968	Round 2
Piqua (Piqua, OH)	Organically Cooled 46	AI	08/1962 01/1966	Round 2
Big Rock Pt (Charlevoix, MI)	BWR 240	GE	05/1964 08/1997	Round 3
Pathfinder (Sioux Falls, SD)	BWR 190	GE	07/1966 08/1967	Round 3
Peach Bottom (Delta, PA)	High Temp Gas Cool 115	General Atomics	01/1966 10/1974	Round 3

Fermi-I, a large-scale (200 MWt) fast breeder reactor operated from May 1963 to September 1972. It did not have serious problems during operation. However, for a number of reasons related to the use of breeder reactors, the technology did not get expanded in the United States.

The Piqua facility, a small-scale (46 MWt) organically cooled and moderated reactor, operated between January 1964 and January 1966 as a demonstration project. During the operation, the plant experienced a number of problems related to in-vessel filters and control rod instability. It ceased operation in 1966. The facility was dismantled in 1967 to 1969 time frame.

Peach Bottom Atomic Power Station, Unit 1 was a 115 MWt, high temperature, gas cooled reactor (HTGR). The heart of the Peach Bottom 1 NSS was a helium-cooled, graphite-moderated reactor operating on a thorium-uranium fuel cycle. The plant was operated from June of 1967 to its final shutdown in October 1974. Later on, the reactor core test samples installed in the reactor were sent to various laboratories for analysis and research.

Fort St. Vrain (not part of the AEC's demonstration program), an 842 MWt HTGR, got an operating license in December 1973 and operated till August 1989.

During the period from 1973 and 1979, the AEC/Nuclear Regulatory Commission (NRC) received a number of construction permit applications for building HTGRs. However, for a number of quality assurance-related and financial reasons, the sponsors withdrew the applications.

By the end of the AEC's demonstration program in 1963, the light water reactor family, i.e., PWR and BWR, emerged as the most promising source of nuclear power in the United States. In a light water reactor, ordinary water is used as the coolant and as the moderator.

1.3 REACTOR CONCEPTS AND CONTAINMENTS

Since January 1974, the U.S. agency that regulates nuclear power plants is the Nuclear Regulatory Commission. It uses defense in depth approach for designing and operating nuclear facilities. The approach has been effective in preventing and mitigating accidents that could release radiation or hazardous materials. The strategy is to create multiple independent and redundant layers of defense to compensate for potential human and mechanical failures so that no single layer, no matter how robust, is exclusively relied upon. Defense-in-depth includes the use of access controls, physical barriers, redundant and diverse key safety functions, and emergency response measures. Containment structure (sometimes termed as reactor building, reactor containment, or simply, containment) is the last barrier against release of radiation and is defined as a gas-tight shell or other enclosure around a nuclear reactor. It confines the fission products that otherwise might be released to the atmosphere in the event of an accident. In the United States, each reactor is enclosed by its own containment structure.

In some designs, a reinforced concrete (RC) shield building encloses the containment structure such as, when the containment structure is fabricated from steel. In some designs, a reinforced concrete reactor building encloses the containment structure and its RC biological shield. Specific designs are discussed later in this chapter.

1.3.1 Pressurized Water Reactors

In a typical pressurized water reactor, the core inside the reactor vessel creates heat (1). The pressurized water in the reactor vessel and primary coolant loop carries the heat to the steam generators (2), which in

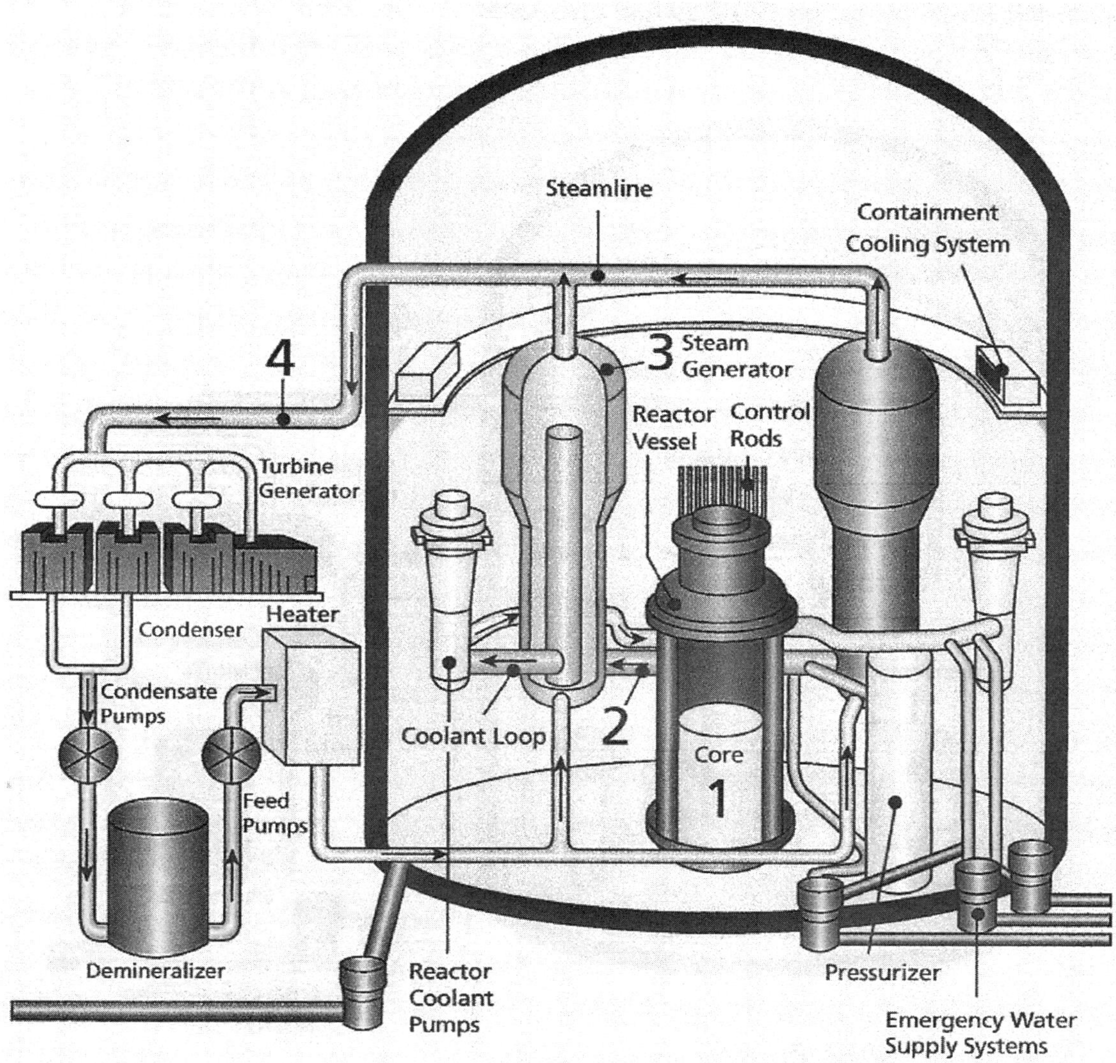

FIG. 1.1
PWR SCHEMATIC [2]

turn vaporize the water in the secondary loop (3). The steam pipes attached to the steam generators carry the steam to the main turbine (4), causing it to turn the turbine generator that produces electricity.

The unused steam is exhausted to the condenser, where it is condensed into water. The resulting water is pumped out of the condenser with a series of pumps, reheated, and pumped back to the steam generator (Fig. 1.1).

The reactor core contains fuel assemblies that are cooled by water circulated using electrically powered pumps. These pumps and other operating systems in the plant receive their power from the electrical grid. If offsite power is lost, emergency cooling water is supplied by other pumps, which can be powered by onsite diesel generators. Other safety systems, such as the containment cooling system, also need electric power.

At present (i.e., as of December 2011), there are three vendors who have supplied the Nuclear Steam Supply Systems (NSSS) for 69 operating PWRs. A typical PWR NSSS would include a reactor pressure vessel (RPV), reactor coolant piping system (RCS), and generally includes pressurizer, reactor coolant pumps, and associated steam generators. Westinghouse (W) has supplied 48 NSSS, Combustion Engineering (CE) has supplied 14 NSSS, and Babcock and Wilcox (B & W) has supplied seven NSSS.

1.3.2 Boiling Water Reactors

In a typical boiling-water reactor (BWR — see Fig. 1.2), the core inside the reactor creates heat (1), and a steam–water mixture is produced. This mixture, in turn, induces very pure water (reactor coolant) to move upward through the core, absorbing heat (2). The steam–water mixture leaves the top of the core and enters the two stages of moisture separation where water droplets are removed before the steam is allowed to enter the steam pipe (3). The steam pipe (4) directs the steam to the main turbine, causing it to turn the turbine generator, which produces electricity.

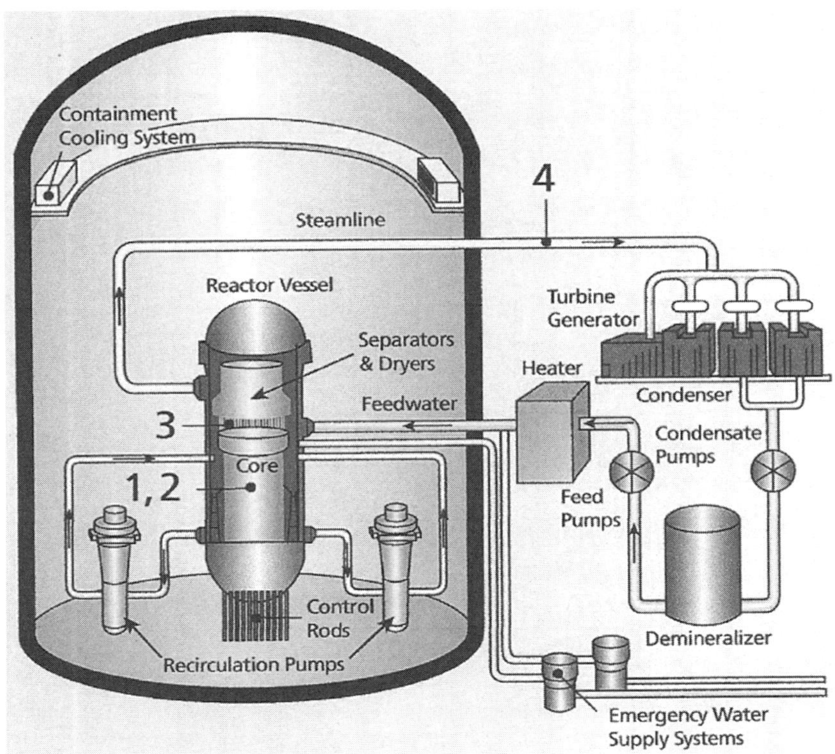

FIG. 1.2
BWR SCHEMATIC [2]

Note: Numbers (1) to (4) in the above paragraph correspond to the numbers 1 to 4 shown in Fig. 1.2. The unused steam is exhausted to the condenser where it is condensed into water. Resulting water is pumped out of the condenser with a series of pumps, reheated, and pumped back to the reactor vessel. The reactor's core contains fuel assemblies that are cooled by water circulated using electrically powered pumps. These pumps and other operating systems in the plant receive their power from the electrical grid.

The drywell, a part of the containment structure (please see Fig. 1.7 as an example), encloses the reactor vessel and recirculation loop. All BWRs except the Big Rock Point plant (not operating) have pressure suppression systems as part of the containment. The steam pipes and feed–water pipes penetrate the drywell of the containment structure.

As of December 2011, there are 35 BWRs. General Electric (GE) is the only supplier of the Boiling Water Reactor NSSS in the U.S. BWR Nuclear Steam Supply System consists of a reactor, reactor core and internal structures, two to five recirculation loops with reactor coolant pumps.

1.4 CONTAINMENT AND CONTAINMENT STRUCTURES

Containment structure is a physical barrier against the release of fission products to the environment following a release from the reactor coolant system (RCS). However, containment (or containment system) includes a number of other safety relevant systems that need to remain functional during normal and accident conditions. Other chapters in this book will describe the loadings generated from the design basis accidents and low probability natural phenomena under which the containment structure and relevant systems have to remain operable. However, a brief description of the critical loads would be useful to understand discussion of containment structures.

One of the most challenging loads that a containment structure has to withstand is the design basis accident (DBA). For a containment structure, the DBA is a loss of coolant accident (LOCA). In hypothesizing a LOCA, it is stipulated that there is a double-ended rupture of the largest pipe in the RCS. The containment structure is designed to withstand the maximum pressure and temperature generated by the LOCA.

1.4.1 PWR Containments

PWR containments can be classified in a number of ways: e.g., based on the material of construction (i.e., concrete, steel, or hybrid) or by the maximum design pressure used. However, it is convenient to classify them by their functional characteristics as follows:

(1) Atmospheric containments: during operation, the containment interior is at normal atmospheric pressure.

(2) Subatmospheric containments: during operation, the containment interior is at pressure lower than the atmospheric.

(3) Ice-condenser containments: ice-baskets, provided in the most of the perimeter in upper area of the containment function as pressure suppression medium during a DBA.

Containments of atmospheric and subatmospheric containments are also termed as large dry containments. Their design pressures are relatively high. Pressure suppression system of ice-condenser containment is designed to keep the maximum design pressure in the containment to a lower level. Regardless of containment class, all PWR containments have auxiliary systems to accomplish (1) containment isolation, (2) containment pressure control, containment heat and fission product removal, and (3) containment combustible gas control. These three systems are briefly described in the following paragraphs as they affect the containment function during normal operation and DBA:

(1) Containment Isolation: during normal operation, containment cooling is typically provided by recirculating ventilation system, not requiring large diameter ventilation pipes penetrating the containment wall to remain open, and the containment remains isolated. Following a LOCA, the containment isolation system causes isolation valves and dampers to close in certain lines that penetrate the containment boundary.

(2) Containment Pressure Control, Heat Removal, and Fission Product Cleanup: in most PWR containments, these functions are integrated in the containment spray system. In most PWRs, the containment spray system initially injects water from the refueling water storage tank (RWST), located outside the containment. The spray system is located in the upper part of the containment. When the RWST is empty, the containment spray pump system is aligned with the containment sump, and the RWST is isolated. A chemical additive is generally added to the spray water to enhance the fission product removable capability.

(3) Containment Combustible Gas Control: as the PWR containment air is not inserted, the post LOCA combustible gas control can be controlled by hydrogen recombiners and igniters, when it is appropriate to do so.

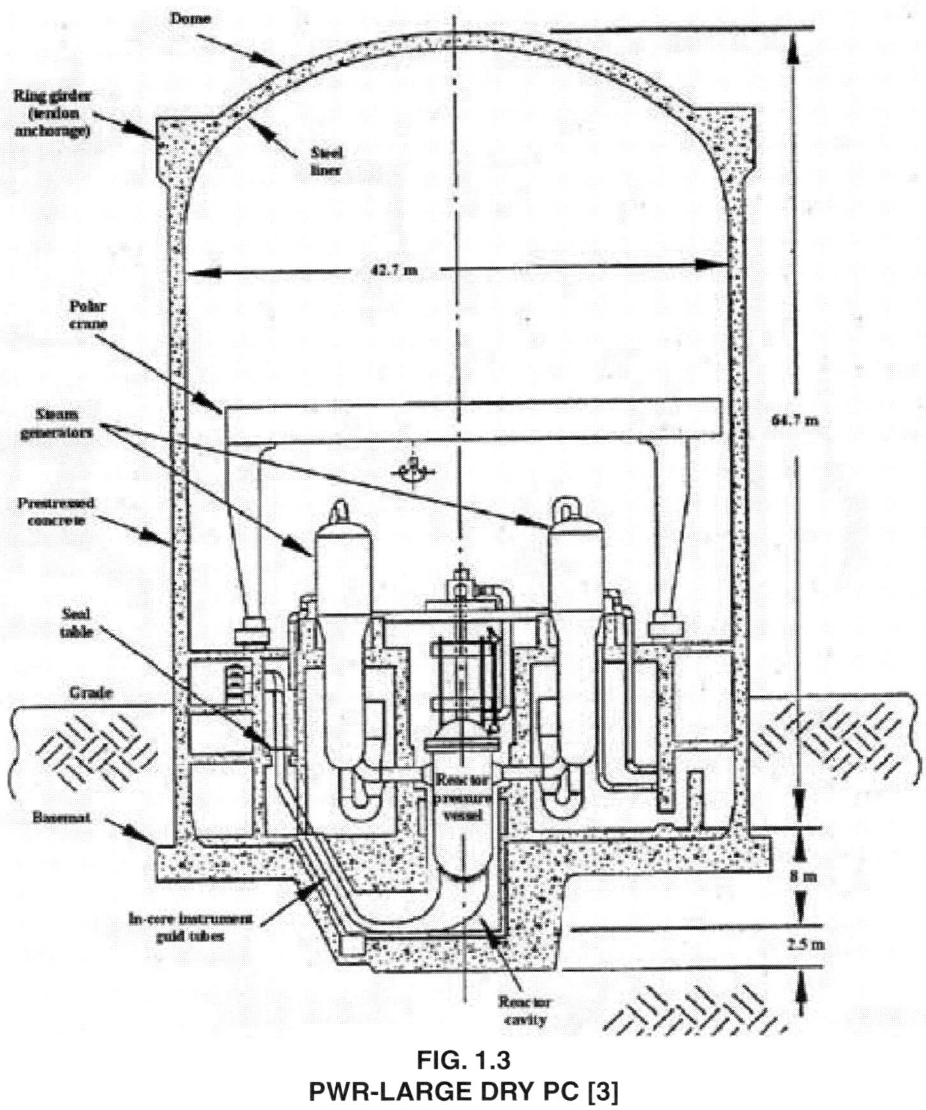

FIG. 1.3
PWR-LARGE DRY PC [3]

1.4.1.1 Large Dry PWR Containments. A large dry containment is constructed from concrete (reinforced or prestressed) or steel. In general, it consists of a 40 m (+) diameter cylindrical shell with a shallow or hemispheric dome, and a reinforced concrete basemat. The large dry concrete containment has an essentially leaktight metal liner plate attached to the inside of the dome and the cylinder, and on the top of the basemat. A large, dry prestressed concrete containment (PC) is shown in Fig. 1.3. In general, a PC consist of (1) a ring girder to anchor vertical and dome prestressing tendons, (2) tendon gallery (not shown in Fig. 1.3) at the bottom to anchor vertical tendons, and (3) buttresses, along the height of the cylindrical shell, to anchor the hoop (or horizontal) tendons.

Figure 1.3 shows a typical large dry prestressed concrete containment. Earlier prestressed concrete containments (Ginna and Robinson) have only prestressing in the vertical direction. They did not have prestressing tendons in the horizontal direction in cylinder and in the dome. Some later designs, used after 1975, have dome and vertical tendons combined to form inverted U-tendons. The evolution of prestressed concrete containments is described in detail by Ashar et al. [4].

In Fig. 1.3, the PC encloses a primary shield wall, a secondary shield wall, and a polar crane support-wall. The primary shield wall encloses the reactor vessel and its supports and supports refueling cavity. The secondary shield wall encloses steam generators, pressurizer, and reactor coolant pumps (not shown in the figure). The polar crane wall supports the polar crane and transfers its load to the containment basemat. In some PCs, the polar crane is supported on heavy steel brackets attached to the locally thickened containment liner that transfers the polar crane load directly to the containment wall.

The large dry reinforced concrete containment has similar configuration, except that there is no need for a ring-girder, buttresses, and a tendon gallery. However, generally, it has hemispherical dome, and the dome and the cylindrical wall are heavily reinforced.

Figure 1.4 shows a large dry steel containment with elliptical bottom head. The steel containment consists of a hemispherical dome, a cylindrical wall, and an elliptical bottom head. The elliptical bottom head is embedded between a reinforced concrete basemat and concrete-fill around the reactor cavity as shown in the figure. A reinforced concrete (RC) shield structure encloses the large dry steel containment. The shield structure serves as a barrier against the external missiles and as a low leakage secondary containment. There is an annular space between the steel containment and the RC shield wall. The annular space could be between 80 and 120 cm wide. The annular space has a number of floors used for supporting equipment. The floors also serve as platforms for maintenance and inspections.

The polar crane is supported by the steel brackets welded to the stiffened steel shell. The basic internal structures in a PWR dry containment consist of a primary shield wall that encloses the reactor pressure vessel (RPV), a secondary shield wall (or crane wall in some designs) that encloses the steam generators, reactor coolant pumps, and a pressurizer. The lower portion of the primary shield is called a reactor cavity, where the reactor pressure vessel is supported by brackets attached to the primary shield wall. In some designs, the RPV is supported by heavy duty nozzles emerging from the RPV. The refueling cavity, where the transfer of new and spent fuel takes place, is just above the primary shield wall (not shown in Fig. 1.4). The operating floor, shown above the steam generators in Fig. 1.4, is generally a reinforced concrete slab supported by steel or concrete girders. This floor (for all large dry PWR containments) is used as a working platform. Most of the maintenance, inspection, and monitoring activities are performed from this operating floor. Normally, in other PWR dry large containments, the bottom of the basemat is flat on sound foundation, and not elliptical, as shown in the figure.

There are 53 PWR Large Dry Containments in the United States. Table 1A.2 of Appendix 1A of this chapter shows the plant-specific information of the Large Dry PWR containments.

1.4.1.2 PWR Subatmospheric Containments. The subatmospheric containments are very similar to the large dry atmospheric containments. For the purpose of this chapter, they are considered as PWR large dry containments. There are seven subatmospheric containments being designed and built by one architect-engineer company. They are all reinforced concrete containments. The subatmospheric containments require frequent venting and purging of the containment air to maintain the inside air pressure below the normal atmospheric pressure. Some owners of this type of containments are considering changing them to atmospheric containments.

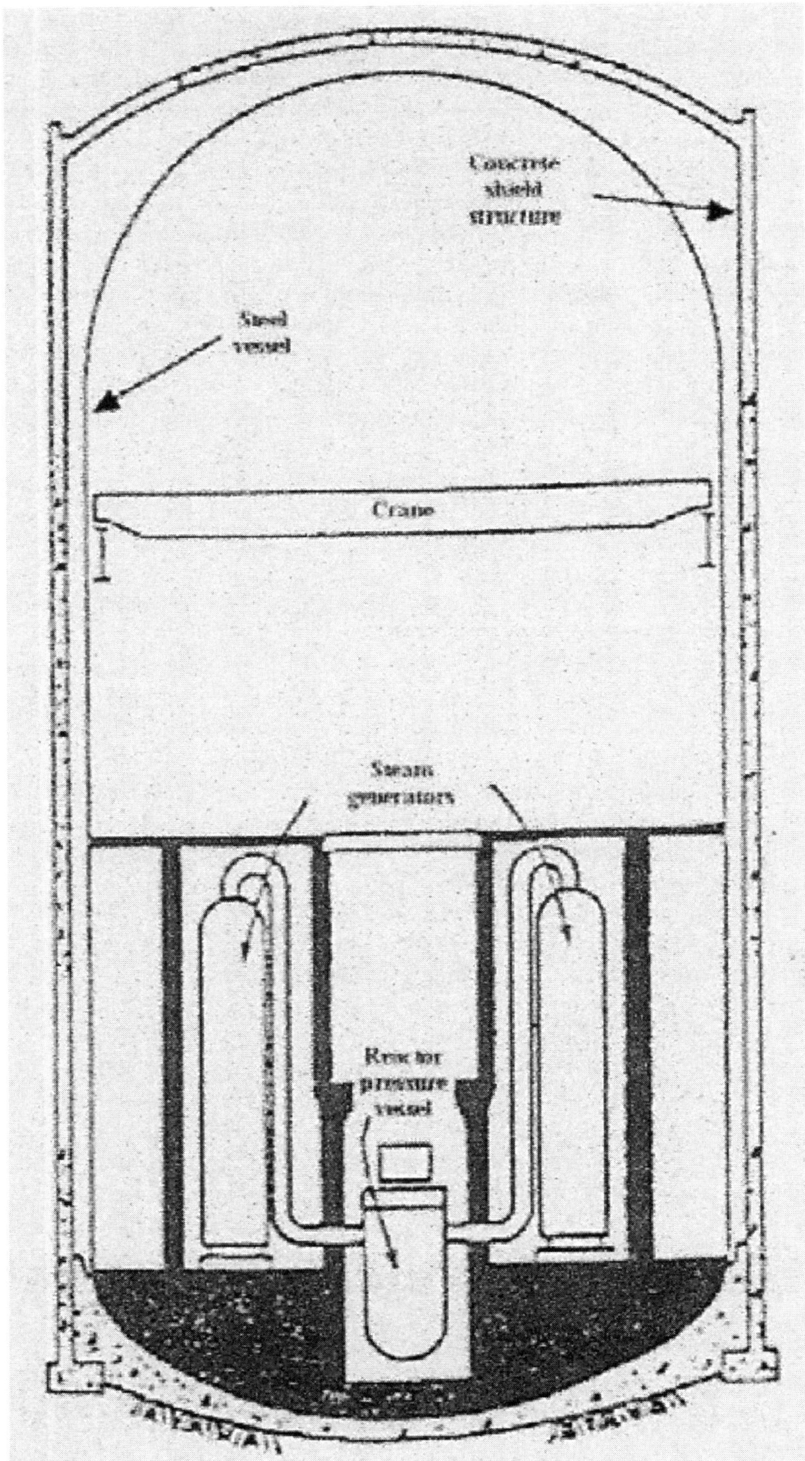

FIG. 1.4
PWR STEEL CONTAINMENT [5]

As shown in Fig. 1.5, the containment consists of a reinforced concrete hemispherical dome, supported by a reinforced concrete cylindrical wall, which in turn is supported by a reinforced concrete basemat. Metal liner is attached to the inside surfaces of the dome and the cylinder by means of steel studs. Metal liner is also installed on the top of the RC basemat. Thus, the carbon steel liner forms a continuous leaktight envelope.

A crane wall supports a polar crane, used for installing and moving heavy equipment. The crane wall structure surrounds steam generators, a pressurizer, and reactor coolant pumps. In some designs, removable block walls are provided to facilitate inspection, maintenance, and replacement, as shown in Fig. 1.5, if needed. A thick primary shield wall surrounds the reactor pressure vessel and supports a refueling cavity.

The refueling cavity is used to transfer the used fuel assemblies from the reactor to the spent fuel pool (SFP) located in the adjoining building. Similarly, the new fuel assemblies are transferred from the SFP to

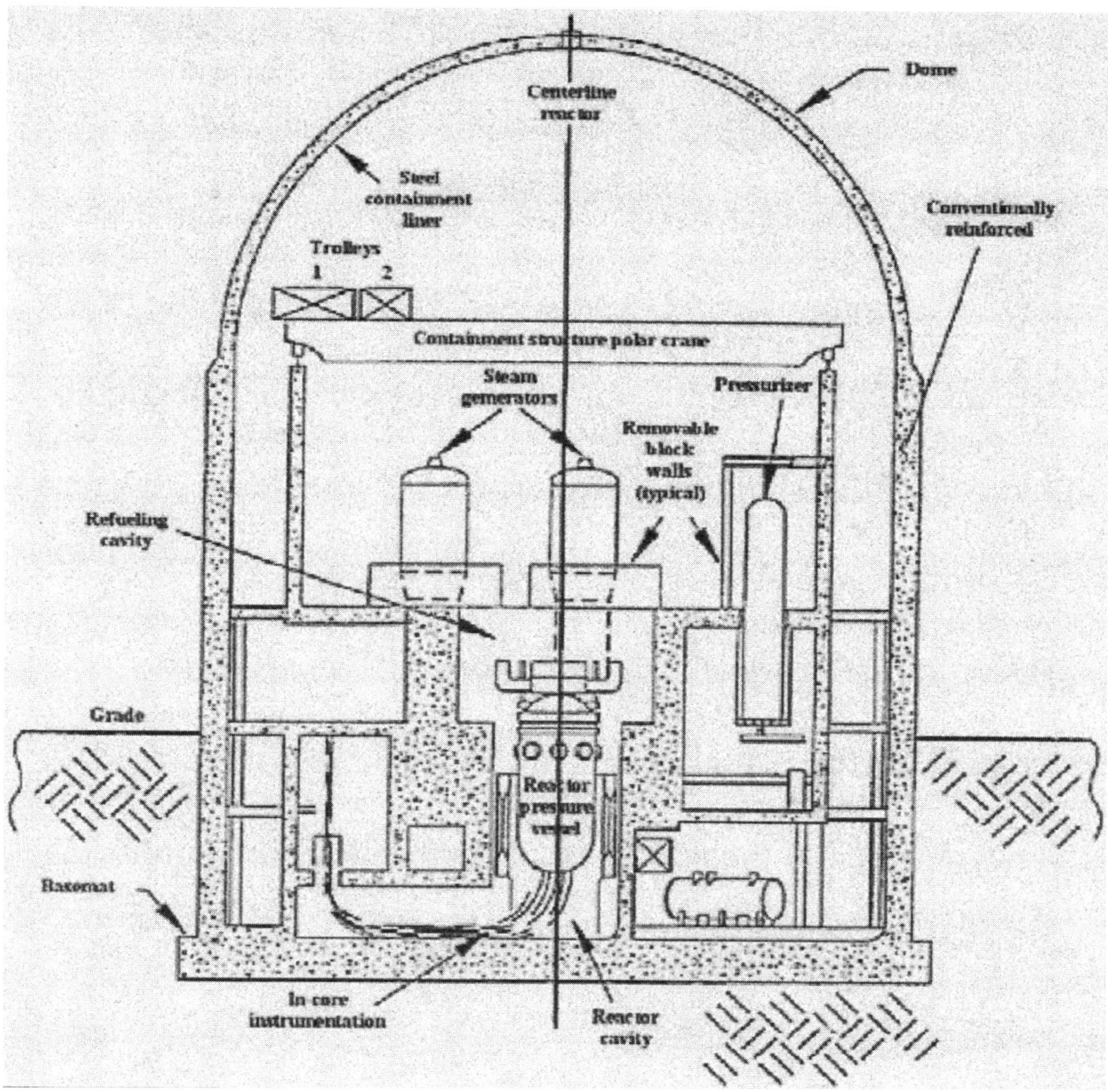

FIG. 1.5
PWR SUBATMOSPHERIC CONTAINMENT [3]

the refueling cavity before being inserted into the reactor. Readers should note that, in general, the arrangement of the refueling cavity and the spent fuel pool and the method of transferring the fuel assemblies are very similar in large dry PWR containments.

There are seven PWR Subatmospheric Containments in the United States. Table 1A.3 of Appendix 1A of this chapter shows the plant-specific information of the PWR subatmospheric containments.

1.4.1.3 PWR Ice Condenser Containments. The third type of PWR containments are the Ice-Condenser Containments. In this type of containments, ice baskets containing ice are provided in the upper portion of the containment (Fig. 1.6). The principal is to direct the flow of steam discharged during the postulated design basis LOCA through the ice baskets, that would suppress the steam pressure to 1/4th to 1/5th of that used in the design of large dry PWR containments. Figure 1.6 depicts a reinforced concrete ice condenser containment. It consists of a reinforced concrete hemispherical dome supported by a reinforced concrete cylinder, which, in turn, is supported by a reinforced concrete basemat. Metal liner plates are attached to the inside of the dome and the cylinder, and installed on the top of the basemat for making the containment essentially leaktight. Compared to the large dry PWR containments, the ice condenser containments have some special features required to ensure proper functioning of the suppression mechanism. At the floor below the ice baskets, there are inlet doors which, under a design basis accident, open up to allow the steam to pass through the ice baskets, which in turn, lowers the pressure in the containment. The thick RC primary shield wall surrounds the reactor pressure vessel, and supports the refueling cavity. The transfer of fuel assemblies from the reactor (through refueling cavity) to the spent fuel pool is similar to that described for other PWR containments. The crane wall supports the polar crane and surrounds the steam generators, pressurizer, and reactor coolant pumps. The operating floor, in general, is a RC slab supported by steel or concrete girders. Most of the movements of the movable equipment, replacement of equipment, and inspection and maintenance activities are performed from this floor. Also, the fuel transfer activities are mostly performed from the operating floor. Just below the ice basket floor, and between the containment liner and the crane wall, there are intermediate floors, which support the high energy pipes (main steam and feed water pipes) and are designed to withstand the pressures and temperatures generated by rupture of these pipes.

There are nine containments with ice-condenser designs in the United States — seven of them are steel containments, and two are of RC construction with the metal liner attached to the inside surfaces of the concrete containments. Figure 1.6 shows a PWR Ice condenser with RC containment. The internal ice-basket layout and the internal structures in the steel ice condenser containments are similar to that shown for the RC containment. However, the free standing steel containment is enclosed by a reinforced concrete shield wall, similar to that shown in Fig. 1.4.

Table 1A.4 of Appendix 1A of this chapter shows the plant-specific parameters of the nine PWR ice condenser containments.

1.4.2 BWR Containments

The NSSS for BWRs are exclusively provided by General Electric. It is most convenient to differentiae the BWR containments by the way they were developed to reflect improvements and efficiency in design, construction, and operation.

1.4.2.1 Early BWR Containments. As shown in Table 1.1, Dresden 1 was part of the Round 1 of the AEC's demonstration program. Big Rock Point was constructed in Round 3 of the AEC's demonstration program. The containments of both these plants can be considered as large dry containments. They did not have the pressure suppression feature, which is the part of the later BWRs. The containment for the two plants consisted of large dry spherical steel containments, comparable to that of the earlier PWR steel containments.

Humboldt Bay was the first U.S. BWR plant with a pressure suppression feature and has been termed as Pre-Mk I containment. It consisted of two chambers; the cylindrical, reinforced concrete drywell enclosing

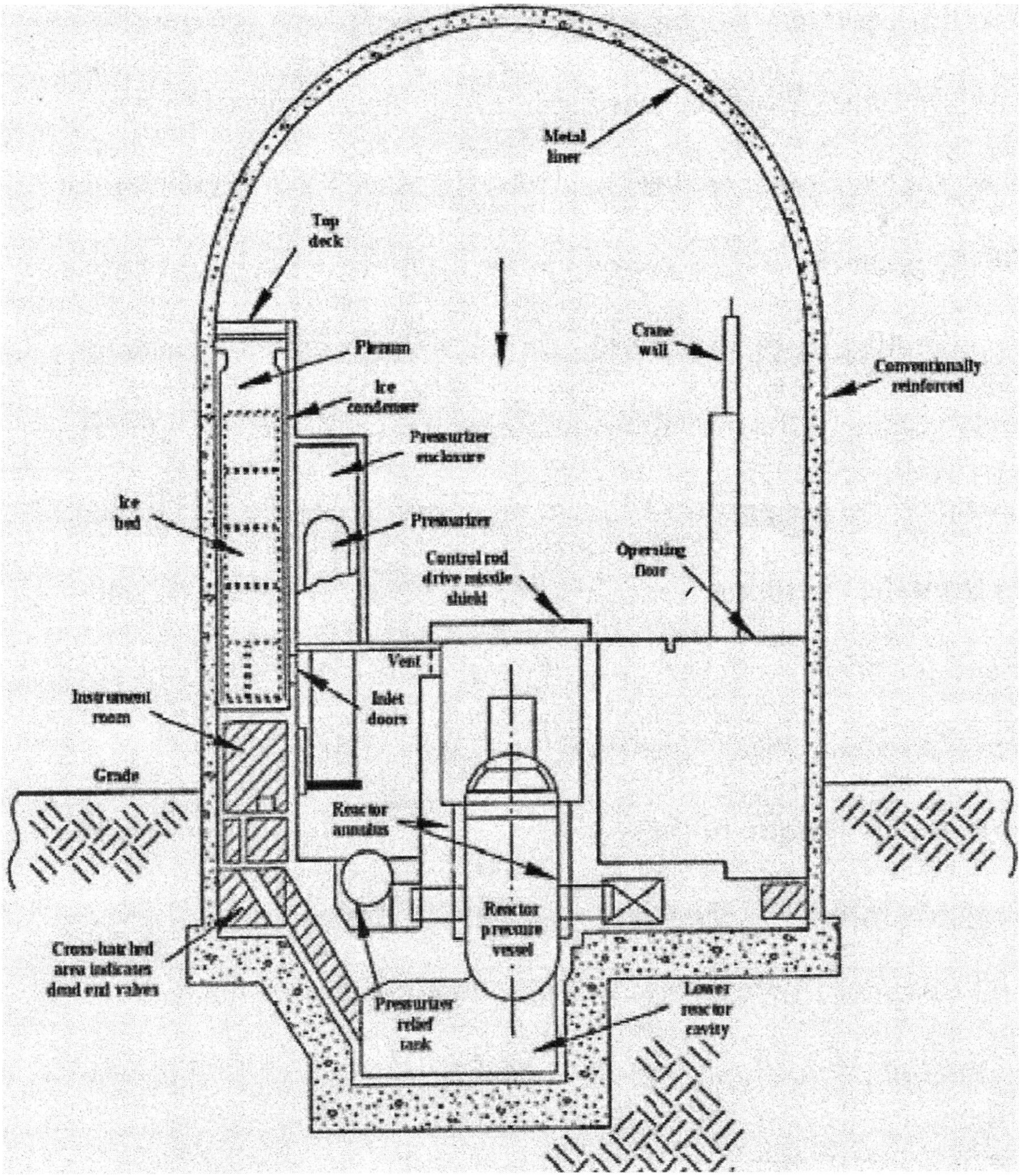

FIG. 1.6
PWR ICE CONDENSER CONTAINMENT [5]

the reactor pressure vessel and primary system components, and the exterior annular cylindrical suppression chamber that contained a large quantity of water used for pressure suppression, and as a heat sink. The drywell vented into the suppression chamber by means of large diameter horizontal vent headers connected to vertical standpipes that were provided to discharge steam below the surface of the suppression pool. NUREG/CR-5640 (1990) [6] provides a detailed discussion of these features.

The vital auxiliary systems of all operating BWR containments are (1) the pressure suppression system, (2) the containment ventilation system, and (3) the standby gas treatment system. The pressure suppression system will be discussed in respective containment category (e.g., BWR Mk I). The other systems are briefly discussed in the following paragraphs.

During operation, the BWR primary containment is closed and isolated with the containment cooling provided by a recirculating ventilation system. The areas enclosed by the reactor building (sometimes termed as secondary containment) are continuously ventilated and exhausted through an elevated stack. Following an accident, the containment isolation system causes isolation valves to close in certain lines penetrating the containment boundary. The secondary containment exhaust is directed to the Standby Gas Treatment System for processing prior to release through the elevated stack.

1.4.2.2 BWR Mk I Containments. BWR Mk I containment consists of a light-bulb shape drywell, a torus (wetwell) containing water required for suppression of the fluid pressure and temperature discharged during the postulated design basis accident. Large diameter steel vent pipes connect the drywell and the torus (see Figs. 1.7 and 1.8). The steel drywell encloses the reactor pressure vessel with removable head, a steel-lined concrete sacrificial shield wall (in Fig. 1.8, it is termed as biological shield), and associated

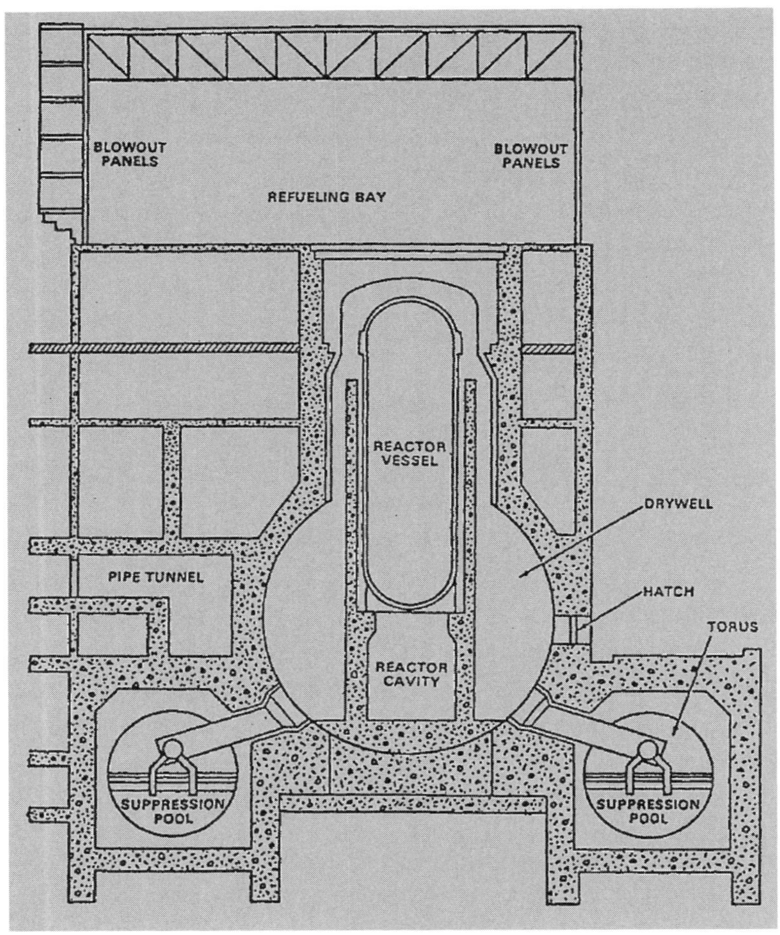

FIG. 1.7
BWR Mk I STEEL CONTAINMENT [7]

pumps and recirculation pipes. The reactor pressure vessel is supported by a pedestal supported on a reinforced concrete floor. The drywell shell at the bottom of the light bulb is embedded between the RC floor and the concrete fill, which is bearing on the reactor building basemat. Ashar and Tan [7] provide the performance experience of MK I containments. The steel drywell encloses the reactor pressure vessel with removable head, a steel-lined sacrificial shield wall (in Fig. 1.8, it is termed as biological shield wall) and associated pumps and recirculation pipes.

The reactor pressure vessel is supported by a pedestal which in turn is supported on a reinforced concrete floor. The drywell shell at the bottom of the light bulb is embedded between the RC floor and the concrete fill, which is bearing on the reactor building basemat. Figure 1.8 shows the components of a typical MK I containment. As Fig. 1.8 is taken from an operating reactor, it shows an area of corrosion found during a routine inspection. Readers can find additional description of this and other events in NUREG-1522 (1995) [8].

If a pipe rupture occurs inside the drywell, the drywell becomes pressurized by steam blowing down from the reactor coolant system. The fluid discharged during an accident has to go through the vent pipes and is collected in the vent header, which in turn distributes it to the downcomers immersed below the water surface. The torus is supported by steel columns, which are bearing on the reactor building basemat. In Fig. 1.8, a sand cushion is shown, just below the vent pipe. The function of the sand cushion is to allow the drywell structure to flex under high temperatures and pressures. Above the operating floor (see Fig. 1.7), the refueling bay area is typically surrounded by a steel structure with blow-out panels. During high winds, the panels are blown out to relieve the pressure on the structure. Figure 1.9 shows the basic layout of an Mk I concrete containment.

The basic function of drywell, suppression chamber, and the vent pipes is the same as that noted for Mk I steel containment. In the case of Mk I concrete containments, the drywell and the torus are steel-lined concrete containments. Most of the other details related to the inside of the drywell and torus are similar to that described for the Mk I steel containment.

In Fig. 1.9, more details are shown related to the structures surrounding the drywell and the torus. The Mk I containment (RC or steel) is enclosed by a low leakage reinforced concrete Reactor Building, which serves as a barrier against external missile, as well as a low leakage secondary containment. Refueling cavity, where the fuel assemblies are transferred from the reactor to the adjoining spent fuel pool is located on the top of the drywell. Out of 22 Mk I containments, 20 are steel and 2 are steel-lined concrete containments. Table 1A.5 of Appendix 1A of this chapter shows the plant-specific parameters of Mk I containments.

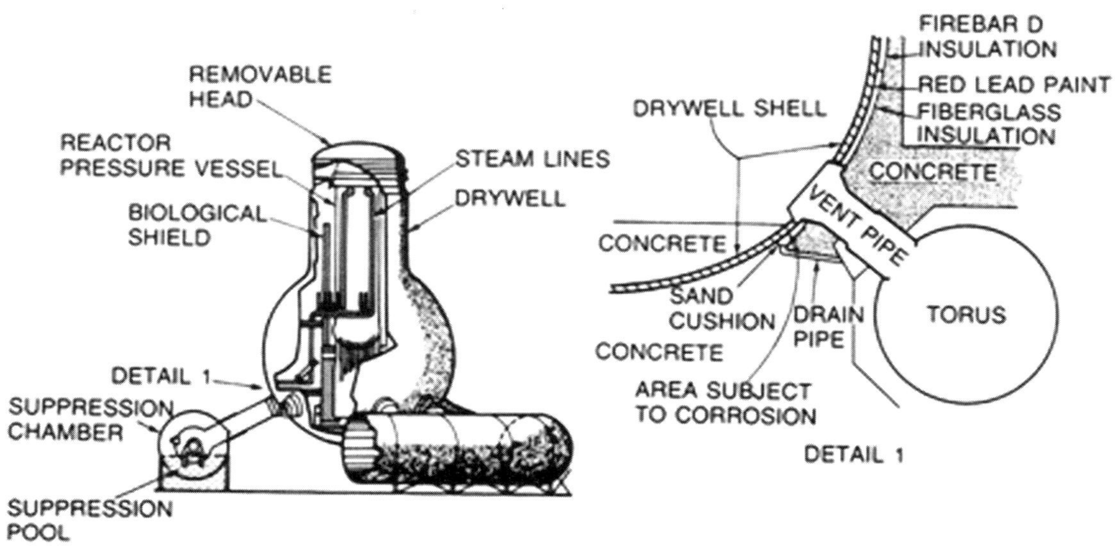

FIG. 1.8
BWR Mk I CONTAINMENT COMPONENTS [7]

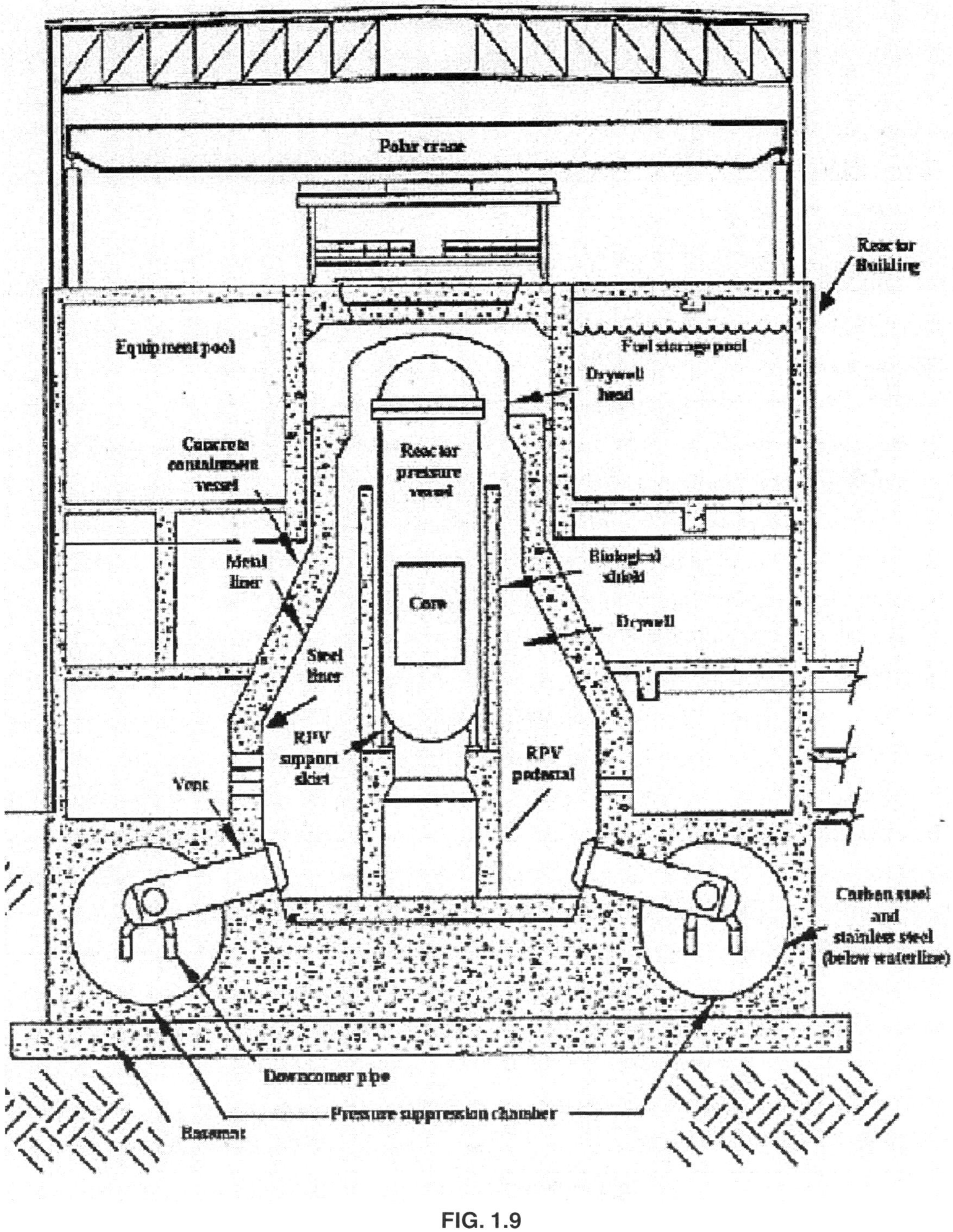

FIG. 1.9
BWR Mk I CONCRETE CONTAINMENT [5]

1.4.2.3 BWR Mk II Containments. Mk II design retained the basic pressure suppression function of the Mk I containment, but rearranged the drywell and suppression pool into an "over/under" configuration as shown in Fig. 1.10. The suppression pool is below the drywell with vertical downcomers, directing the LOCA-induced steam to below the surface of the suppression pool water. This simpler vent configuration results in lower head loss in the downcomers, resulting in a lower containment design pressure than that in the Mk I design. There are eight Mk II containments in the United States. One is constructed of steel, five are reinforced concrete containments, and two are prestressed concrete containments.

Figure 1.10 shows a typical concrete containment configuration. The conical upper portion of the containment is the drywell, and the cylindrical lower portion is the suppression chamber or wet-well. The downcomer pipes penetrate the operating floor of the drywell and are immersed below the water surface in the

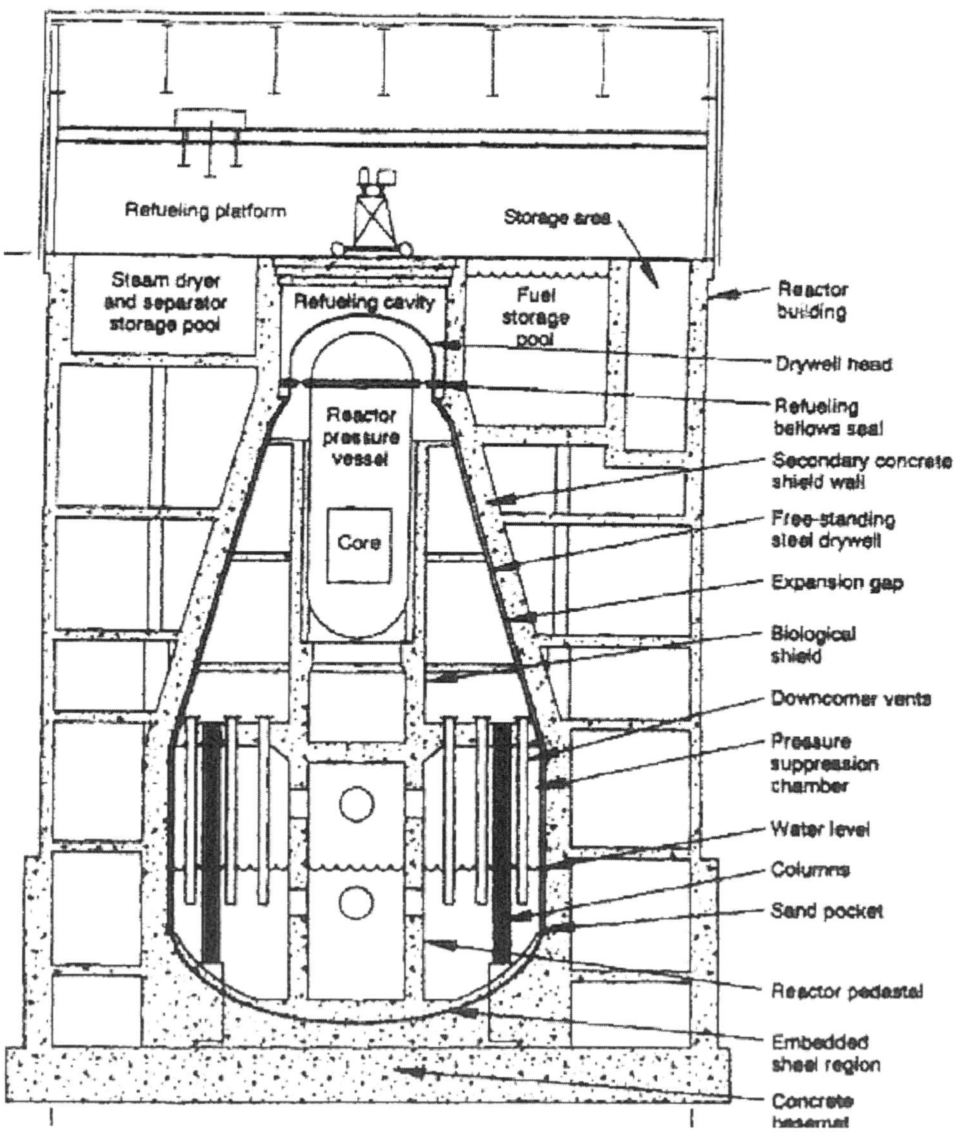

FIG. 1.10
BWR Mk II CONTAINMENT [5]

suppression pool. Metal liner (carbon steel or stainless steel) is attached to the inside of the drywell and the wetwell, as well as to the basemat floor. The reactor vessel is supported by a pedestal that bears down on the top of the reinforced concrete basemat. The reactor vessel is surrounded by a biological shield. The reactor vessel head, as well as the drywell head, is removable for maintenance and refueling activities. The layout of the Mk II steel containment is similar to that of the concrete containment, except that reinforced concrete shield building encloses the steel containment. A low leakage reactor building encloses the entire containment (drywell and the wetwell) and protects the safety-related equipment against external missiles (see Fig. 1.10). In Mk I and Mk II containments, the spent fuel pools are located adjacent to the refueling cavity and are located inside the reactor building.

Table 1A.6 of Appendix 1A of this chapter shows the information related to all Mk II containments.

1.4.2.4 BWR Mk III Containments. Compared to BWR Mk II containment, the Mk III containment is simpler structural design with the reactor placed lower in the containment resulting in lower seismic response and better construction schedule. In the Mk III containment design, the downcomers of the Mk I and Mk II design have been replaced by a weir wall with horizontal vent that facilitates the suppression of the steam pressure generated during the postulated DBA.

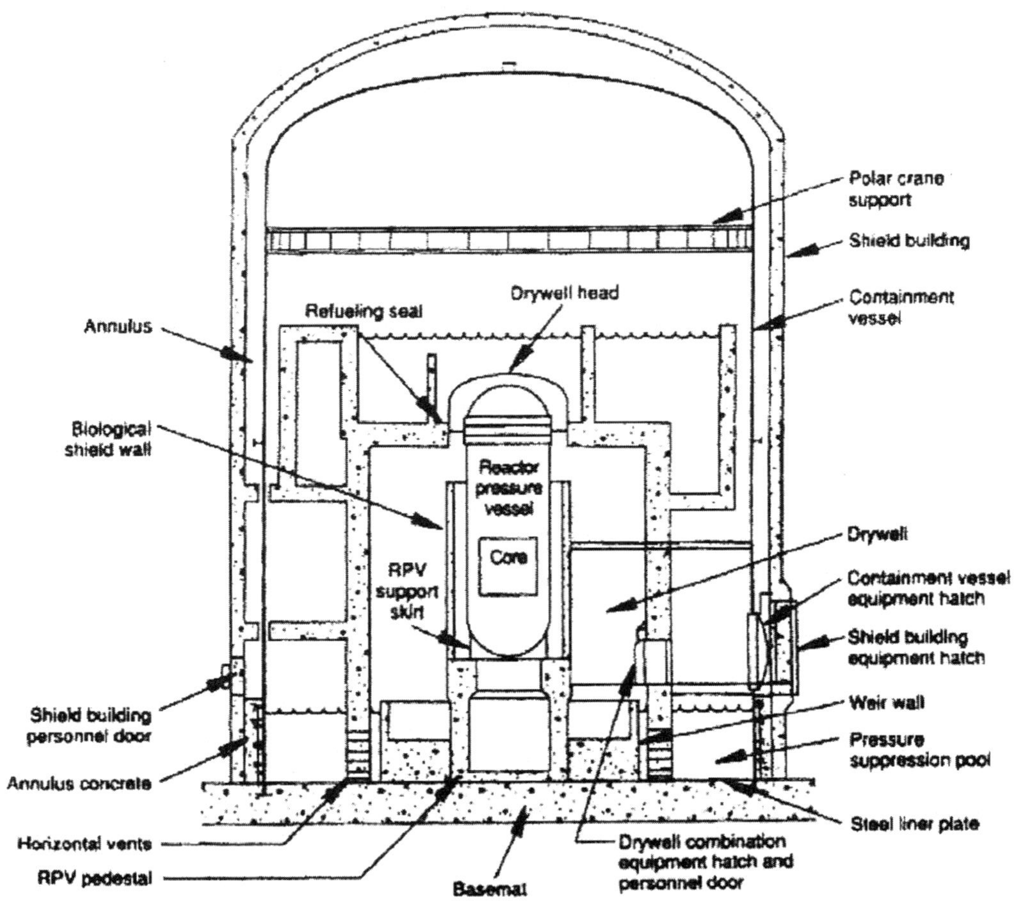

FIG. 1.11
BWR Mk III STEEL CONTAINMENT [5]

Mk III containment consists of a drywell and pressure suppression chamber located between the RC weir wall and the primary containment shell, as shown in Fig. 1.11. The primary containment can be reinforced concrete structure with a metallic liner (carbon steel, stainless steel) attached to the inside of the concrete surfaces in the dome and the cylindrical area and installed over the basemat. Figure 1.11 shows a steel primary containment with an RC shield wall around the containment. In the case of a steel primary containment, the annulus concrete is placed between the steel shell and the concrete shield wall in the suppression chamber area as shown in the figure. One Mk III containment has steel stiffening rings around the steel shell to withstand the hydrostatic pressure of the water in the suppression chamber.

There are four plants with Mk III designs, two are of RC construction, and two are primarily steel shell construction. The volume and arrangement of the Mk III containment provides lower containment design pressure and better accessibility of the RPV and other equipment and components enclosed by the drywell. In Mk I and Mk II designs, the spent fuel pool was outside the drywell and was located in the reactor building close to the refueling cavity (see Figs. 1.7, 1.9, and 1.10). In Mk III containment design, the SFP is located in a separate fuel handling building (not shown in the figure).

Table 1A.7 of Appendix 1A of this chapter shows the information related to all Mk III containments.

APPENDIX 1A — PLANT SPECIFIC INFORMATION

Tables 1A.2 to 1A.7 provide plant specific containment-related information. We have taken great care in making sure that the information comes from the publicly available reliable sources, such as NRC's NUREG Reports. However, it should be noted that they are for relative information of the plant specific containment parameters and not to be used as design- or construction-related details.

For the readers interested in more plant specific information, publicly available NRC Web Site: http://www.nrc.gov/info-finder/reactor/.html may be useful.

Abbreviations used in the tables.

TABLE 1A.1 ABBREVIATIONS

MWt — Megawatt thermal	kPa — kilo Pascal
BW — Babcock & Wilcox	PC — Prestressed Concrete
CE — Combustion Engineering	W — Westinghouse Electric
SL — Sargent & Lundy	PG & E — Pacific Gas & Electric Company
SSI — Southern Services Incorporated	GHDR — Gibbs & Hill, & Girham & Richardson
UE &C — United Engineers & Constructors	PSE — Pioneer Services Engineers
DBDP — Duke & Bechtel	FLUR — Flur Pioneer
PSE&G — Public Service Electric & G	GA — Gilbert Associates
S&W — Stone & Webster	Duke — Duke Power Company
AEP — American Electric Power	RC — Reinforced concrete
B&R — Burns & Roe	GE — General Electric
TVA — Tennessee Valley Authority	PC/RC — Prestressed vertically and reinforced horizontally and in dome

TABLE 1A.2
LARGE DRY PWR CONTAINMENTS

Plant	NSSS Vendor	Eng-Architect	Power (MWt)	Start of Operation	Containment Design Pressure (kPa)	Material of Construction	Containment Air Volume (m^3)
Arkansas 1	BW	Bechtel	2568	1974	406	PC with steel liner	59,183
Arkansas 2	CE	Bechtel	3026	1980	372	PC with steel liner	50,404
Braidwood 1/2	W	SL	3586/3586	1988/1988	421	PC with steel liner	82,119
Byron 1/2	W	SL	3586/3586	1985/1987	421	PC with steel liner	82,119
Callaway	W	Bechtel	3565	1984	414	PC with steel liner	70,793
Calvert Cliffs 1/2	CE	Bechtel	2737/2737	1975/1976	345	PC with steel liner	56,630
Comanche Pk 1	W	Bechtel	3612	1990	345	RC with steel liner	84,382
Comanche Pk 2	W	Bechtel	3458	1993	345	RC with steel liner	84,382
Crystal River 3	BW	GA	2609	1977	379	PC with steel liner	56,630
Davis Besse	BW	Bechtel	2817	1978	276	Steel shell with RC shield bldg	81,260
Diablo Canyon 1/2	W	PG & E	3411/3411	1985/1986	324	RC with steel liner	74,470
Farley 1/2	W	SSI	2775/2775	1977/1981	372/372	PC with steel liner	57,480
Fort Calhoun	CE	GHDR	1500	1973	414	PC with steel liner	29,730
Ginna	W	GA	1775	1970	414	PC/RC with steel liner	28,230
Indian Pt 2/3	W	UE & C	3216/3216	1974/1976	372/372	RC with steel liner	73,907
Kewaunee	W	PSE	1772	1974	317	Steel shell with RC shield bldg	37,348
Millstone 2	CE	Bechtel	2700	1975	372	PC with steel liner	54,370
Oconee 1/2/3	BW	DBDP	1568/2568/2568	1973/1974/1974	407/407/407	PC with steel liner	54,080
Palisades	CE	Bechtel	2565	1971	379	PC with steel liner	45,310
Palo Verde 1/2/3	CE	Bechtel	3990	1986/1987/1988	414/414/414	PC with steel liner	73,620
Point Beach 1/2	W	Bechtel	1540/1540	1970/1972	414/414	PC with steel liner	28,317
Prairie Island 1/2	W	FLUR	1650/1650	1973/1974	283/283	Steel shell w RC shield bldg	62,298
Robinson 2	W	Ebasco	2339	1971	293	PC/RC with steel liner	59,468
Salem 1/2	W	PSE&G	3459/3459	1977/1981	324/324	RC with steel liner	74,190
San Onofre 2/3	CE	Bechtel	3438/3438	1983/1984	414/414	PC with steel liner	66,260
Seabrook 1	W	UE & C	3648	1990	448	RC with steel liner	76,460
Shearon Harris 1	W	Ebasco	2900	1987	310	RC with steel liner	70,793
S. Texas 1/2	W	Bechtel	3853/3853	1988/1989	386/386	PC with steel liner	93,446
St. Lucie 1/2	CE	Ebasco	2700/2700	1976/1983	303/303	Steel shell with RC shield bldg	70,793
Three Mile Island 1	BW	UE & C	2568	1974	379	PC with steel liner	56,634
Turkey Pt 3/4	W	Bechtel	2300	1972/1973	345/345	PC with steel liner	50,971
Virgil Summer 1	W	GA	2900	1984	379	PC with steel liner	53,802
Vogtle 1/2	W	Bechtel	3625/3625	1987/1989	359/359	PC with steel liner	76,456
Waterford 3	CE	Ebasco	3716	1985	303	Steel shell with RC shield bldg	75,890
Wolf Creek 1	W	Bechtel	3565	1985	414	PC with steel liner	70,793

TABLE 1A.3
PWR SUBATMOSPHERIC CONTAINMENTS

Plant	NSSS Vendor	Eng-Architect	Power (MWt)	Start of Operation	Containment Design Pressure (kPa)	Material of Construction	Containment Air Volume (m^3)
Beaver Valley 1/2	W	S&W	2900/2900	1976/1987	372/372	RC with steel liner	50,971
Millstone 3	W	S&W	3650	1986	310	RC with steel liner	291,665
N. Anna 1/2	W	S&W	2940/2940	1978/1980	310/310	RC with steel liner	50,971
Surry 1/2	W	S&W	2546/2546	1872/1973	414/414	RC with steel liner	50,971

TABLE 1A.4
PWR ICE-CONDENSER CONTAINMENTS

Plant	NSSS Vendor	Eng-Architect	Power (MWt)	Start of Operation	Containment Design Pressure (kPa)	Material of Construction	Containment Air Volume (m^3)
Catawba 1/2	W	Duke	3411/3411	1985/1986	207/207	Steel shell with RC shield bldg	33,980
D.C. Cook 1/2	W	AEP	3304/3468	1975/1978	83/83	RC with steel liner	35,151
McGuire 1/2	W	Duke	3411/3411	1981/1984	193/193	Steel shell with RC shield bldg	33,980
Sequoyah 1/2	W	TVA	3455/3455	1981/1982	75/75	Steel shell with RC shield bldg	33,700
Watts Bar 1	W	TVA	3459	1996	103/103	Steel shell with RC shield bldg	32,410

TABLE 1A.5
BWR MK I CONTAINMENTS — VENDOR GE

Plant	Eng-Architect	Power (MWt)	Start of Operation	Containment Design Pressure (kPa)	Material of Construction	Containment Air Volume (m^3)	
						Drywell	Wetwell
Browns Ferry 1/2/3	TVA	3458/3458/3458	1974/1975/1977	386/386/386	Steel shell with concrete shielding	4502	3370
Brunswick 1/2	UE&C	2923/2923	1977/1975	427/427	RC with steel liner	4647	3511
Cooper	B&R	2419	1974	386	Steel shell with concrete shielding	4115	3112
Dresden 2/3	S&L	2957/2957	1970/1971	427/427	Steel shell with concrete shielding	4481	3320
Duane Arnold	Bechtel	1912	1975	386	Steel shell with concrete shielding	3708	2680
Firmi 2	S&L	3430	1988	400	Steel shell with concrete shielding	4636	3618
Fitzpatrick	S&W	2536	1975	386	Steel shell with concrete shielding	4248	3228
Hatch 1/2	Bechtel	2804/2804	1975/1979	400/400	Steel shell with concrete shielding	4141	3142
Hope Creek 1	Bechtel	3840	1986	427	Steel shell with concrete shielding	4786	3780
Monticello 1	Bechtel	1775	1971	379	Steel shell with concrete shielding	3800	3065
Nine Mile Pt 1	NIAG	1850	1969	427 Drywell 241 Wetwell	Steel shell with concrete shielding	5097	3398
Oyster Creek	B & R	1930	1969	527 Drywell 241 Wetwell	Steel shell with concrete shielding	5097	3625
Peach Bottom 2/3	Bechtel	3514/3514	1974/1974	386/386	Steel shell with concrete shielding	4978	3616
Pilgrim 1	Bechtel	2028	1972	386	Steel shell with concrete shielding	4163	3175
Quad Cities 1/2	S & L	2957/2957	1973/1973	386	Steel shell with concrete shielding	4481	3313
Vermont Yankee	Ebasco	1912	1972	386	Steel shell with concrete shielding	3800	3065

TABLE 1A.6
BWR MK II CONTAINMENTS — VENDOR GE

Plant	Eng-Architect	Power (MWt)	Start of Operation	Containment Design Pressure (kPa)	Material of Construction	Containment Air Free Volume (m^3) Drywell	Wetwell
Columbia	B & R	3486	1984	310	RC drywell, steel wetwell + concrete shielding	5679	4083
LaSalle 1/2	S & L	3489/3489	1984/1984	310	PC/RC dry & wetwell + concrete shielding	5927	4658
Limerick 1/2	Bechtel	3458/3458	1986/1990	386	RC with steel liner & concrete shielding	6897	4182
Nine Mile Point 2	S & W	3467	1988	310	RC with steel liner & concrete shield	5437	4383
Susquehanna 1/2	Bechtel	3952	1983/1985	365	RC with steel liner & concrete shield	6784	4151

TABLE 1A.7
BWR MK III CONTAINMENTS — VENDOR GE

Plant	Eng-Architect	Power (MWt)	Start of Operation	Containment Design Pressure (kPa)	Material of Construction	Containment Air Free Volume (m^3) Drywell	Wetwell
Clinton 1	S & L	3478	1987	103	RC drywell and steel lined RC containment	7079	30,449
Grand Gulf 1	Bechtel	3998	1985	103	RC Drywell and steel lined RC containment	7646	35,793
Perry 1	GA	3758	1987	103	RC drywell & steel containment	7872	22,710
Riverbend 1	S & W	3091	1986	103	RC drywell & steel containment	7108	20,190

REFERENCES

[1] Mazutan G., Walker J., 1997, "Controlling the Atom" U.S. Nuclear Regulatory Commission (US NRC), Reprinted in April 1997.

[2] NUREG 1350, 2010, "NRC Information Digest," U.S. NRC, Washington, DC.

[3] NUREG/CR 6424, 2007, "Report on Aging of Nuclear Power Plant Reinforced Concrete Structures," Prepared by D. J. Naus, C. Oland, B. Ellingwood, Oak Ridge National Laboratory, for US NRC.

[4] Ashar H., Tan C. P., Naus D., 1994, "Prestressing in U.S. Nuclear Power Plants," Published in two parts by Concrete International, American Concrete Institute, Detroit, Michigan.

[5] EPRI TR-103835, Rev. 1, 1994, "License Renewal Industry Report," Electric Power Research Institute, Palo Alto, California.

[6] NUREG/CR 5640, 1990, "Overview and Comparison of U.S. Commercial Nuclear Power Plants," prepared by F. Lobner, C. Danahoe, C. Cavallin, Science Application International Corporation, for U.S. NRC.

[7] Ashar H., Tan C. P., 1994, "Inservice Performance of Containment Structures – U.S. Experience, Presented at the 11th International Conference on Structural Mechanics in Reactor Technology, Tokyo, Japan.

[8] NUREG 1522, 1995, "Assessment of Inservice Condition of Safety-Related Nuclear Power Plant Structures," prepared by H. Ashar and G. Bagchi, U.S. NRC, Washington, DC 20555.

CHAPTER

2

REGULATORY REQUIREMENTS AND CONTAINMENTS

Hansraj Ashar

Chapter 2 of the book provides an in-depth historical background of regulations related to the licensing of nuclear power plants. At the time (circa 1952), only experience the AEC had was related to the small-scale government reactors and the demonstration reactors (see Table 1.1 of Chapter 1). With this knowledge and help from the research conducted at the national laboratories (e.g., Argon, Idaho), the AEC embarked on developing regulations that could apply to the commercial power reactors. One of the constraints imposed by the U.S. Congress was that the regulations should not be so restrictive that they would discourage the development of a private nuclear industry. At the same time, the regulations should be adequate to protect the public health and safety. During the period 1954 to 1962, the AEC's review of the construction permit applications and development of regulations, were conducted side by side. By 1962, the AEC developed workable regulations that were used by the applicants of construction permit for nuclear power reactors.

The chapter, moreover, describes the regulatory framework, a description of non-mandatory guidance documents and technical parts of Chapter 1 of Title 10 of the Code of Federal Regulation (CFR). Section 2.5 describes the containment-related regulations, and how these regulations are implemented using national standards and NRC's guidance documents.

Keywords: Code of Federal Regulation (CFR), Regulatory Guides, Standard Review Plan, NUREG (related to nuclear regulations) Series Reports, Steel Containment, Steel Lined Concrete Containment

2.1 INTRODUCTION (HISTORICAL BACKGROUND)[1]

As discussed in Chapter 1, the Atomic Energy Commission (AEC or the Agency) licensed the first batch of demonstration reactors under the utilization regulation. In 1955, the Agency spent considerable time on two sets of rules that applied to power reactors. Regulation on special nuclear materials (principally, plutonium, uranium-233, or uranium enriched in the isotopes uranium-233 or uranium-235) and regulations on operator's licensees. These regulations were necessary to implement the 1954 Atomic Energy Act. Under the law, the government retained title to all special nuclear material because of its strategic importance in making atomic bombs. Congress imposed upon the AEC the responsibility to protect special nuclear material against loss, diversion, and unauthorized use. The Atomic Energy Act specifically told the AEC that special nuclear material need to be distributed in a way that no user would be "permitted to construct an atomic bomb." Moreover, the Act emphasized the development of standards that would "protect health and minimize danger to life and property."

[1]I have extracted the historical information mainly from Mazuzan and Walker [1] and Walker [2].

A Power reactor owner (an applicant) applying for a fissionable material license can concurrently apply for an application for a reactor facility. The AEC insisted on this procedure to assure that the licensee had adequate supplies of fuel for duration of its 40-year license period. The key problem in developing the regulations was to strike a balance between the strict accountability rules and extent of burden on the licensees. There were those in the Congress and in the AEC who wanted strict procedures for accountability, and those who wanted the licensees to implement accountability rules that would be monitored by the AEC. Similar debates took place on the subject of "operator's licensees." The industry advisory groups agreed with the original AEC position that the Commission should not regulate the adequacy of Operator's license. However, the task force assigned to develop the regulation came out with a simple rule. The rule required the applicant (for operator's license) to take an operating test and pass a medical examination. The rule also applied to the people who were trained in government contracts and navy reactor programs by waiving the requirements for operators who could prove their qualifications to the AEC. The simplicity of the process stipulated that the selection of reliable operators is the responsibility of the management of the nuclear facility.

The Atomic Energy Act of 1954 had outlined a two-step procedure for granting licenses: (1) issuance of construction permit and (2) issuance of operating license. Even while the regulations were being developed, the AEC was processing the applications for construction permits. Thus, the regulations were being developed in parallel with the processing of reactor licenses. With the implementation of this arrangement, the applicants were able to informally discuss their proposals with the AEC. These earlier rapports determined whether the applications included the necessary technical and financial information that would incorporate necessary safety provisions required under the existing (or under development) regulations at the time.

During these rapports, the AEC staff emphasized that the responsibility for the safety of the reactor rested with the applicant. The AEC would issue a construction permit if it found the safety analysis submitted by a utility for a proposed reactor to be acceptable. After the utility completed the construction and the AEC determined that the plant fully met safety requirements, the applicant would receive a license to load fuel and begin operation.

The preliminary discussions highlighted matters, such as the relationship between the site and the reactor containment. The applicant had to select a site on the basis of complete knowledge of all factors that would affect radiological consequences. The applicant had to consider site-specific hydrology, meteorology, and seismology. The applicant had to consider population density of the surrounding areas and the potential population growth in the future years. In addition, the applicant had to determine the land use of the surrounding areas for industrial, agricultural, and residential purposes and assess the potential for surface and groundwater contamination as a result of operation of the proposed nuclear facility. These considerations resulted in identification of outstanding technical problems that the applicant had to resolve before submitting a formal request for a license. The AEC staff provided guidance, as well as the information from the state-of-the-art research conducted by the National Laboratories. All these information resulted in a workable licensing process and was instrumental in the development of nuclear industry.

In spite of this workable licensing process, the evaluation of applications was considered as a difficult task. Everyone in the nuclear field recognized that a reactor accident could destroy the infant industry, or at least set it back many years. Excessive caution could overload the structures, systems, and components (SSC) with unnecessary and expensive safeguard requirements that could have negative impact on the development of the industry. The solution was to find the appropriate balance despite many unanswered questions related to the nuclear energy in general, and reactor facility in particular. Both, the AEC regulators and the applicants lacked definitive safety standards and criteria against which to judge a reactor application. A phase of the industry marked by a number of varied reactor designs (see Table 1.1 of Chapter 1), all in the developmental stage, made it more difficult to apply uniform standards. Even with the experience gained during mid-1950s, many gaps in technical knowledge remained to be resolved before the regulators could develop standards with greater certainty. The issues of major uncertainties could be summarized as follows: (1) the properties of steel and other metals under stress in a reactor vessel, (2) means of quantifying the effects of radiation on reactor materials, (3) reaction of water with aluminum, zirconium, uranium, thorium, and (4) the measures required to decontaminate a large area in the event of a major accident. The absence of definitive knowledge of such issues required the AEC staff to pay special attention for each application. In spite of the lack of codes, specifications, and formulas, the AEC believed it could move forward in the safe licensing of reactors in parallel with the results of research findings at various National Laboratories. The research programs analyzed and tested, as necessary, metal–water reaction, metal ignition, reactor fuses, containment viabilities, and reactor instrumentation and controls.

2.2 DEVELOPMENT OF REGULATIONS

The 1954 Atomic Energy Act explicitly incorporated a goal of minimum regulation for the new nuclear industry. Subsequently, the AEC developed regulations and regulatory organizations, which implemented the national policy, i.e., minimum regulations that would protect national security and public health and safety. The rule that the AEC finally adapted in 1962 was based on (1) experience with the government owned reactors, (2) case-by-case evaluations of several reactor applications in the demonstration program (see Table 1.1 of Chapter 1), (3) the demand for guidance from nuclear industry and the congressional committee, and (4) collective knowledge about elements of nuclear technology. The discussion of site adequacy of the reactors resulted in a position that each reactor must have an essentially leak tight barrier, i.e., a containment structure, that would enclose the reactor vessel, and nuclear steam supply system components. Thus, the containment structure was defined as one of the vital engineered safety features of a power reactor.

The AEC's regulatory staff, which was created soon after the passage of the Atomic Energy Act of 1954, confronted the task of writing regulations and devising licensing procedures rigorous enough to ensure safety but flexible enough to allow for new findings and rapid changes in nuclear technology. Within a short period of time, the staff drafted rules and definitions on radiation protection standards, the distribution and safeguarding of fissionable materials, and the qualifications of reactor operators.

Between 1954 and 1962, the AEC initiated a regulatory organization and developed a regulatory process to evaluate safety issues, published regulations, reviewed license applications, and verified that its rules and license conditions were being followed. Agency officials believed that the regulatory framework they had established was sufficiently rigorous to ensure that the reactors are operated safely, but not so inflexible that it discouraged the growth of the technology. Striking a balance between those objectives was s difficult task, and it became more so after the surge in reactor applications in mid-1960s. Also, the size of the plants under review dramatically increased.

2.3 REGULATORY FRAMEWORKS

The AEC reexamined and revised its regulatory procedures to meet the new demands, and the regulatory requirements continued to evolve.

The readers should note that after a number of debates and discussions on the subject of the AEC's role as the proponent of nuclear power, and the regulator of nuclear power, in January 1974, under Energy Reorganization Act of 1974, the U.S. Congress established a new independent organization as Nuclear Regulatory Commission (NRC). The regulatory framework formed by the AEC remained intact for a while, however, was evolved later to incorporate the experiences gained from the operating reactors.

Though this chapter of the book focuses on the containment structures of the operating reactors, the following discussion briefly describes the overall regulatory framework under which the operating reactors have been reviewed and licensed.

The readers can find the NRCs regulations in Chapter I of Title 10, "Energy," of the Code of Federal Regulations (CFR). Chapter I is divided into Parts 1 through 199. The relevant parts are briefly discussed in Section 2.4.

2.3.1 Guidance Documents and Reports

To explain how containment-related General Design Criteria (GDCs), as well as the relevant parts of the regulation, are applied, it is necessary to discuss the NRC-generated guidance documents and reports that go hand in hand with the implementation of the regulation.

2.3.1.1 Regulatory Guides (RGs). Regulatory guides are issued in 10 divisions and are intended to aid licensees in implementing regulations, in general, and GDCs, in particular. The guides most applicable to nuclear reactors are in Division 1, Power Reactors. Division 2 guides address Research and Test Reactors, and Division 4 guides address Environmental and Siting.

I would like to point out that the regulatory guides are the documents that provide the (NRC) staff position on specific subject, and their use is not mandatory. The applicants/licensees can propose another acceptable alternative that would meet the basic GDC requirements.

Each regulatory guide is listed by Division, Title, Date issued, and Revision No. In this discussion, I am only discussing the regulatory guides related to power reactors. There are over 220 regulatory guides in the Division that provide guidance on implementing various design criteria. I will discuss only the containment- and quality assurance-related guides corresponding to the applicable GDC. Request for single copies of draft or active regulatory guides (which may be reproduced) should be made to the U.S. Nuclear Regulatory Commission, Washington, DC 20555, Attention: Reproduction and Distribution Services Section, or by fax to (301) 415-2289; or by email to Distribution@nrc.gov.

2.3.1.2 Standard Review Plan (SRP). Standard Review Plan — SRP {NUREG-800 (2007) [3]} provides information on the procedure that the NRC staff would use when reviewing the license applications. SRP consists of 19 chapters.

The following is a list of SRP chapters pertinent to Containments.

Chapter 2: Site Characteristics and Site Parameters
Chapter 3: Design of Structures, Components, Equipment, and Systems
Chapter 6: Engineered Safety Features
Chapter 14: Initial Test Program and ITAAC-Design Certification (see Chapter 9 of this book for more information)
Chapter 16: Technical Specifications
Chapter 17: Quality Assurance
Chapter 19: Severe Accidents

The technical positions built into SRP review procedures are not mandatory requirements. The applicants can propose alternatives to specific technical issue that can meet the relevant GDC.

2.3.1.3 NUREG-Series Reports. NRC regularly publishes reports or brochures on regulatory decisions, results of research, results of incident investigations, and other technical and administrative information in NUREG-Series Publications. The reports are designated as follows:

- Publications Prepared by NRC Staff NUREG-(nnnn) — four numerical units
- Brochures Prepared by NRC Staff NUREG/BR-(nnnn) — four numerical units
- Conference Proceeding Prepared by NRC Staff or Contractors NUREG/CP-(nnnn) — four numerical units
- Publications Prepared by NRC Contractors NUREG/CR-(nnnn) — four numerical units
- Publications Resulting from International Agreements NUREG/IA-(nnnn) — four numerical units

For example, NUREG-1430 is an NRC staff prepared document describing the Standard Technical Specification (STS) related to Babcock & Wilcox (B&W) plants.

Web (HTML) versions of some NUREG-series publications are available in NRC's Formal Publication Collection.

2.3.1.4 Other NRC Documents. In addition to the guidance documents, NRC issues, as needed, Generic Letters (GL), Information Notices (INs), Bulletins (B), and Regulatory Issue Summaries (RIS). These documents are issued based on the experiences gained related to the condition and performance of SSCs during operation of nuclear power Plants. For example, RIS are issued to clarify the NRC staff position on the implementation of codes and standards.

2.4 TECHNICAL PARTS OF CHAPTER 1 OF TITLE 10

The parts governing the licensing and operation of nuclear power plants, together with a short summary (not as precise as the regulation), are described below for easy reference. Part 52 is not directly applicable to the operating reactors. However, it has been summarized here because Chapter 9 of the book discusses the regulation in context with advanced reactors, where Part 52 is applicable.

2.4.1 Requirements of Parts 20 and 21

Part 20 — Standards for Protection Against Radiation

The regulations in this part establish standards for protection against ionizing radiation resulting from activities conducted under licenses issued by the Nuclear Regulatory Commission. These regulations are issued under the Atomic Energy Act of 1954 and the Energy Reorganization Act of 1974.

Part 21 — Reporting of Defects and Non-Compliance

This part applies to any individual director or responsible officer of a firm constructing, owning, operating, or supplying the components of any facility or activity which is licensed or otherwise regulated pursuant to the Atomic Energy Act of 1954 (as amended) or the Energy Reorganization Act of 1974. If such an individual obtains information indicating: (a) that the facility, activity, or basic component supplied to such facility or activity fails to comply with the Atomic Energy Act of 1954, as amended, or any applicable rule, regulation, order, or license of the Commission relating to substantial safety hazards or (b) that the facility, activity, or basic component supplied to such facility or activity contains defects, which could create a substantial safety hazard, to immediately notify the Commission of such failure. Under this regulation, the vendors of the structures, systems, and components, as well as the designers of the nuclear plant's SSCs are required to report to the NRC potential defects in the hardware and/or calculations that would have long-term effects on the safety of the plants.

2.4.2 Requirements of Part 50 and Its Subsections

Part 50 — Domestic Licensing of Production and Utilization Facilities

The regulations in this part are promulgated by the Nuclear Regulatory Commission pursuant to the Atomic Energy Act of 1954, and the Energy Reorganization Act of 1974, to provide for the licensing of production (reprocessing facility) and utilization facilities (reactors). This part also gives notice to all persons who knowingly provide to any licensee, applicant, contractor, or subcontractor, components, equipment, materials, or other goods or services, that relate to a licensee's or applicant's activities subject to this part, that they may be individually subject to NRC enforcement action for violation. The following are some of the technical subsections and appendices of this part:

Part 50.34 — Contents of Applications; Technical Information

Each application for a construction permit is required to include a preliminary safety analysis report (PSAR). The PSAR is required to include, as a minimum, the following technical information related to the proposed reactor facility:

- assessment of the site on which the nuclear reactor facility is to be located, including site evaluation factors identified in Part 100 of Chapter 1 (10 CFR Part 100);
- assessment to contain an analysis and evaluation of the major structures, systems, and components of the facility which bear significantly on the acceptability of the site under the site evaluation factors identified in Part 100 of this chapter;
- the extent to which generally accepted engineering standards are applied to the design of the reactor, and its safety-related SSCs;
- the extent to which the reactor incorporates unique, unusual or enhanced safety features having a significant bearing on the probability or consequences of accidental release of radioactive materials;
- the evaluation must determine that the dose limits prescribed in 10 CFR Part 100.11(a) are met;
- the safety features that are to be engineered into the facility and those barriers that must be breached as a result of an accident before a release of radioactive material to the environment can occur. Special attention must be directed to plant design features intended to mitigate the radiological consequences of accidents. In performing this assessment, an applicant shall assume a fission product release from the core into the containment assuming that the facility is operated at the ultimate power level contemplated.

Based on the comments and questions raised during the review of a PSAR, the applicant is required to develop a final safety analysis report (FSAR) that would include information describing the facility, the design bases and the limits on its operation, and describes a safety analysis of the structures, systems, and components and of the facility as a whole. The FSAR is required to address the following (a partial listing of significant requirements):

- all current information, such as the results of environmental and meteorological monitoring programs, which has been developed since issuance of the construction permit, relating to site evaluation factors identified in Part 100 of this chapter;
- a description and analysis of the structures, systems, and components of the facility, with emphasis upon performance requirements, the bases, with technical justification upon which such requirements have been established;
- For nuclear reactors, such items as the reactor core, reactor coolant system, instrumentation and control systems, electrical systems, containment system, other engineered safety features, auxiliary and emergency systems, power conversion systems, radioactive waste handling systems, and fuel handling systems need to be discussed to the extent they are pertinent;
- for facilities other than nuclear reactors, such items as the chemical, physical, metallurgical, or nuclear process to be performed, instrumentation and control systems, ventilation and filter systems, electrical systems, auxiliary and emergency systems, and radioactive waste handling systems need to be discussed, to the extent they are pertinent;
- the kinds and quantities of radioactive materials expected to be produced in the operation and the means for controlling and limiting radioactive effluents and radiation exposures within the limits set forth in Part 20 of this chapter;
- analysis and evaluation of ECCS cooling performance following postulated loss-of-coolant accidents shall be performed in accordance with the requirements of Section 50.46 for facilities for which a license to operate may be issued.

The above listing describes the essential technical information needed in the PSAR and FSAR. In addition, the regulation requires complete information regarding the operation of the facility; plans for preoperation and inservice testing including the acceptance criteria, description of accidents considered, and the final technical specifications (see Subsection 50.36).

Part 50.36 — Technical Specifications (TS)

Each applicant for a license authorizing operation of a production (reprocessing) or utilization facility (reactor) is required to include in his/her application proposed technical specifications in accordance with the requirements of this section. A summary statement of the bases or reasons for such specifications, other than those covering administrative controls, needs to be included in the application, but does not become part of the technical specifications.

To aid the licensees and applicants to develop plant-specific technical specification, NRC has developed standard technical specification (STS) corresponding to each Nuclear Steam Supply System vendor. Thus, NUREG-1430 contains STS for Babcock & Wilcox Plants, NUREG-1431 for Westinghouse Plants, NUREG-1432 for Combustion Engineering Plants, NUREG-1433 for General Electric BWR 4 Loop Plants, and NUREG-1434 for BWR 6 Loop Plants.

Part 50.44 — Combustible Gas Control for Nuclear Power Reactors

This regulation principally applies to pressure suppression containment structures. However, as the containments of BWR MK I and BWR Mk II (see Chapter 1 of this book for description of these containments) are inserted during operation, the BWR Mark III containments, and PWR Ice Condenser containments are affected by this regulation. In case of a loss of coolant accident (LOCA), there is likelihood for generation of combustible gases (e.g., hydrogen), which could result in hydrogen detonation. Means (hydrogen recombiners, igniters) have to be provided to alleviate the possibility of detonation.

Part 50.46 — Acceptance criteria for Emergency Core Cooling Systems (ECCS) for LWRs

The following is a brief description of the intent of this regulation:

Each boiling or pressurized light-water nuclear power reactor fueled with uranium oxide pellets within cylindrical zircaloy or ZIRLO cladding must be provided with an emergency core cooling system (ECCS) that must be designed so that its calculated cooling performance following postulated loss-of-coolant accidents conforms to the criteria set forth in this section of the regulation. ECCS cooling performance must be calculated in accordance with an acceptable evaluation model and must be calculated for a number of postulated loss-of-coolant accidents of different sizes, locations, and other properties sufficient to provide assurance that the most severe postulated loss-of-coolant accidents are calculated. In general, the evaluation model must include sufficient supporting justification to show that the analytical technique realistically describes the behavior of the reactor system during a loss-of-coolant accident. Comparisons to applicable experimental data must be made and uncertainties in the analysis method and inputs must be identified and assessed so that the uncertainty in the calculated results can be estimated. This uncertainty must be accounted for, so that, when the calculated ECCS cooling performance is compared to the criteria set forth in this section of the regulation, there is a high level of confidence that the criteria would not be exceeded.

Appendix K, Part II Required Documentation, sets forth the documentation requirements for each evaluation model.

Part 50.48 — Fire Protection

This regulation requires the licensees of operating plants to have fire protection plan. This fire protection plan must (i) describe the overall fire protection program for the facility; (ii) identify the various positions within the licensee's organization that are responsible for the program; (iii) state the authorities that are delegated to each of these positions to implement those responsibilities; and (iv) outline the plans for fire protection, fire detection and suppression capability, and limitation of fire damage.

Additionally, the plan must also describe specific features necessary to implement the program such as, (i) administrative controls and personnel requirements for fire prevention and manual fire suppression activities; (ii) automatic and manually operated fire detection and suppression systems; and (iii) the means to limit fire damage to structures, systems, or components important to safety so that the capability to shut down the plant safely is ensured.

Part 50.55a — Codes and Standards

This regulation requires the applicants of construction permit and licensees of operating reactors to properly use the industry Codes and Standards related to design, construction, inspection, quality assurance, preoperational testing, inservice inspections and testing of safety-related structures, systems, and components. The requirements of the codes and standards are incorporated by reference in this regulation. The regulation also includes modifications and limitations of the requirements of the referenced codes and standards.

By general reference to ASME Code, Subsection NE (2007) [4] of Section III of the ASME Boiler & Pressure Vessel Code is used for design, construction, inspection, and preoperational testing of steel containments. However, Subsection NE does not explicitly describe the loads, load combinations, and acceptance criteria for steel containments. Thus, subsection NE is supplemented with the staff position in Regulatory Guide 1.57 (2007) [5]. Also, Subsections IWE (2007) [6] and IWL (2007) [7] for inservice inspection requirements of steel and concrete containments are incorporated by reference in Subsection 50.55a.

Part 50.60 — Acceptance Criteria for Fracture Prevention Measures for LWRs for Normal Operation

This section provides reference to Appendices G and H of this part of the regulation. The Appendices, essentially, require the information related to fracture toughness and material surveillance program.

Part 50.65 — Requirements for Monitoring the Effectiveness of Maintenance at NPPs

The requirements of this section are applicable during all conditions of plant operation, including normal shutdown operations. The requirements of this section are applicable to all safety-related structures, systems, and components (SSCs), as well as to some essential SSCs characterized as none-safety as defined in this regulation. Ashar and Bagchi [8] discuss how the regulation is implemented for structures of nuclear facilities.

Part 50.69 — Risk-Informed Categorization and Treatment of SSCs for Components for Power Reactors

The categorization of NPP structures, systems and components has been based on deterministic classification of NRC's Regulatory Guide (RG) 1.26 (2007) [9], and RG 1.29 (2007) [10]. In 1990s, NRC initiated a program for using risk based or risk-informed criteria for a number of requirements, including that for categorization of SSC. As of June 2011, the use of these criteria is *optional*. Under this regulation, SSCs are categorized in four categories:

- Risk-Informed Safety Class (RISC)-1 structures, systems, and components (SSCs) means safety-related SSCs that perform safety significant functions.
- Risk-Informed Safety Class (RISC)-2 structures, systems and components (SSCs) means nonsafety-related SSCs that perform safety significant functions.
- Risk-Informed Safety Class (RISC)-3 structures, systems and components (SSCs) means safety-related SSCs that perform low safety significant functions.
- Risk-Informed Safety Class (RISC)-4 structures, systems and components means nonsafety-related SSCs that perform low safety significant functions.

Safety significant function means a function whose loss could result in a significant adverse effect on defense-in-depth, safety margin, or risk.

Part 50.72 — Immediate Notification Requirements for Operating NPPs

By this regulation, all licensees of operating reactors are required to notify the NRC Operations Center via the Emergency Notification System. The regulation requires the licenses to follow the categorization of 1-hour report, 4-hour report, and 8-hour report, as specified in the Regulation. The licensee shall also notify the NRC immediately after notification of the appropriate State or local agencies and not later than 1 hour after the time the licensee declares one of the Emergency Classes.

Eight-hour reports: If an event or a finding is not reported under other paragraphs of this regulation, the licensee is required to notify the NRC as soon as practical and in all cases within 8 hours of the occurrence of any of the following:

Any event or condition that results in the condition of the nuclear power plant, including its principal safety barriers (e.g., Containment), being seriously degraded; or the nuclear power plant being in an unanalyzed condition that could significantly degrade plant safety.

2.4.3 Requirements of Part 50 Appendices

Appendix A — General Design Criteria for Nuclear Power Plants

Under the provisions of Subsection 50.34, an application for a construction permit is required to include the principal design criteria for a proposed facility. The principal design criteria establish the necessary design, fabrication, construction, testing, and performance requirements for SSCs important to safety; that is, to ensure that the performance of the SSCs will provide reasonable assurance that the facility can be operated without undue risk to the health and safety of the public.

These General Design Criteria establish minimum requirements for the design, construction, inspection, preoperational testing, periodic inspection and testing of the water-cooled nuclear power plant SSCs similar in design and location of plants for which construction permits have been issued by the Commission. The General Design Criteria are also considered to be generally applicable to other types of nuclear power reactors (e.g., High Temperature Gas Cooled Reactors — HTGRs) and are intended to provide guidance in establishing the principal design criteria for such other reactors.

Readers should note that the development of these General Design Criteria is not yet complete. For example, some of the definitions need further amplification. Also, some of the specific design requirements for the SSCs important to safety have not as yet been suitably defined. Their omission does not relieve any applicant from considering these matters in the design of a specific facility and satisfying the necessary safety requirements. Such matters include:

- Consideration of the need to design against single failures of passive components in fluid systems important to safety;
- Consideration of redundancy and diversity requirements for fluid systems important to safety. A system could consist of a number of subsystems each of which is separately capable of performing the specified system safety function. The minimum acceptable redundancy and diversity of subsystems and components within a subsystem, and the required interconnection and independence of the subsystems have not yet been developed or defined;
- Consideration of the type, size, and orientation of possible breaks in components of the reactor coolant pressure boundary in determining design requirements to suitably protect against postulated loss-of-coolant accidents;
- Consideration of the possibility of systematic, non-random, concurrent failures of redundant elements in the design of protection systems and reactivity control systems.

It is expected that the criteria will be augmented and changed from time to time as important new requirements for these and other features are developed.

There will be some water-cooled nuclear power plants for which the General Design Criteria may not be sufficient, and for which additional criteria must be identified and satisfied in the interest of public safety. In particular, it is expected that additional or different criteria will be needed to take into account unusual sites and environmental conditions, and for water-cooled nuclear power units of advanced design. Also, there may be water-cooled nuclear power units for which fulfillment of some of the General Design Criteria may not be necessary or appropriate. For plants such as these, departures from the General Design Criteria need to be identified and justified.

Appendix B — Quality Assurance Criteria for Nuclear Power Plants and Fuel Reprocessing Plants

Every applicant for a construction permit is required by the provisions of Subsection 50.34 to include in its PSAR a description of the quality assurance program to be applied to the design, fabrication, construction, and testing of the structures, systems, and components of the facility. Every applicant for an operating license is required to include, in its FSAR, information pertaining to the managerial and administrative controls to be used to assure safe operation.

The pertinent requirements of this appendix apply to all activities affecting the safety-related functions of those structures, systems, and components. These activities include designing, purchasing, fabricating, handling, shipping, storing, cleaning, erecting, installing, inspecting, testing, operating, maintaining, repairing, refueling, and modifying.

As used in this appendix, "quality assurance" comprises all those planned and systematic actions necessary to provide adequate confidence that a structure, system, or component will perform satisfactorily in service. Quality assurance includes quality control, which comprises those quality assurance actions related to the physical characteristics of a material, structure, component, or system which provide a means to control the quality of the material, structure, component, or system to predetermined requirements.

The essential elements of the quality assurance program are briefly described here for ready reference.

- Organization — The applicant shall be responsible for the establishment and execution of the quality assurance program.
- Quality Assurance Program — The applicant shall establish at the earliest practicable time, consistent with the schedule for accomplishing the activities, a quality assurance program which complies with the requirements of this appendix.
- Design Control — Measures shall be established for the identification and control of design interfaces and for coordination among participating design organizations. These measures shall include the establishment of procedures among participating design organizations for the review, approval, release, distribution, and revision of documents involving design interfaces.
- Procurement Document Control — Measures shall be established to assure that applicable regulatory requirements, design bases, and other requirements which are necessary to assure adequate quality are suitably included or referenced in the documents for procurement of material, equipment, and services, whether purchased by the applicant or by its contractors or subcontractors.
- Instrumentation, Procedures, and Drawing — Activities affecting quality shall be prescribed by documented instructions, procedures, or drawings, of a type appropriate to the circumstances and shall be accomplished in accordance with these instructions, procedures, or drawings.
- Document Control — Measures shall be established to control the issuance of documents, such as instructions, procedures, and drawings, including changes thereto, which prescribe all activities affecting quality.

- Control of Purchased Material, Equipment, and Services — Measures shall be established to assure that the purchased material, equipment, and services, whether purchased directly or through contractors and subcontractors, conform to the procurement documents.
- Identification and Control of Materials, Parts, and Components — Measures shall be established for the identification and control of materials, parts, and components, including partially fabricated assemblies. These measures shall assure that identification of the item is maintained by heat number, part number, serial number, or other appropriate means, either on the item or on records traceable to the item, as required throughout fabrication, erection, installation, and use of the item.
- Inspection — A program for inspection of activities affecting quality shall be established and executed by or for the organization performing the activity to verify conformance with the documented instructions, procedures, and drawings for accomplishing the activity.
- Test Control — A test program shall be established to assure that all testing required to demonstrate that structures, systems, and components will perform satisfactorily in service is identified and performed in accordance with written test procedures which incorporate the requirements and acceptance limits contained in applicable design documents.
- Corrective Action — Measures shall be established to assure that conditions adverse to quality, such as failures, malfunctions, deficiencies, deviations, defective material and equipment, and nonconformance are promptly identified and corrected. In the case of significant conditions adverse to quality, the measures shall assure that the cause of the condition is determined and corrective action taken to preclude repetition.
- Quality Assurance Records — Sufficient records shall be maintained to furnish evidence of activities affecting quality. The records shall include at least the following: Operating logs and the results of reviews, inspections, tests, audits, monitoring of work performance, and materials analyses.
- Audits — A comprehensive system of planned and periodic audits shall be carried out to verify compliance with all aspects of the quality assurance program and to determine the effectiveness of the program.

The expanded version of the quality assurance requirements has been published by the ASME as NQA-1 (2008) [11] and its Addenda NQA-1a (2009) [12]. They are considered as complying with the requirements of 10 CFR Part 50 Appendix B, to the extent they are endorsed in RG 1.28 (2010) [13].

Appendix E — Emergency Planning and Preparedness for Production and Utilization Facilities

This appendix establishes minimum requirements for emergency plans for use in attaining an acceptable state of emergency preparedness. These plans shall be described generally in the PSAR for a construction permit and submitted as part of the FSAR for an operating license. These plans, or major features thereof, may be submitted as part of the site safety analysis report for an early site permit.

Appendix G — Fracture Toughness Requirements

This appendix specifies fracture toughness requirements for ferritic materials of pressure-retaining components of the reactor coolant pressure boundary of light water nuclear power reactors to provide adequate margins of safety during any condition of normal operation, including anticipated operational occurrences and system hydrostatic tests, to which the pressure boundary may be subjected over its service lifetime. The requirements of this appendix apply to the following materials:

- Carbon and low-alloy ferritic steel plate, forgings, castings, and pipe with specified minimum yield strengths not over 345 MPa (50,000 psi), and to those with specified minimum yield strengths greater than 345 Mpa (50,000 psi) but not over 621 MPa (90,000 psi) if qualified by using methods equivalent to those described in paragraph G-2110 of Appendix G of

Section XI of the latest edition and addenda of the ASME Code incorporated by reference into Subsection 50.55a (b) (2).

- Welds and weld heat-affected zones in the materials specified in the bullet above.
- Materials for bolting and other types of fasteners with specified minimum yield strengths not over 896 Mpa (130,000 psi).

The definitions of various technical terms such as ferritic material, the required fracture toughness tests, and requirements related to the fracture toughness of reactor pressure vessel are provided in this regulation.

Appendix H — Reactor Vessel Material Surveillance Program Requirements

This regulation describes the material surveillance program to monitor changes in the fracture toughness properties of ferritic materials in the reactor vessel beltline region of light water nuclear power reactors which result from exposure of these materials to neutron irradiation and the thermal environment. Under the program, fracture toughness test data are obtained from material specimens exposed in surveillance capsules, which are withdrawn periodically from the reactor vessel.

Appendix J to Part 50 — Primary Reactor Containment Leakage Testing for Water-Cooled Power Reactors

Under this regulation, the licensees of all operating light water cooled reactors are required to perform integrated and local leak rate testing of their primary containment structures and of the systems and components penetrating through the containment structures. The test requirements provide for preoperational and periodic verification of the leak-tight integrity of these components. The purpose of the tests are to assure that (a) leakage through the primary reactor containment and systems and components penetrating primary containment shall not exceed allowable leakage rate values as specified in the technical specifications or associated bases; and (b) periodic surveillance of reactor containment penetrations and isolation valves is performed so that proper maintenance and repairs are made during the service life of the containment, and of the systems and components penetrating primary containment. These test requirements may also be used for guidance in establishing appropriate containment leakage test requirements in technical specifications or associated bases for other types of nuclear power reactors. These performance criteria are the essential attributes of the containment structures, penetrating pipes, access provisions, and isolation systems; and they are discussed in detail in Chapter 4 of the book.

Appendix S — Earthquake Engineering Criteria for Nuclear Power Plants

This appendix applies to applicants for a construction permit or operating license under Part 50 (10 CFR Part 50), or a design certification, combined license, design approval, or manufacturing license under Part 52 (10 CFR Part 52) of this chapter based on the applicability criteria defined in the regulation. These criteria implement General Design Criterion 2 to the extent the criterion requires structures, systems, and components important to safety to withstand the effects of earthquakes.

2.4.4 Requirements of Part 52 to Part 100

Part 52 — Licenses, Certifications, and Approvals for Nuclear Power Plants

This part governs the issuance of early site permits, standard design certifications, combined licenses, standard design approvals, and manufacturing licenses for nuclear power facilities licensed under Atomic Energy Act of 1954 (as amended), and Title II of the Energy Reorganization Act of 1974. This regulation is applicable to the licensing of advanced reactors. However, unless otherwise specifically provided for in this part, the regulations in 10 CFR Chapter I apply to a holder of or applicant for an approval, certification, permit, or license. The implementation of this regulation is discussed in detail in Chapter 9 of this book.

Part 54 — Requirements for Renewal of Operating Licensees for Nuclear Power Plants

This part governs the issuance of renewed operating licenses and renewed combined licenses for nuclear power plants licensed pursuant to the Atomic Energy Act of 1954 (as amended), and Title II of the Energy Reorganization Act of 1974. These requirements, as related to containment structures, are discussed in detail in Chapter 5 of the book.

Part 73 — Physical Protection of Plants and Materials

This part prescribes requirements for the establishment and maintenance of a physical protection system which will have capabilities for the protection of special nuclear material at fixed sites and in transit and of plants in which special nuclear material is used. To achieve the general performance objective, a licensee shall establish and maintain, or arrange for, a physical protection system that:

- provides the performance capabilities for in-transit protection or for fixed site protection;
- is designed with sufficient redundancy and diversity to ensure availability of the capabilities;
- includes a safeguards contingency capability that can meet the criteria in Appendix C to this part;
- includes a testing and maintenance program to assure control over all activities and devices affecting the effectiveness, reliability, and availability of the physical protection system, including a demonstration that any defects of such activities and devices will be promptly detected and corrected for the total period of time they are required as a part of the physical protection system.

Part 100 — Reactor Site Criteria

The purpose of this part is to establish approval requirements for proposed sites for stationary power and test reactors subject to Part 50 or Part 52 of Chapter 1 of Title 10. There exists a substantial base of knowledge regarding power reactor design, construction, and operation. This base reflects that the primary factors that determine public health and safety are the reactor design, construction, and operation.

Siting factors and criteria are important in assuring that (1) radiological doses from normal operation and postulated accidents will be acceptably low, (2) that natural phenomena and potential man-made hazards will be appropriately accounted for in the design of the plant, (3) that site characteristics are such that adequate security measures to protect the plant can be developed, and (4) that physical characteristics unique to the proposed site that could pose a significant impediment to the development of emergency plans are identified.

2.5 CONTAINMENT-RELATED REGULATIONS

The following discussion describes the existing regulatory framework, in context of reactor site criteria, general design criteria, quality assurance criteria, and other regulatory provisions applicable to assure the adequacy of various aspects of site location, design, construction, inspections, and testing of containment structure.

2.5.1 Reactor Site Criteria

Selection of site is one of the most important endeavors for locating a reactor facility. The 10 CFR Part 100 provides detailed requirements and guidance for selecting a site for a proposed reactor facility. The relevant parts of the requirements are discussed in the following paragraphs.

The basic characteristics, that the reactor facility applicant has to address are; (1) population density and the usage characteristics of the site environs, including the exclusion area, low population zone, and population center distance, (2) physical characteristics of the site, including seismology, meteorology, geology, and hydrology. Specifically, the following attributes need to be considered:

- Appendix A of Part 100, "Seismic and Geologic Siting Criteria for Nuclear Power Plants," describes the nature of investigations required to obtain the geologic and seismic data necessary to determine site suitability and to provide reasonable assurance that a nuclear power plant can be constructed and operated at the proposed site without undue risk to the health and safety of the public. It describes procedures for determining the quantitative vibratory ground motion design basis at a site due to earthquakes and describes information needed to determine whether and to what extent a nuclear power plant need be designed to withstand the effects of surface faulting.
- Meteorological conditions at the site and in the surrounding area should be considered.
- Geological and hydrological characteristics of the proposed site may have a bearing on the consequences of an escape of radioactive material from the facility. Special precautions should be planned if a reactor is to be located at a site where a significant quantity of radioactive effluent might accidentally flow into nearby streams or rivers or might find ready access to underground water tables.
- Where unfavorable physical characteristics of the site exist, the proposed site may be found to be acceptable, if the design of the facility includes appropriate and adequate compensating engineering safeguards.

As an aid in evaluating a proposed site, an applicant should assume a fission produce release from the core, the expected leak rate from the containment, and the meteorological conditions pertinent to the proposed site to derive an exclusion area, a low population zone, and population center distance. For the purpose of this analysis, which shall set forth the basis for the numerical values used, the applicant should determine the following:

- An exclusion area of such size that an individual located at any point on its boundary for 2 hours immediately following onset of the postulated fission product release would not receive a total radiation dose to the whole body in excess of 0.25 Sievert (Sv) (25 rem), or a total radiation dose in excess of 3.0 Sv (300 rem) to the thyroid from iodine exposure.
- A low population zone of such size that an individual located at any point on its outer boundary who is exposed to the radioactive cloud resulting from the postulated fission product release (during the entire period of its passage) would not receive a total radiation dose to the whole body in excess of 0.25 Sv (25 rem) or a total radiation dose in excess of 3.0 Sv (300 rem) to the thyroid from iodine exposure.
- A population center distance of at least one and one-third times the distance from the reactor to the outer boundary of the low population zone. In applying this guide, the boundary of the population center shall be determined upon consideration of population distribution. Political boundaries are not controlling in the application of this guide. Where very large cities are involved, a greater distance may be necessary because of total integrated population dose consideration.

When a site is proposed for construction of multiple reactor facilities, the applicant is expected to show that the simultaneous operation of multiple reactors at a site will not result in total radioactive effluent releases beyond the allowable limits of applicable regulations. For further information in developing the exclusion area, low population zone, and population centers, the readers should refer to Technical Issue Document (TID) 14844 (1962) [14]. The calculations described in TID 14844 may be used as a point of departure for consideration of particular site requirements which may result from evaluation of the characteristics of a particular reactor, its purpose, and method of operation.

In addition to the above relevant information, Part 100 provides (1) factors to be considered when evaluating sites, (2) non-seismic site criteria, and (3) geologic and seismic site criteria.

2.5.2 General Design Criteria

Appendix A of Part 50 of Title 10 (written as 10 CFR Part 50, Appendix A) of the NRC regulation provides general design criteria (GDCs) for nuclear power plants. The criteria are broadly categorized in six categories. Each category has between 5 and 16 design requirements. I have reproduced here the categories relevant to containment.

Category I: Overall Requirements — Consist of Five Criteria:

- Quality Standards and Records (GDC 1)
- Design Bases for Protection Against Natural Phenomena (GDC 2)
- Fire Protection (3)
- Environmental and Dynamic Effects Design Bases (4)
- Sharing of Structures, Systems, and Components (5)

Category II: Protection by Multiple Fission Product Barriers — Consists of Ten Criteria:

- Reactor Design (6)
- Reactor inherent Protection (7)
- Suppression of Reactor Power Oscillations (8)
- Instrumentation and Control (9)
- Reactor Coolant Pressure Boundary (10)
- Reactor Coolant System Design (11)
- Containment Design (12)
- Electric Power Systems (13)
- Inspection and Testing of Electric Power Systems (14)
- Control Room (15)

Category V: Reactor Containment Design Basis — Consists of Eight Criteria:

- Containment Design Basis (50)
- Fracture Prevention of Containment Pressure Boundary (51)
- Capability for Containment Leakage Rate Testing (52)
- Provisions for Containment Testing and Inspection (53)
- Systems Penetrating Containment (54)
- Reactor Coolant Pressure Boundary Penetrating Containment (55)
- Primary Containment Isolation (56)
- Closed Systems Isolation Valves (57)

The relevant underlined and italicized criteria are discussed in Section 2.5.3.

2.5.3 Implementation of Containment-Related GDC

GDC 1 Quality Standards and Records

"Quality standards and records: Structures, systems, and components important to safety shall be designed, fabricated, erected, and tested to quality standards commensurate with the importance of the safety functions to be performed. Where generally recognized codes and standards are used, they shall be identified and evaluated to determine their applicability, adequacy, and sufficiency and shall be supplemented or modified as necessary to assure a quality product in keeping with the required safety function. A quality assurance program shall be established and implemented in order to provide adequate assurance that these structures, systems, and components will satisfactorily perform their safety functions. Appropriate records of the design, fabrication, erection, and testing of structures, systems, and components important to safety shall be maintained by or under the control of the nuclear power unit licensee throughout the life of the unit."

This is a general criterion applicable to all safety-related structures, systems, and components. ASME NQA-1 (2008) [11], and NQA-1a (2009) [12] QA requirements are applicable to the extent they are endorsed by NRC in RG 1.28 (2010) [13]. Generally recognized Code for design, construction, and testing of steel containments is Subsection NE [4] of Section III, Division 1 of the Boiler and Pressure Vessel (B & PV) Code of the American Society of Mechanical Engineers (ASME), as endorsed by RG 1.57 [5]. Generally recognized Code for concrete containments is Subsection CC [15] of Section III, Division 2 of the ASME B & PV Code, as endorsed by RG 1.136 (2007) [16]. The general requirements for structures and components of both the divisions are specified in Subsection NCA [17] of Section III of the ASME B & PV Code.

GDC 2 Design Bases for Protection against Natural Phenomena

"Design bases for protection against natural phenomena: Structures, systems, and components important to safety shall be designed to withstand the effects of natural phenomena such as earthquakes, tornadoes, hurricanes, floods, tsunami, and seiche without loss of capability to perform their safety functions. The design bases for these structures, systems, and components shall reflect: (1) Appropriate consideration of the most severe of the natural phenomena that have been historically reported for the site and surrounding area, with sufficient margin for the limited accuracy, quantity, and period of time in which the historical data have been accumulated, (2) appropriate combinations of the effects of normal and accident conditions with the effects of the natural phenomena and (3) the importance of the safety functions to be performed."

To comply with the requirements of this GDC, the applicants are required to consider the effects of extreme natural phenomena as summarized below:

- Earthquakes: criteria for defining safe shutdown earthquake (SSE) are developed from the site-related factors (see Appendix A of 10 CFR Part 100 and SRP Section 2.5.2), and Criteria provided in Appendix S of 10 CFR Part 50. The staff reviews the site-related factors in Section 2 of Standard Review Plan (Ref. 3), and Design methodology in Section 3.7 of SRP. The staff has developed several regulatory guides [e.g., RG 1.60 (2010) [18], RG 1.61 (2007) [19] to help the applicants and licensees to develop the plant-specific design and analysis.
- Hurricanes, Tornadoes: the safety-related plant structures (including containment structures) are designed to withstand hurricane-related winds and low probability tornado effects as discussed in SRP Sections 3.3.1, 3.3.2, and in RG 1.76 (2007) [20]. The safety-related structures are designed to withstand a spectrum of tornado generated missiles, as stipulated in RG 1.76 [20].
- Probable Maximum Flood (PMF): the applicants are required to consider the effects of PMF as stipulated in SRP Sections 2.4.3, and 3.4.2, and potential for upstream dam failure as stipulated in SRP Section 2.4.4.
- Probable Maximum Surge (PMS) and Seiche Flooding: the applicants are required to consider the potential effects of PMS and Seiche flooding as stipulated in SRP Section 2.4.5.
- Probable Maximum Tsunami (PMT): The applicants are required to consider the coastal sites likely to be subjected to PMT based on the historical records with sufficient margins as discussed in SRP Section 2.4.6.

Containment structures are appropriately designed for the applicable natural phenomena in combination with the applicable design basis loadings. Detailed discussion of the loads and load-load combinations is provided in Chapter 3 of this book. Detailed discussion of "Beyond Design Basis Natural Phenomena" is provided in Chapter 8 of this book.

GDC 4 Environmental and Dynamic Effects Design Bases

"Environmental and dynamic effects design bases: structures, systems, and components important to safety shall be designed to accommodate the effects of and to be compatible with the environmental conditions associated with normal operation, maintenance, testing, and postulated accidents, including loss-of-coolant accidents. These structures, systems, and components shall be appropriately protected against dynamic effects, including the effects of missiles, pipe whipping, and discharging fluids, that may result from equipment failures and from events and conditions outside the nuclear power unit. However, dynamic effects associated with postulated pipe ruptures in nuclear power units may be excluded from the design basis when analyses reviewed and approved by the Commission (i.e., NRC) demonstrate that the probability of fluid system piping rupture is extremely low under conditions consistent with the design basis for the piping."

This criterion is met when an applicant has considered the potential for internally generated missiles originated outside the containment structure, inside the containment structure, turbine generated missiles, site proximity missiles, as discussed in SRP Sections 3.5.1.1, 3.5.1.2, 3.5.1.3, and 3.5.1.4. Barrier designs have been performed as stipulated in SRP Section 3.5.2. Concrete containments are designed to withstand these types of missiles. Free-standing steel containments cannot withstand the externally generated missiles, such as turbine or tornado generated missiles. However, the free-standing steel containments are protected from such effects by the concrete shield building(s) surrounding the containment structures. In the later portion of this GDC, in lieu of designing for a pipe rupture loadings, NRC allows the licensees to demonstrate that the probability of pipe rupture is very low {also called leak before break (LBB) evaluation. SRP Section 3.6.3 and NUREG-1061 [21] provide evaluation procedure for demonstrating the validity for LBB consideration.

GDC 16 Containment Design

"Containment design: Reactor containment and associated systems shall be provided to establish an essentially leak-tight barrier against the uncontrolled release of radioactivity to the environment and to assure that the containment design conditions important to safety are not exceeded for as long as postulated accident conditions require."

In the United States, either essentially leak-tight steel, or a steel lined concrete containment surrounds each nuclear power reactor. The inside surfaces of the concrete containments are lined with between 6.3 mm (1/4 inch) and 12.6 mm (1/2 inch) steel liner plates. Around the large containment penetrations, or near the crane supports, the plates are locally thickened, as necessary. The piping systems penetrations, as well as the containment structure as a whole, are periodically tested as required by Appendix J of 10 CFR Part 50. The leak rate test pressure is derived from the conservatively calculated design basis peak pressure. More information on this subject is provided in Chapter 4 of this book. Additional requirements related to containment isolation provisions for fluid systems penetrating the containment structure are provided in the discussion of GDCs 54 to 57.

GDC 50 Containment Design Basis

"Containment design basis: The reactor containment structure, including access openings, penetrations, and the containment heat removal system shall be designed so that the containment structure and its internal compartments can accommodate, without exceeding the design leakage rate and with sufficient margin, the calculated pressure and temperature conditions resulting from any loss-of-coolant accident. This margin shall reflect consideration of (1) the effects of potential energy sources which have not been included in the determination of the peak conditions, such as energy in steam generators and as required by Section 50.44 energy from metal–water and other chemical reactions that may result from degradation but not total failure of emergency core cooling functioning, (2) the limited experience and experimental data available for defining accident phenomena and containment responses, and (3) the conservatism of the calculational model and input parameters."

The essential requirements of this GDC are briefly stipulated in SRP Section 6.2.1, "Containment Function Design," as follows:

The staff reviews information regarding the functional capability of the reactor containment presented in Section 6.2.1 of the applicant's safety analysis report (SAR or FSAR). The containment encloses the reactor system and in the event of an accident, it is the final barrier against the release of significant amounts of radioactive fission products. The containment structure must be capable of withstanding, without loss of function, the pressure and temperature conditions resulting from postulated loss-of-coolant accident (LOCAs), and steam line or feed water line break accidents. The containment structure must also maintain functional integrity, as needed, following a postulated accident; i.e., it must remain a low leakage barrier against the release of fission products for as long as the postulated accident conditions require.

The design and sizing of containment systems are largely based on the pressure and temperature conditions which result from release of the reactor coolant in the event of a LOCA. The containment design basis includes the effects of stored energy in the reactor coolant system, decay energy, and energy from other sources, such as the secondary system, and metal–water reactions, including the recombination of hydrogen and oxygen. The containment system is not required to be a complete and independent safeguard against a LOCA by itself, but it functions to contain any fission products released while the ECCS (see Section 50.46) cools the reactor core.

GDC 51 Fracture Prevention of Containment Pressure Boundary

"Fracture prevention of containment pressure boundary: The reactor containment boundary shall be designed with sufficient margin to assure that under operating, maintenance, testing, and postulated accident conditions (1) its ferritic materials behave in a non-brittle manner, and (2) the probability of rapidly propagating fracture is minimized. The design shall reflect consideration of service temperatures and other conditions of the containment boundary material during operation, maintenance, testing, and postulated accident conditions, and the uncertainties in determining (1) material properties, (2) residual, steady state, and transient stresses, and (3) size of flaws."

To meet the requirements of this GDC, ferritic containment pressure boundary materials are required to meet the fracture toughness criteria for testing as identified in Article NE-2300 [4] of Section III, Division 1, or Article CC-2520 [15] of Section III, Division 2 of the ASME Code.

GDC 52 Capabilities for Containment Leak Rate Testing

"Capability for containment leakage rate testing: The reactor containment and other equipment which may be subjected to containment test conditions shall be designed so that periodic integrated leakage rate testing can be conducted at containment design pressure."

All containment structures and systems penetrating through them are provided with provisions so that they can be tested for leak rates at the peak calculated pressure, as required by Appendix J of 10 CFR Part 50. Detailed discussion of Leak Rate Testing is provided in Chapter 4 of this book.

GDC 53 Provisions for Containment Testing and Inspection

"Provisions for containment testing and inspection: The reactor containment shall be designed to permit (1) appropriate periodic inspection of all important areas, such as penetrations, (2) an appropriate surveillance program, and (3) periodic testing at containment design pressure of the leaktightness of penetrations which have resilient seals and expansion bellows."

All containment structures are designed to allow for testing at the containment design pressure, and they are tested at 1.10 (steel containments) to 1.15 (concrete containments) of the design pressures before the plant become operational. They are inspected at periodic interval in accordance with the requirements of Subsections IWE [6] and IWL [7] of Section XI of the ASME code, as required by 10 CFR Part 50.55a. Penetrations with resilient seals and expansion bellows are periodically tested as required by Appendix J of 10 CFR 50.

GDC 54 Piping Systems Penetrating Containment

"Piping systems penetrating containment: Piping systems penetrating primary reactor containment shall be provided with leak detection, isolation, and containment capabilities having redundancy, reliability, and performance capabilities which reflect the importance to safety of isolating these piping systems. Such piping systems shall be designed with a capability to test periodically the operability of the isolation valves and associated apparatus and to determine if valve leakage is within acceptable limits."

The industry standard ANSI/ANS N260 (1976) [22] addresses the requirements of GDC 54 to GDC 57 related to the containment isolation provisions for fluid systems. The NRC staff has endorsed these provisions in RG 1.141 [23] with certain additional provisions. In combination, the standard and the RG provide adequate requirements that the piping system penetrating containment are designed with capability to periodically test the operability, and acceptability of valve leakage.

GDC 55 Reactor Coolant Pressure Boundary Penetrating Containment

"Reactor coolant pressure boundary penetrating containment: Each line that is part of the reactor coolant pressure boundary and that penetrates primary reactor containment shall be provided with containment isolation valves as follows, unless it can be demonstrated that the containment isolation provisions for a specific class of lines, such as instrument lines, are acceptable on some other defined basis:

(1) one locked closed isolation valve inside and one locked closed isolation valve outside containment; or

(2) one automatic isolation valve inside and one locked closed isolation valve outside containment; or

(3) one locked closed isolation valve inside and one automatic isolation valve outside containment. A simple check valve may not be used as the automatic isolation valve outside containment; or

(4) one automatic isolation valve inside and one automatic isolation valve outside containment. A simple check valve may not be used as the automatic isolation valve outside containment.

Isolation valves outside containment shall be located as close to containment as practical and upon loss of actuating power, automatic isolation valves shall be designed to take the position that provides greater safety.

Other appropriate requirements to minimize the probability or consequences of an accidental rupture of these lines or of lines connected to them shall be provided as necessary to assure adequate safety. Determination of the appropriateness of these requirements, such as higher quality in design, fabrication, and testing, additional provisions for inservice inspection, protection against more severe natural phenomena, and additional isolation valves and containment, shall include consideration of the population density, use characteristics, and physical characteristics of the site environs."

As discussed under GDC 54 explanation, the cited isolation provisions also apply to the requirements of this GDC. The staff review process in SRP 6.2.4 would require the licensees to address the following specific provisions:

1. The design of containment isolation provisions, including:
    - The number and location of isolation valves (i.e., the isolation valve arrangements and their physical locations as to the containment);
    - The actuation and control features for isolation valves;
    - The positions of isolation valves for normal plant operating conditions (including shutdown), post-accident conditions, and valve operator power failures;
    - The valve actuation signals;
    - The basis for selection of closure times of isolation valves;
    - The mechanical redundancy of isolation devices;
    - The acceptability of closed piping systems inside containment as isolation barriers.

2. The protection of containment isolation provisions against loss of function from missiles, pipe whip, and earthquakes.
3. The environmental conditions inside and outside the containment considered in the design of isolation barriers.
4. The design criteria applied to isolation barriers and piping.
5. The provisions for detecting needs to isolate remote manual-controlled systems like engineered safety feature systems.
6. The design provisions and technical specifications for testing of isolation barrier operability and leakage rate.
7. The calculation of containment atmosphere released prior to isolation valve closure for lines that provide direct paths to the environs.
8. The containment purging/venting design features minimizing purging time consistent with as low as reasonably achievable (ALARA) principles for occupational exposure.
9. The reliability of the purge system in isolating under accident conditions.
10. The containment isolation and valve indication provisions for station blackout (SBO).

GDC 56 Primary Containment Isolation

"Primary containment isolation: Each line that connects directly to the containment atmosphere and penetrates primary reactor containment shall be provided with containment isolation valves as follows, unless it can be demonstrated that the containment isolation provisions for a specific class of lines, such as instrument lines, are acceptable on some other defined basis:

(1) one locked closed isolation valve inside and one locked closed isolation valve outside containment; or

(2) one automatic isolation valve inside and one locked closed isolation valve outside containment; or

(3) one locked closed isolation valve inside and one automatic isolation valve outside containment. A simple check valve may not be used as the automatic isolation valve outside containment; or

(4) one automatic isolation valve inside and one automatic isolation valve outside containment. A simple check valve may not be used as the automatic isolation valve outside containment.

Isolation valves outside containment shall be located as close to the containment as practical and upon loss of actuating power, automatic isolation valves shall be designed to take the position that provides greater safety."

SRP 6.2.4 uses the following acceptance criteria to address the requirements of GDCs 55 and 56: Isolation valves for lines penetrating (GDC 55) the primary containment boundary, as parts of the reactor coolant pressure boundary, or as direct connections to the containment atmosphere (GDC 56) are as follows:

- one locked-closed isolation valve, one inside and one outside containment; or
- one automatic isolation valve inside and one locked-closed isolation valve outside containment; or
- one locked-closed isolation valve inside and one automatic isolation valve
- outside containment; or
- one automatic isolation valve inside and one outside containment.

GDC 57 Closed Systems Isolation Valves

"Closed system isolation valves: Each line that penetrates primary reactor containment and is neither part of the reactor coolant pressure boundary nor connected directly to the containment atmosphere shall have at least one containment isolation valve which shall be either automatic, or locked closed, or capable of remote manual operation. This valve shall be outside containment and located as close to the containment as practical. A simple check valve may not be used as the automatic isolation valve."

For the requirement that lines penetrating the primary containment boundary is neither part of the reactor coolant pressure boundary nor connected directly to the containment atmosphere have at least one locked-closed, remote-manual, or automatic isolation valve outside containment.

The implementation of these regulatory requirements and provisions stipulated in the guidance documents ensures that the primary containment and the associated systems would function effectively during the normal and postulated accidents and natural phenomena.

REFERENCES

[1] Mazuzan G., and Walker S. (1997): "Controlling the Atom," U.S. Nuclear Regulatory Commission, Washington DC 20555. Last Printing 1997.
[2] Walker S. (2010): "Containing the Atom," U.S. Nuclear Regulatory Commission, Washington DC 20555. January 2010.
[3] NUREG-0800 (2007): "Standard Review Plan," U.S. Nuclear Regulatory Commission, Washington DC 20555, 2007.
[4] ASME B & PV, Section III, Div. 1, Subsection NE (2007): "Class MC (Metal Components)," published by the American Society of Mechanical Engineers, Park Avenue, New York, NY 10016-5990, 2007 Edition.
[5] Regulatory Guide 1.57 (2007): "Design Limits and Load combinations for Metal Primary Reactor Containment System Components," U.S. Nuclear Regulatory Commission, Washington DC, 20555. 2007.
[6] ASME B & PV, Section XI, Div. 1, Subsection IWE (2007): "Requirements for Inservice Inspection of Class MC and Metallic Liners of Class CC components of LWR Plants," published by the American Society of Mechanical Engineers, 3 Park Avenue, New York, NY 10016-5990. 2007.
[7] ASME B & PV, Section XI, Div. 1, Subsection IWL (2007): "Requirements for Inservice Inspection of Class CC Concrete Components of LWR Plants," published by the American Society of Mechanical Engineers, 3 Park Avenue, New York, NY 10016-5990, 2007 Edition.
[8] Ashar H., and Bagchi G. (1999): "Implementation of Maintenance Rule for Structures," Nuclear Engineering and Design, Elsevier Science, New York, NY 10159. 1999.
[9] Regulatory Guide 1.26 (2007): "Quality Group Classifications and Standards for Water, Steam, and Radioactive-Waste-Containing Components of Nuclear Power Plants," U.S. Nuclear Regulatory Commission, Washington DC, 20555.
[10] Regulatory Guide 1.29 (2007): "Seismic Design Classification," U.S. Nuclear Regulatory Commission, Washington DC, 20555, 2007.
[11] ASME NQA-1-2008, "Quality Assurance Requirements for Nuclear Facility Applications," American Society of Mechanical Engineers, 3 Park Avenue, New York, NY 10016. 2008.
[12] ASME NQA-1a-2009 Addenda to ASME NQA-1-2008, "Quality Assurance Requirements for Nuclear Facility Applications," American Society of Mechanical Engineers, 3 Park Avenue New York, NY 10016, 2008–2009.
[13] Regulatory Guide 1.28 (2010): "Quality Assurance Program criteria (Design and Construction)," U.S. Nuclear Regulatory Commission, Washington DC, 20555, 2010.
[14] Technical Information Document (TID)-14844) (1962): "Calculation of Distance Factors for Power and Test Reactor Sites," U.S. Nuclear Regulatory Commission, Washington DC, 20555. 1962.
[15] ASME B & PV, Section III, Division 2, Subsection CC (2007): "Class CC (Concrete Containments)," published by the American Society of Mechanical Engineers, 3 Park Avenue, New York, NY 10016-5990, 2007.

[16] Regulatory Guide 1.136 (2007): "Design Limits, Loading Combinations, Materials, Construction and Testing of Concrete Containments," U.S. Nuclear Regulatory Commission, Washington DC, 20555, March 2007.
[17] ASME B & PV, Section III, Subsection NCA (2007): "General Requirements for Division 1 and Division 2," published by the American Society of Mechanical Engineers, 3 Park Avenue, New York, NY 10016-5990, 2007.
[18] Regulatory Guide 1.60 (2010): "Design Response Spectra for Seismic Design of Nuclear Power Plants," U.S. Nuclear Regulatory Commission, Washington DC, 20555, 2010.
[19] Regulatory Guide 1.61 (2007): "Damping Values for Seismic Design of Nuclear Power Plants," U.S. Nuclear Regulatory Commission, Washington DC, 20555, 2007.
[20] Regulatory Guide 1.76 (2007): "Design-Basis Tornado and Tornado Missiles for Nuclear Power Plants," U.S. Nuclear Regulatory Commission, Washington DC, 20555, 2007.
[21] NUREG-1061, Volume 3 (1984): "Report of the U.S. Nuclear Regulatory Commission Piping Review Committee," U.S. Nuclear Regulatory Commission, Washington DC 20555, 1984.
[22] ANSI/ANS N271 (1976): "Containment Isolation Provisions for Fluid Systems," American Nuclear Society, La Grange Park, IL, 1976.
[23] Regulatory Guide 1.141 (2010): "Containment Isolation Provisions for Fluid Systems," U.S. Nuclear Regulatory Commission, Washington DC, 20555. 2010.

CHAPTER

3

DESIGN, CONSTRUCTION, INSPECTION AND TESTING OF CONTAINMENT STRUCTURES

Javeed Munshi, Shen Wang and Abdul Sheikh

This chapter provides a summary of analysis and design and pre-operational inspection and testing requirements for concrete and steel containments. Both conventionally reinforced as well as prestressed concrete containments are included. Containments are intended to provide a leak-tight boundary against accident pressure loading which essentially dictates the generally used cylindrical shape. The size of the containment and available free volume depends upon equipment layout and operational requirements. The available free volume within the containment affects the magnitude of pressure that can be generated during an accident. In general, conventionally reinforced concrete and steel containments have been used for small containments with low to moderate internal design pressure. But large containments with relatively high internal design pressure require either a prestressed concrete or a steel containment. In general, selection of containment type (conventional concrete, prestressed concrete or steel) depends upon considerations for the equipment size and layout, design pressure and other loadings, construction cost and schedule.

General regulatory recommendations for containment design are given in Standard Review Plan SRP 3.8, Regulatory Guide 1.136, and in RG 1.57. Other relevant SRP Sections include Section 3.3 (Wind Loading), Section 3.4 (Flood Protection), Section 3.5 (Missile Impact and Aircraft Hazards) and Section 3.7 (Seismic Design). Other Regulatory Guides of interest are RG 1.29: Seismic Design Classification, RG 1.60: Design Response Spectra for Seismic Design of Nuclear Power Plants, RG 1.61: Damping Values for Seismic Design of Nuclear Power Plants and RG 1.76: Design Basis Tornado and Tornado Missiles for Nuclear Power Plants, and RG 1.142 Safety Related Concrete Structures other than containment structures.

Keywords: Concrete containment, steel containment, prestressed containment, ASME Code, Finite element modeling, design of containment, impact design, aircraft impact, Inspection, testing

3.1 CONCRETE CONTAINMENTS

3.1.1 Introduction

Design of concrete containments in the United States is now carried out under ASME Section III, Div 2 Code [1] for Concrete Containments—Joint ACI/ASME Technical Committee on Concrete Pressure Components for Nuclear Service under the sponsorship of the American Concrete Institute and the American Society of Mechanical Engineers. The two Societies have agreed that requirements for concrete containments will be published as Section III, Division 2, of the ASME Boiler and Pressure Vessel Code. The older containments were designed in accordance with NRC SRP section 3.8.1 and the American Concrete Institute (ACI) Building Code Requirements for Structural Concrete (318). Some licensees have incorporated some of the requirements of the ASME III-Div. 2 Code, as part of reconstitution efforts during license renewal. The following paragraphs provide a summary of various section of the current Section III, Div 2 code:

Section CC 2000 provides material requirements for concrete constituents, mix design, grout, reinforcement, prestressing steel, liner plate, liner anchor and welding. This section also covers the required certifications, identification, testing, performance requirements and examinations and supplier quality requirements.

Section CC 3000 covers design of reinforced and prestressed concrete containments for various load conditions and allowables. This Section also includes design requirements for liner plate and liner anchors, fatigue and impulse and impact loads.

Section CC 4000 provides requirements for fabrication and construction of concrete materials, components and systems for concrete containments. This section also includes required certifications and traceability.

Section CC 5000 provides requirements for construction testing and examination for materials and components used in concrete containments. This section also provides requirements for procedures for testing and qualification, certification, and evaluation requirements.

Section CC 6000 provides the requirements for structural integrity testing, Section 7000 provides requirements for overpressure protection and Section 8000 provides requirements for nameplates, stamps, and reports.

3.1.2 Conventionally Reinforced Concrete Containments

Conventionally reinforced containments are generally cylindrical structures with a hemispherical dome supported on a flat slab [2]. Conventional bonded reinforcement is provided in the structure to carry all applicable load actions including tension, shear, and flexure. The design and detailing provisions are given in ASME Section III, Div 2 Code and generally follow the basic principles of reinforced concrete design given in older editions ACI 318. The ASME Code is currently being updated to the latest editions of ACI 318 [3]. Special detailing is provided around boundaries, discontinuities, and openings. The concrete cylindrical shell is usually between 42 and 52 in. thick, and the dome is 30 to 36 in. thick. The base slab is usually 8 to 10 ft thick. The concrete sections are, generally, heavily reinforced with up to 8 layers of No. 11 to No. 18 rebars with a reinforcement ratio of 2% to 4%. The inside surface of the concrete is lined with carbon steel liner plate. The liner plate in the base slab is covered with an additional 2 foot thick concrete slab.

3.1.3 Prestressed Concrete Containments

Prestressed concrete containments are cylindrical structures with a hemispherical or sphere torus dome and a flat basemat slab with a tendon gallery running under the cylindrical wall [2]. As indicated above, prestressing is required when containment pressure requirements become large, and the consequent design loads cannot be handled by conventional reinforced concrete. Early prestressed containments in the United States, such as, Palisades and Turkey Point were designed for 1.5 times the accident design pressure. They consisted of a cylinder, shallow dome with a ring girder and six vertical buttresses. The ring girder is provided to accommodate vertical and dome prestressing tendons end anchorages. The buttresses are used to accommodate hoop prestressing tendons end anchorages. The prestress level was subsequently reduced to 1.2 Pa with three buttresses and about 1000 ton capacity tendons. The shallow dome was replaced with a hemisphere. The vertical inverted U-tendons were used which start in the tendon gallery below the base slab ran over the dome and anchored at the tendon gallery on the opposite side. This eliminates the need for the ring girder and number of tendons is significantly reduced. However, loss of prestress in the longer inverted U-tendons due to friction and elastic shortening is higher than those in the earlier containments. Use of inverted U-tendons also eliminates potential corrosion of prestressing anchors due to limited exposure to the environment. The tendons comprised of wire with buttonheads or strands. But due to high loss of prestress due to relaxation of steel, especially at higher temperatures, more recent plants use low relaxation prestressing wires or strands. Except for a few earlier plants (i.e., Robinson and TMI-2), all US nuclear power plant containments use unbounded post-tensioning system. Unbonded post-tensioning allows periodic surveillance and inspection of the tendons over the life of the plant for corrosion and degradation. References [2] and [4–8] give a detailed discussion on various aspects of prestressed containment design and testing.

Because of the level of prestressing, the concrete strength used in prestressed containments generally range from 5000 psi to 6000 psi and should exhibit low creep properties. Creep and shrinkage of concrete and relaxation of prestressing steel are important aspects of prestressed concrete design as they affect the magnitude of prestress loss in the long-term.

3.1.4 Modeling and Analysis Considerations

SRP 3.7.2 [9] provides general outline for the required analysis approach. In general, elastic structural finite element analysis is used for containment analysis. The analysis should include cracking that is likely to soften the structure and change its dynamic characteristics and damping. In case of post-tensioned containments, it is possible that cracking may not be significant due to level of prestress present during, for example, a safe shutdown earthquake. Since the analysis also needs to include the soil structure interaction (SSI) effects, a two-step analysis approach is generally used one for static loading (non seismic loads) using commercial software programs like ANSYS or GTDTRUDL and other for a separate analysis involving SSI using, for example, using SASSI program. The static model may include soil springs as approximate and simplified assumptions for soil boundary effects. The SSI model, on the other hand, includes detailed soil impedence and embedment effects. In order to minimize differences between the two models, a SASSI model is usually developed from the static ANSYS or GTSTRUDL model. To ensure that SASSI model represents the static model without the soil boundary effects, a correlation study assuming fixed base boundary conditions for both models is carried out using the two programs (for example, ANSYS and SASSI) to ensure that two models have the same global static and dynamic characteristics. The SASSI analysis is carried out for a suite of site specific ground motions representing various soil conditions that may be present. The results of the SASSI seismic analyses are then combined with analysis of the structure for other non seismic loading using appropriate load combinations given in ASME Code (Table CC 3230-1).

Note that since the current approach involves discrete and detailed finite element modeling of the containment and its internal structures, the analysis results produce a significant amount of element output results that need to processed for appropriate combinations and subsequent design. Therefore, a robust post-processing approach is required to handle the data in order to automate the process and eliminate human error. As a result, such post processer becomes part of the analysis and design process and needs to be independently verified and appropriately documented. The reduction of thermal forces is allowed due to cracking. The thick basemat is modeled through several layers of brick or solid elements while the containment may be modeled using shell elements. The containment shell elements need to be appropriately connected to solid basemat elements at the interface. In order to get a more accurate picture of force resultants around discontinuities, boundaries and openings and major penetrations (such as equipment hatch), finer mesh submodels of these regions may be used. A mesh sensitivity study is carried out to ensure accuracy of the models before detailed analysis is carried out. See more discussion in References [10–11].

For prestressed containments, the post-tensioning effect is modeled either directly by discrete modeling of the tendons or indirectly through temperature or pressure loading. In case of discrete modeling, the tendon slip/movement may also be captured for an unbounded tendon.

Note also that in prestressed concrete containments, there is a beneficial effect of pressurization that can be accounted for in design. As the containment is pressurized under accident pressure load, it tends to grow in volume. This tendency to grow reduces the effect of precompression in concrete. But at the same time, it tends to stretch the tendons and thus cause a slight increase in the prestressing force.

3.1.5 Concrete Containment Design Criteria

The loads and load combinations are given in CC-3230 of ASME Section III, Div 2 for concrete containments. The design philosophy for concrete containments is based on ensuring that concrete under any set of load conditions does not reach its nominal or crushing capacity and there is no significant yielding of steel to result in any general yield condition. To accomplish this, appropriate safety margins against these conditions are built into the Code provisions in terms of allowable stresses for various service and factored load conditions. The material allowable stresses are given in the Code for both service and factored loads and provide a healthy margin of safety in design of concrete containments. Chapter 1 provides some additional details on concrete containments.

For service loads, the load factors are unity since the design allowables are well below concrete compressive strength and steel yield strength. Therefore, the structural response under service load conditions is kept well within the material capacities with adequate safety margins. For factored loads, the load factors may be greater than one depending on the probability of load occurrence, confidence in load magnitude and the safety margins required for a guaranteed performance.

The containment is designed for all of the applicable load combinations with SSE applied simultaneously in three orthogonal directions generally combined using the square root sum of the squares (SRSS) method or 100:40:40 rule as recommended in NRC Regulatory Guide RG 1.92. In most cases, the most critical load combination for concrete containments results due to the combined effect of the design accident pressure and SSE at the base of the containment shell in the vertical direction or 1.5 times the accident pressure (Pa) load combination in the hoop direction at mid height of the containment shell.

The allowable stresses discussed above are given in ASME Section III, Div 2 for various load effects such as primary and secondary or a combination thereof. This Code categorized membrane forces, moments and shears as either primary or secondary depending upon their location in the containment and the source that the forces come from or origin. Primary mode behavior is the ability of internal forces to equilibrate applied loads. Secondary resistance is exhibited when internal forces are not required to balance external forces, or the external loads are self-limited. For example, membrane forces, moments, and shears are all considered primary for externally applied loads at critical locations of the containment. On the other membrane, forces and moments from volume change effects are considered secondary.

Secondary force may be due to (a) a local, internal, force, or moment that results from applied loads, but is not required to equilibrate such loads; or (b) a local, internal force, or moment that results from nonload, volume change effects, such as shrinkage strain and thermal strain.

Analyses is carried out in stages for primary membrane, primary membrane plus bending, primary plus secondary membrane and primary plus secondary membrane and bending of critical containment sections. For each of these analysis steps there is an allowable concrete and reinforcement stress/strain that needs to be satisfied as given in Tables CC 3421-1 and CC 3431-1 of ASME Section III, Div 2 code.

The Code requires an allowable stress design approach for both service as well as factored loads. The intent is to ensure that concrete and steel both remain within their respective capacity limits for all load situations in order to safeguard the containment integrity. Although allowable stress for reinforcement for load resisting purpose should not exceed 0.9fy, some limited and local yielding (or partial yield) is allowed in some situations, for example, where multiple layers of reinforcement are provided for bending, in diagonal reinforcement and in reinforcement adjacent to large openings. These exceptions do not include design for shear in which case reinforcement should always remain below yield.

There is no guidance in the Code [1] on how to treat thermal effects. ACI 349-06 [12] may be used for guidance; however the requirements of ASME Section III, Division 2 must be satisfied.

Gurfinkel [13] gives detailed discussion on thermal effects in reinforced concrete containments.

Note that because of lack of data, the development lengths in areas of bi-axial tension are increased by 25%. As a result of this requirement, couplers or Cadwelding is generally used to avoid use of excessively long splices.

3.1.6 Design Considerations for Prestressed Containments

The level of prestress is generally designed to provide sufficient balancing force to resist all of the accident design pressure load and in some cases also some or all of the tangential shear loads due to SSE. Thus, depending upon the level of prestress, the following three design situations may arise in design of post-tensioned containments:

1. A containment with enough prestressing to cover both design pressure and tangential shear so no conventional reinforcement is needed for these actions.
2. A containment which relies on conventional reinforcement to carry some of the tangential shear while some of the shear is carried by concrete in compression as a result of prestressing.
3. A containment where all tangential shear is carried by conventional reinforcement and concrete does not carry any shear.

In cases 1 and 2, concrete is expected to carry shear which may be affected by cracking as a result of accident thermal forces. In order to reduce effects of thermal cracking, higher level of prestress may be required which in turn will result in thicker walls and thus induce higher earthquake forces. Because of this thermal cracking issue, this type of design is both complicated and uneconomical. In Case 3, the design is simple as it requires designing prestress for accident pressure and provide conventional reinforcement for seismic loads. Under thermal loads, concrete will crack and thus relieve some of the thermally induced forces. Since concrete is considered cracked, it cannot carry any shear or tension for which conventional reinforcement is provided.

The tangential shear reinforcement is generally provided to resist lateral loads due to seismic or wind effects in two orthogonal directions. If this orthogonal reinforcement is not sufficient, additional layers inclined at 45° may be required for reinforced concrete containments. For detailed discussion on design for tangential shear effects see Oesterle [14].

Prestress losses need to be conservatively computed in order to ensure that sufficient prestress force remains available to counter accident pressure loads during the life of the containment [6]. A detailed discussion of prestressing losses is provided in RG 1. 35.1 [15] and in Appendix 5A of this book. Appropriate design and reinforcement detailing is required in the tendon anchorage region to make sure that there is no excessive bursting cracking that could jeopardize the design prestress level in the tendon. Adequate corrosion protection is also required to ensure long-term durability of the prestressing system. For more discussion on tendon anchorage design and testing see References [16–17] (Fig. 3.1).

Prestressed containments also require careful consideration of radial tension reinforcement because of several instances of delaminations [18]. It is recommended that prestressing tendons should not be placed near a free edge. In addition, radial tension reinforcement be provided if detailed analysis is not carried out to confirm that delamination is not likely to occur.

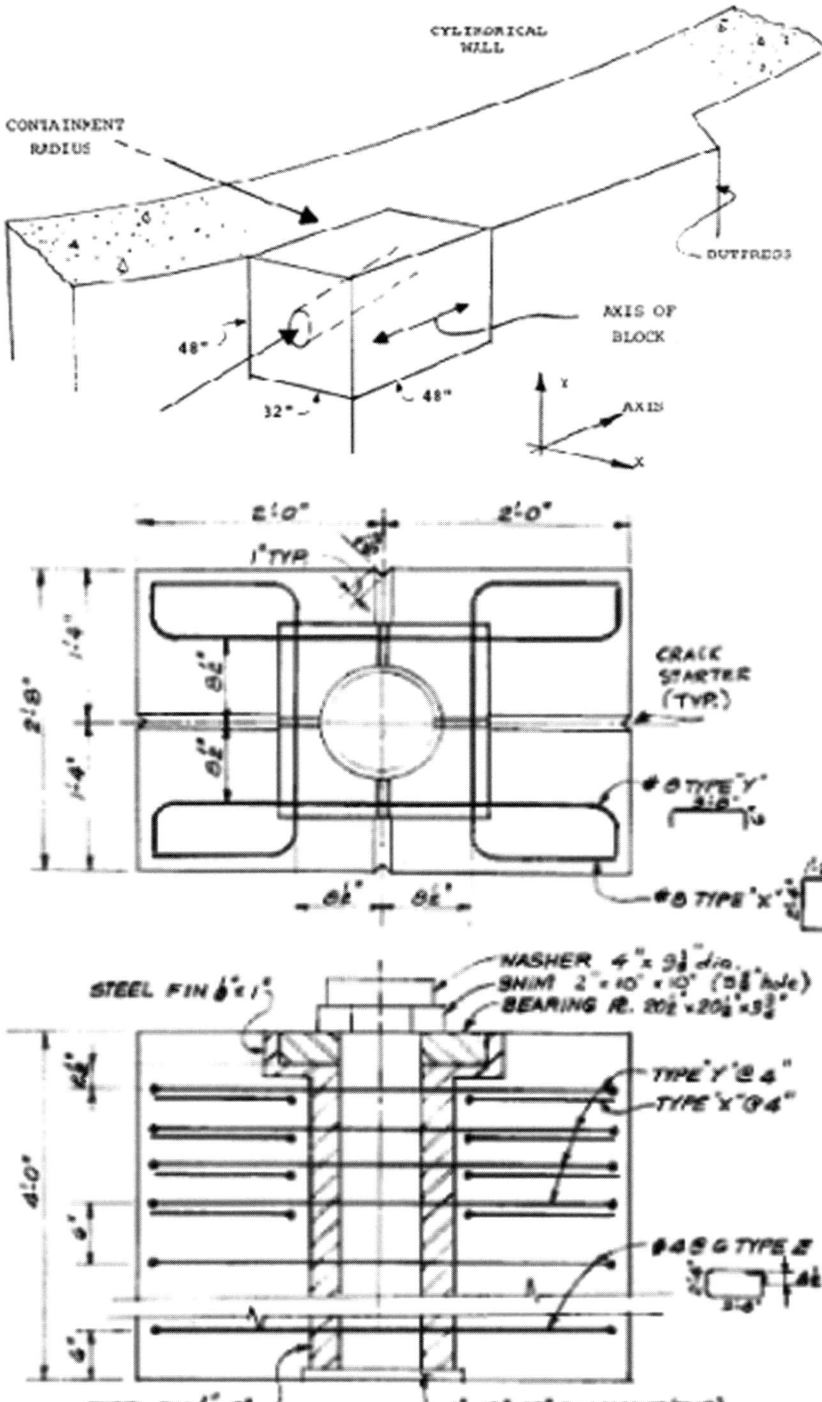

FIG. 3.1
PRESTRESSED ANCHORAGE BLOCK TESTING AT A BUTTRESS [16]

3.1.7 Liner and Liner Anchor Design

A liner plate, generally 6 mm (1/4 in.) to 12 mm (1/2 in.) thick, is attached to inside face of concrete containment is for leak protection. The liner is erected first during construction process and serves as a form for concrete placement. The liner is braced with structural steel sections (channels and angles at regular spacings) or welded studs which serve as embeds in the concrete. Although the Code does not allow the liner to be considered as a stiffness or a strength element contributing to capacity, it should be ductile enough to accommodate the stresses/strains due to various imposed load conditions including accident thermal without jeopardizing the leak tightness. Under accident thermal and pressure loads, it is important to include the liner — concrete interaction to adequately determine stresses/strains in concrete and liner. The liner can, however, be used as a structural element to resist construction loads during concrete placement and small attachment loads from cable tray, conduit support, and ladders. For large attachments, equipments and personal hatches, and piping penetrations, the liner plate is thickened locally to transfer loads to the concrete. References [19–22] provide a comprehensive background on liner behavior, design, detailing, and testing (Fig. 3.2).

The limiting condition for liner is the strain and its impact on anchorage in concrete. There are no strength or buckling stability issues with liner as the liner is restrained by the concrete and anchors. However, buckling needs to be considered for temporary loads during construction and installation prior to concrete placement. So as long as the liner is able to perform its leak-tight boundary function, there is no limit on its strength or stability. The only issue relates to the effect of imposed deformations of the liner on anchors. The integrity of anchorage system must be maintained during the design. The liner has to be a ductile element so that it can exhibit good deformation capacity beyond yield. Consequently, full penetration butt welds are necessary to ensure leak tightness and adequate ductility of the liner plate.

The liner anchors are generally tested for the imposed ultimate deformation from the liner plate. The liner may assume an initial inward curvature during thermal accident load which is acceptable as there is no structural consequence of this deformation or resulting buckling. The force induced in the liner depends upon the liner strength and strain hardening. Therefore, liner strength is kept to a minimum with well-known yield and post-yield behavior, if possible. Since it is practically difficult to get a well-defined yield for the plate material, higher anchor loads are induced due to higher actual yield for which high strength ductile anchors need to be used. Note that local crushing around the anchors is considered acceptable during a rare accident condition. Anchors also need to be designed for mechanical loads that may come from small attachments to the liner plate during construction or operation.

3.1.8 Pre-Service Inspection and Testing (Concrete)

The pre-service (CC-5000) and in-service inspection and testing requirements for concrete containments (both reinforced and unbonded post-tensioned) are given in Subsection IWL-1000 of ASME Section XI of the Code. This section does not include requirements for bonded (grouted) tendons. This section should be used together with Subsection IWA for General Requirements to ensure structural and leak-tightness of the containment. Subsection IWE applies to containment liner (see discussion on steel containments in Section 3.2.7). USNRC RG 1.90 provides guidance for in-service inspection of prestressed concrete containments with grouted tendons (Fig. 3.3).

The pre-service inspection and testing is used to establish a successful start and to serve as a baseline for inspection and testing during operational service of the containment. The pre-service examination includes 100% of the concrete surfaces of the containment and post-tensioning system based on plant construction record. Pressure integrated leak rate testing (ILRT) or local leak rate testing (LLRT) is carried out per Appendix J of 10CFR Part 50 which provides containment leakage testing in terms of Type A, B, and C tests. The Type A test is a measurement of the overall integrated leakage rate of the primary containment; whereas Type B and C tests are local leak rate tests designed to detect and measure local leakage across each pressure-containing or leakage-limiting boundary for primary containment. See Chapter 4 of this Book for detailed discussion of ISI and periodic leak rate testing of containments.

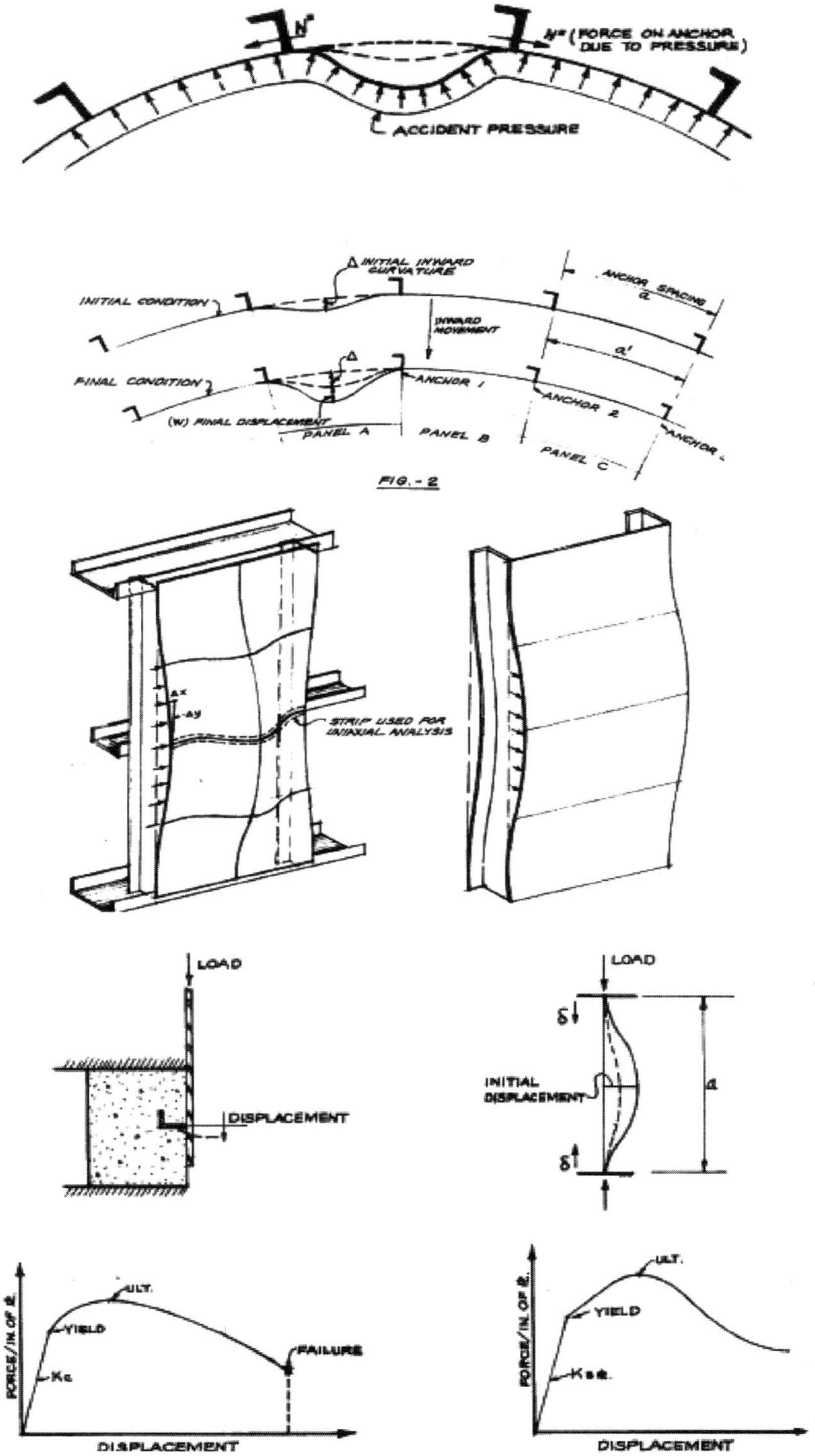

FIG. 3.2
LINER AND ANCHOR BEHAVIOR AND TEST SIMULATION [19–20]

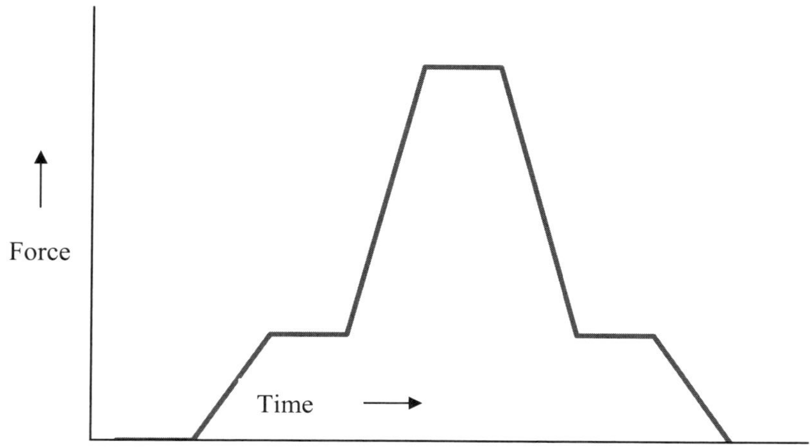

FIG. 3.3
TYPICAL RIERA LOADING CURVE FOR AIRCRAFT IMPACT [58]

A professionally registered Responsible Engineer is assigned the responsibility of evaluating the concrete and post-tensioning system to ensure the structural and leak-tightness of the concrete containment.

The visual examination should include inspection of concrete surfaces for any apparent distress/cracks/voids, detailed condition of concrete around anchorages and detailed inspection of strands and anchorage hardware. ACI 349.3R provides detailed guidance and quantitative inspection criteria for concrete surface examination.

The structural integrity test (SIT) requirements given in Section CC-6000 of ASME Section III, Div 2 Code [1] are a pre-requisite for Code acceptance and stamping before start of operation. The test is performed at 1.15 times the containment design pressure to evaluate design compliance and quality of construction. A detailed test procedure is required that uses calibrated instrumentation to measure strains/displacements, crack widths, and ambient conditions. Cracks exceeding 0.01 inches in width are recorded before the test, at peak pressure and after depressurization. These and other representative measurements are compared against the acceptance criteria given in CC-6400.

3.1.9 Severe Accident Analysis

Containment performance at beyond design basis accident internal pressure and temperature is required as an input for determining the offsite consequences and accident progression of the containment during a severe accident. Extensive research and scale model testing of reinforced and prestressed concrete containments to determine behavior at beyond design basis accident pressure has been performed in the last 25 years at Sandia National Laboratories (SNL) [23] and Central Electricity Generating Board (CEGB), England [24]. Concrete containments start to leak before a complete rupture or failure. It is extremely difficult to accurately predict the location and leakage rate of the concrete containment due to beyond design basis internal pressure and temperature. Hessheimer and Dameron [23], and Dameron, Rashid, and Tang [25] provide guidance for predicting leak area and leak rate in containments. Hessheimer and Dameron [23] recommend a non-linear finite element analysis of the concrete containment to predict containment performance and leakage.

Hessheimer and Dameron [23] have concluded that global, free field strain of 1.5% to 2.0% for reinforced and 0.5% to 1.0% for prestressed concrete can be achieved before failure or rupture. In addition, leakage in concrete containment increases appreciably after the rebars and liner plate yield. Furthermore, under gradual increase in internal pressure, containment leakage continues to grow without failure and rupture. Sheikh

[26] provides the following simplified approach for predicting containment performance during a severe accident. This approach has been used in the NRC's State-of-the-Art Consequence Analyses Project [27]

$$P_{fail} = (A_{hoop} * Y_{rebar@2\%} + A_{liner} * Y_{liner@2\%}) / R$$
$$P_{yield} = (A_{hoop} * Y_{rebar} + A_{liner} * Y_{liner}) / R$$

where:

P_{fail} = Containment failure pressure (containment leakage greater than 100%)
P_{yield} = Containment pressure at which hoop rebars and liner plate yield
A_{hoop} = Area of the hoop rebars
A_{liner} = Area of the liner plate
Y_{rebar} = Yield stress of the rebar
Y_{liner} = Yield stress of the liner plate
$Y_{rebar@2\%}$ = Stress in the rebar at 2% strain
$Y_{liner@2\%}$ = Stress in the rebar at 2% strain
R = Radius of the containment

The results using this simple approach are quite consistent with detailed finite element analyses using state of the art computer codes and test data [26].

3.2 STEEL CONTAINMENTS

3.2.1 Introduction

The steel containment vessel is an integral part of the containment system, which serves to limit releases in the event of an accident and also to provide the ultimate heat sink. Current steel containment vessels in United States generally fall into three categories [2]: 1) cylindrical steel containments, 2) spherical steel containments, and 3) BWR Mark I/II containments which typically have a drywell with the shape of an inverted light bulb or a cone, connected to a wetwell with the shape of a torus or a cylinder. Vertical or circumferential stiffeners may be attached to the exterior of the vessel. The steel containment is generally protected by a 24 in to 36 in reinforced concrete cylindrical shield building in order to provide shielding against release of radioactive material and for external missile protection. Chapter 1 provides some additional details on steel containments.

Cylindrical steel containment vessels are common due to relative ease of fabrication and erection. Most cylindrical steel containments have a hemispherical, torispherical, or ellipsoidal dome made of steel. The lower section of such containment could either be a hemispherical, torispherical, or ellipsoidal steel head similar to its dome, or a hybrid steel cylinder with concrete basemat with a leak tight steel liner.. The main advantage of a spherical containment is that it requires only about one-half the wall thickness of a cylindrical vessel for the same internal pressure loading. However, fabrication and erection of a spherical containment vessel is relatively costly and time consuming, compared to a single-curved cylindrical containment. The thickness of cylindrical steel containment generally varies between 2.5 and 3.7 cm (1 to 1.5 in.) which may be thickened around the transition regions at the top and bottom head and around openings and penetrations to account for stress concentrations.

Penetrations through the steel containment shell are provided for piping, instrumentation, control lines and electrical power. Large diameter openings such as equipment hatches and personnel access should be evaluated for buckling, which will be further discussed in Section 3.2.4. The reactor building polar crane can either be supported by the locally thickened containment vessel, or by an interior structure.

Controlling loads for the design of a steel containment vary depending upon its reactor type (such as BWR vs. PWR) and the seismic region where the nuclear plant is located. But because of their relatively low mass, steel containments induce less seismic loads as compared to concrete containments. However, localized pressure transient loads and buckling failure modes are much more critical in the design of a steel containment than in the case of a concrete containment, which has significantly higher built in structural stiffness.

Design of steel containments in the United States is carried out under ASME BPV Code Section III, Div 1 Subsection NE — Class MC Components [28] as supplemented by RG 1.57 [30]. For the aforementioned hybrid steel cylinder and concrete basemat, the design provisions for the lower (concrete) portion of the containment is governed by ASME BPV Code Section III, Div 2 [1] for concrete containments.

3.2.2 Modeling and Analysis Considerations

NUREG-0800 SRP 3.7.2 provides general outline for the required analysis approach. In general, elastic structural finite element analysis is used for containment analysis. Global models are developed as main tools to perform analysis as needed. Since the analysis also needs to include the soil structure interaction (SSI) effects, it generally requires two separate global analyses, one for static loading (non seismic loads) using commercial software programs like ANSYS and a separate analysis involving SSI using, for example, using SASSI program [31–33]. In order to minimize differences between the two global models, a SSI model is usually developed based on the static model and correlation study for fixed base boundary conditions is performed to ensure that two models have the same global and dynamic characteristics. The SSI analysis is then conducted for a suite of site specific ground motions representing various soil conditions that may be present. The results of the SSI seismic analyses are then combined with analysis of the structure for other non-seismic loading using appropriate load combinations given in design codes.

Nonlinear finite element model may be developed using software program such as BOSOR-5 [34], for example, to perform buckling analysis. The nonlinear buckling analysis has been further discussed in Section 3.2.4.

A lumped-mass stick model may be used as a valuable tool in the early stages of the seismic analysis. But with the advent of computer technology and finite element methods, three-dimensional (3-D) model utilizing shell or solid elements are now generally carried out to meet the current expectations. Due to the homogeneous material characteristic of steel containments, shell elements are generally adequate for the analysis and design purpose. If solid elements are used, nonlinear through thickness stress may result. In this case, such stresses may be linearized in order to be compared with the code allowables. The concrete basemat which provide anchorage for the steel containment is modeled through several layers of brick or solid elements while the containment may be modeled using shell elements. The containment shell elements need to be appropriately connected to solid basemat elements at the interface.

Besides the global model, local models with geometry details and refined mesh may be developed separately in order to evaluate effects of discontinuities or large penetrations such as an equipment hatch or a personnel access. Boundaries of a local model need to be established at adequate distance from the subject "geometry disturbs" so that disturbing effects on the boundary are negligible. The results from the analysis using the global model are applied as the boundary conditions for the local model.

The detailed finite element analysis results produce a significant amount of element output results that need to be processed for appropriate combinations and subsequent design. Therefore, a robust post-processing approach is required to handle the data in order to automate the process and eliminate human error. But such post processer becomes part of the analysis and design process and needs to be independently verified and appropriately documented.

3.2.3 Steel Containment Design Criteria

The material requirements for steel containment vessels are specified in ASME BPV III Code NE-2000 and RG 1.57. Table NE-2121(a)-1 lists the materials that are allowed for steel containment vessel construction. Additional requirements regarding certification and marking of materials and certification of materials suppliers are provided in NCA-3800 [35].

The design loadings for steel containment vessel are specified in NE-3110 of Section III of ASME B & PV Code. In particular, each service loading to which the containment may be subjected shall be categorized as one of service levels A, B, C, and D. These four service levels are to represent the four component conditions "Normal," "Upset," "Emergency," and "Faulted" which were used in earlier versions of ASME BPV III Code. These service levels are established to permit different levels of allowable stresses. In general, the less likely a given condition, the higher the permitted allowable stresses. It remains the responsibility of the owner to determine corresponding service level for each postulated loading condition. If the containment vessel is expected to operate "normally" under a given set of conditions, then service level A is specified for those conditions. On the other hand, if a given set of loads is considered to be an unlikely event for the containment, a higher service level is specified. For example, the internal pressures and temperatures associated with the design basis accident must be accommodated with service level A allowable stress intensities, while service level C allowable are permitted for these pressures acting in combination with the safe shutdown earthquake (SSE). Load combinations for each of the ASME NE service levels are given in RG 1.57 [30].

Rules for "design by analysis" are specified in NE-3200. The theory of failure used in NE-3200 is the maximum shear stress theory, in which the term "stress intensity" is defined as twice that of the maximum shear stress, and equals the difference between the algebraically largest and smallest principal stresses at a given point [34]. The basic ASME BPV Code, Section III criteria are expressed in terms of allowable stress intensities. The code criteria are further complicated by the fact that the ASME BPV III defines different categories of stresses. These are denoted as [35]:

1) General primary membrane stress
2) Local primary membrane stress
3) Primary bending stress
4) Secondary stress
5) Peak stress

As implied by the name, a primary stress (categories 1, 2, and 3) is one which results from an imposed loading and one which is required to satisfy the law of equilibrium. In particular, a general membrane stress is defined as a global stress which extends in such a large region that an effective redistribution is not possible if the material yields. On the other hand, a local primary membrane stress is the one which results at major structural discontinuities such as the boundary between a cylinder and the dome. Such localized stresses are able to be redistributed to the adjacent areas when material yields. A primary bending stress is the through thickness bending caused by external loads. Primary bending stresses may appear in such geometries as flat heads and stiffeners, but do not exist in axisymmetric thin shell structures subject to axisymmetric loadings such as internal pressure. A secondary stress is generally extreme fiber stress developed by self-constraint of the structure, thus it is self-limiting. That is, the conditions that cause the stress will be satisfied when the material yields or distorts locally, and one application of the load will not result in failure. A peak stress is the increment of stress addition to the primary plus secondary stresses, due to local discontinuities or local thermal, including stress concentration. The peak stress is mainly for the purpose of fatigue analysis, and no limit is specified on it [35]. Some relaxation of allowable stress limits is allowed if a plastic analysis is performed. Rules of such relaxation related to plastic analysis are presented in NE-3228.

Details of deriving aforementioned stress categories and detailed stress limits for various service levels are presented in ASME Code Subsection NE and the Chapter 9 of Companion Guide to the ASME Boiler & Pressure Vessel Code [35]. The stresses calculated using a 3-D finite element model mentioned in Section 3.2.2 need to be categorized before compared with allowable stresses. Table NE-3217-1 in ASME Code is provided to assist the designer for this purpose.

Besides the option of "design by analysis," subsection NE also specifies rules for "design by formula" in NE-3300. However, application of this option in steel containment design is limited only to service levels A and B without substantial mechanical or thermal load. For service levels C and D, and also for service levels A and B with significant mechanical or thermal load, the design by analysis option must be used.

3.2.4 Buckling Analysis

Unlike cases of concrete containments, shell buckling is a significant design consideration for steel containment vessels. There are generally three methods allowed to calculate the critical buckling stresses [2, 35]:

1) To perform a rigorous nonlinear analysis, considering geometric imperfection and large deformation.
2) To carry out a classic linear analysis, then reduce margins of its results by applying a "knockdown factor." The "knockdown factors" are used to reflect the difference in buckling stress of a perfect shell and an actual shell, due to imperfection.
3) To test the physical model under the same conditions which the containment is expected to be subjected.

The third option is not only expensive, but is also likely to raise questions regarding the size and details of the model. The first option is a comprehensive analytical method to determine the buckling loads for steel containment vessels. However, it could also be expensive and time consuming, and the analyst needs in-depth knowledge and experience in the subject matter. With the advance of computer technology, this option appears to be more attractive especially as a supplement to the second option. The second option is the easiest and by far the most widely adopted method for the buckling evaluation of steel containment vessels. Global or local finite element models mentioned in Section 3.2.2 could be used to carry out the classic linear analysis. Then the knockdown factors are applied to results in order to establish the critical buckling allowable for the containment vessel assuming its imperfection is within the tolerances of the Code. The knockdown factors are generally obtained from tests of specimens, which are fabricated to tolerance allowed by Code. Interaction equations are developed to account for biaxial stress conditions and for multiple loads.

Most steel containments in United States are designed using the second option, either following ASME BPV III, NE-3133, or following alternative rules of ASME Code Case N-284. This Code Case was developed in order to supplement NE-3133 which is recognized to be insufficient in terms of covering geometries and loadings varieties for steel containment design. The methodology of the Code Case N-284 has been extensively used not only on containment vessels, but also on other components such as storage tanks. Due to the more rigorous approach of N-284 compared to NE-3133, lower design margins are recommended. Acknowledging that the knockdown factors are based on the lower bound of the test data, N-284 introduces some additional safety margin into the rules. Use of Code Case N-284 in lieu of NE-3133, especially when these design margins are used, will lead to a relatively economical design. It has been argued by some that the use of aforementioned design margins is considered somewhat controversial and has not been universally accepted [35]. However, in RG 1.57 issued in 2007, Code Case N-284-2 was allowed to be considered for complex geometries and loading conditions which is not covered by NE-3133. A flow chart is provided at the end of N-284 as a guideline for the design process.

As additional data sources, several test programs have been conducted since the Code Case N-284 was written, and considerable information with respect to the knockdown factor for containment shell and its components are available in the literature [7, 36–39].

3.2.5 Severe Accident Evaluation

Previous sections cover the analysis and design of steel containment vessels to withstand the design bases accident in combination with other design loadings, such as seismic. However, there is also an interest in the ultimate capability of the containment vessel to resist internal pressure at certain specified temperature, which is originated from postulated severe accident scenarios [40]. This is often referred as the probabilistic risk assessment analysis or the severe accident evaluation. The design codes utilized in such type of analysis are the same as used for general design of steel containment, i.e., ASME BPV III Subsection NE [29] and ASME Code Case N-284 [41–42].

A severe accident scenario is considered as an unlikely event for the containment vessel thus it is logical to only consider service levels C or D. For example, the NRC document SECY-93-087 [43], which addresses the design of revolutionary Advanced Light-Water Reactor (ALWR), stipulates that as a deterministic containment performance goal, the containment should maintain a reliable leak-tight barrier approximately 24 hours following the onset of core damage under severe accident challenges, by ensuring that ASME service level C limits are not exceeded.

For a steel containment design process described in Sections 3.2.3 and 3.2.4, design basis load combinations are applied to the containment and responses of the containment are evaluated against code allowable, in terms of stress intensities (NE 3200) and buckling safety factor (NE 3133 or Code case N-284). For a severe accident evaluation, usually only dead load and internal pressures due to accident need to be considered, and ultimate capacities of the steel containment are presented in terms of critical allowable internal pressures at certain specified temperatures. Elements of the containment, such as cylindrical shell, top and bottom heads, equipment hatches and personnel accesses, mechanical and electrical penetrations, are evaluated individually. The maximum allowable internal pressure of each element is determined by calculating the pressure at which criteria in terms of stress intensities or buckling are reached. Thereafter, the critical internal pressure of the steel containment is determined as the smallest allowable internal pressure of all elements. Finite element models specified in Section 3.2.2 and methodologies described in Sections 3.2.3 and 3.2.4 are utilized for above evaluations. For small mechanical and electrical penetrations, additional test data of allowable pressure are available in the literatures [44–47]. Additional information regarding the steel containment tests under severe accident loading is provided in Chapter 6 of this Book.

3.2.6 Fabrication and Installation

The fabrication and installation requirements of steel containment vessels are specified in ASME BPV III Code, Article NE-4000, which covers certification of materials and fabrication, forming welding, heat treatments, tests, repairs, and examinations.

Material traceability is specified in detail in this sub section. In particular, identification markings should be carried by all pressure-retaining materials and remain distinguishable until assembly. Forming processes generally should be qualified by conducting a procedure test using specimens with the same material, specification, grade or class. The tolerances for forming cylindrical, conical, or spherical shells are presented in NE-4221. If a parameter (dimension, size, etc) is not specified with tolerance, the value should be considered nominal and the designer can determine allowable tolerance using engineering judgment and standard practice [35].

NE-4240 specifies fabrication requirements for weld joints in terms of categories A, B, C, and D. Welding qualifications are covered in NE-4300.

3.2.7 Pre-Service Inspection and Testing

The pre-service and in-service inspection and testing requirements for steel containments are given in Subsection IWE of ASME Section XI of the Code. Subsection IWE applies to Class MC metal containments and their attachments. It also applies to metallic shell and penetration liners of Class CC components and their attachments. This section should be used together with Subsection IWA for General Requirements to ensure structural and leak-tightness of the containment.

ASME III, NE-5000 provides the requirements for pre-operational testing of Steel containments. IWE-1000 provides the pre-service inspection requirements to the extent it is required for performing inservice inspections. More details regarding the ISI and ILRT are provided in Chapter 4 of this Book.

The pre-service inspection and testing is used to establish a successful start and to serve as a baseline for inspection and testing during operational service of the containment. The pre-service examination includes 100% of the containment/liner. Integrated leak rate testing (ILRT) or local leak rate testing (LLRT) is carried out per Appendix J of 10CFR Part 50. A Registered Professional Engineer or equivalent is assigned the responsibility of evaluating the structural and leak-tightness of the containment/liner.

The pre-service examination per IWE 2200 should be met before start of plant operation.

3.3 CONTAINMENT EVALUATION FOR IMPACT AND IMPULSE

The containment must be designed for impulse and impact from accident loading, tornado missile and aircraft impact. Impact loading results from tornado missile, aircraft, pipe whip or other missiles. Impulsive loads result from jet impingement, blast pressure, compartment pressurization, and pipe whip restraint reactions.

ASME BPV III Codes do not give detailed guidance on design for impact and impulse loads. The Code is silent on the utilization of the dynamic increase factor (DIF) for impact and impulse loading. Therefore, the design used need to be justified through detailed analysis and/or testing. For concrete containments, ACI 349-06 may be referred to for some basic guidance on the subject. Recognizing the extreme nature of these loads, local damage/yielding and plastic behavior of the containment is allowed in order to dissipate the energy of the impact and impulse loads. For concrete, the level of plasticity is controlled through allowable ductility factors for impulsive and impactive loads prescribed in ASME Section III, Div 2.

For a containment vessel of steel or concrete, a shield building (or secondary containment) is provided to serve as a missile barrier against external impact and impulse, thus the containment vessel itself need only to be designed for internal impact and impulse load. The analysis and design rules for impact and impulse loads are essentially same for both the shield building and the primary containment. Herein, the term "containment" is used in this section for both primary containments and shield buildings unless specified otherwise.

Missile impact effects are generally assessed in terms of local damage and global response. Local damage (damage that occurs in the immediate vicinity of the impact area) is assessed in terms of perforation and spalling. Spalling (or scabbing) is defined as displacement of material from the back face of the containment. Evaluation of local effects is essential to ensure that components within the containment would not be damaged directly by a missile perforating the containment, or by secondary missiles such as spall particles. Empirical formulae are used to assess local damage.

Evaluation of global response is essential to ensure that protected items are not damaged or functionally impaired by deformation or collapse of the impacted structure. Global response is assessed in terms of deformation limits, strain energy capacity, structural integrity, and structural stability. Structural dynamics principles are used to predict structural global response.

The general procedures for analysis and design of containments for missile impact effects include:

(a) Defining the missile properties (such as type, material, deformation characteristics, geometry, mass, trajectory, strike orientation, and velocity).
(b) Defining the containment and its properties, (such as geometry, section strength, deformation limits, strain energy absorption capacity, stability characteristics, and dynamic response characteristics).
(c) Determining impact location, material strength, and thickness required to preclude local failure (such as perforation for steel containments and spalling for concrete containments).
(d) Determining global response of the subject containment considering other concurrent loading conditions.
(e) Checking adequacy of structural design (stability, integrity, deformation limits, etc.) to verify that local damage and global response (e.g., maximum deformation) will not impair the function of safety-related items.

3.3.1 Evaluation of Local Effect

Evaluation of local effects consists of estimating the extent of local damage and characterization of the interface force-time function. Local damage is confined to the immediate vicinity of the impact location on the containment shell and consists of missile deformation, penetration of the missile into the shell, possible perforation of the element, and in the case of concrete containment, production of secondary missiles by the dislodging of concrete particles from the back face of the containment (spalling). Due to the complex physical processes associated with missile impact, local effects are evaluated primarily by application of empirical relationships based on missile impact test results.

Concrete containments subject to missile impact are generally required to have a thickness sufficient to preclude spalling. Steel containments, on the other hand, are generally designed to preclude perforation.

There is a significant amount of data on various types of missiles impact tests on different targets [2, 48–54]. This information is useful for local damage assessments for postulated missile impacts on containments as well as for determining interface forcing functions which are utilized in global response evaluations as described below.

3.3.2 Evaluation of Global Response

When a missile strikes a structure, large forces develop at the missile-structure interface, which decelerate the missile and accelerate the structure. The response of the structure depends on the dynamic properties of the structure and the time-dependent nature of the applied loading (interface force-time function). The force-time function is, in turn, dependent on the type of impact (elastic or plastic) and the nature and extent of local damage.

In an elastic impact, the missile and the structure deform elastically, remain in contact for a short period of time (duration of impact), and subsequently disengage due to the action of elastic interface restoring forces.

In a plastic impact, the missile or the structure (or both) may deform plastically or sustain permanent deformation or damage (local damage). Elastic restoring forces are small, and the missile and the structure tend to remain in contact after impact. Plastic impact is much more common in nuclear plant design than elastic impact (which is rarely encountered). For example, test data [51–52] indicate that the impact from all postulated tornado-generated missiles can be characterized as a plastic collision.

If the interface forcing function can be defined or conservatively idealized (from empirical relationships as described in Section 3.3.1 or from theoretical considerations), the structure can be modeled mathematically, and conventional analytical or numerical techniques can be used to predict global structural response. If the interface forcing function cannot be defined, the same mathematical model of the structure can be used to determine structural response by application of conservation of momentum and energy balance techniques with due consideration for type of impact (elastic or plastic).

In either case, in lieu of a more rigorous analysis, a conservative estimate of structural response can be obtained by first determining the response of the impacted structural element and then applying its reaction forces to the supporting structure. The predicted structural response enables assessment of structural design adequacy in terms of strain energy capacity, deformation limits, stability and structural integrity. Classical structure dynamic concepts and methodologies such as response chart solution and energy balanced solution [55–56] are then adopted in assessment of nuclear containments for global response.

In lieu of a comprehensive finite element analysis, a dynamically similar lumped mass spring model of the structure can be utilized to facilitate determination of structural response due to missile impact loading. For dynamic similarity and calculation convenience, the model is defined so that the following conditions are satisfied.

(a) The deflection, internal strain energy, and kinetic energy of the model must be the same as that of the structure.
(b) The external work done on the model by the missile and by other concurrent loads must be equal to (or greater than) the external work done on the structure.

3.3.3 Finite Element Analysis

Finite element models can be used to analyze containment for impact or impulse loadings. With availability of the current finite element tools, a rigorous impact analysis using a nonlinear finite element model can be carried out to simulate the impact event. Such a model, however, is complicated by system geometric and material nonlinearities, modeling issues of contact interface and the local triaxiality effects on the response ductility limit. The explicit solving technique is generally necessary to ensure convergence of the analysis results. However, such an analysis can be expensive and time consuming. Commercial software programs such as ABAQUS and LS-DYNA are used to facilitate this evaluation.

An alternative approach to utilize finite element method in containment impact load analysis is to decouple the complicated process into the following steps: 1) determine interface forcing function; 2) local effect evaluation; 3) global effect evaluation, and using customized models to deal with individual step separately. For example, the local damage may be evaluated using an equivalent static analysis with the interface forcing function being determined by tests data as described in Section 3.3.1. On the other hand, a linear elastic finite element model with adequate geometric detail can be utilized in the evaluation of the global effect to achieve more accurate results, in lieu of the lumped mass spring model used in Section 3.3.2. An excellent example implementing such concept is presented by Wang et al. [57] in the design method of nuclear safety related pipes for missile impact.

3.3.4 Special Consideration for Aircraft Impact Assessment

The events of Sept. 11, 2001 have drawn public attention to the potential impact of a large aircraft crashing into a nuclear power plant structures. As a result, The Nuclear Regulatory Commission (NRC) issued a revision to Code of Federal Regulations 10 CFR Part 50. The new rule in 10 CFR 50.150 requires that a design-specific assessment of the effects of impact of a large commercial aircraft be performed on the containment structure. This requirement is applicable only to the new reactor containments for which the construction permit or operating license or standard design certification is issued after July 13, 2009. The NRC staff has previously evaluated the impact of a large aircraft on the existing nuclear power plant containments.

The impact of a large aircraft is considered as a beyond design basis event. The NRC has developed a load-time function, also called Riera loading [58] function that provides a simple method to approximate the load-versus time history of normal impact of a large aircraft. This load time-function is based on extensive studies, computer modeling, and impact tests performed by the Sandia National Laboratories. In addition, the containment structures are required to be evaluated for impact of aircraft's rigid components such as engine and landing gear. The load-time function and characteristics of the rigid missiles are provided by the NRC to the nuclear power plant designers on as need basis.

The containment is evaluated for the load-time function using a non-linear explicit analysis. The LSDYNA and ABAQUS computer codes are normally used for this analysis. The containment structure is modeled using individual concrete, rebar, steel plate, and prestressing steel elements as applicable for different type of structures. The concrete constitutive models used in the analysis need special consideration. Winfrith concrete constitutive model in LSDYNA has been found to predict good results. It is essential that the concrete constitutive model used is bench marked to the existing impact test data such as water slug impact tests performed at Sandia National Laboratories [59]. The Nuclear Energy Institute's Report 07-013 [60], has been approved by the NRC to provide guidance on the selection of constitutive models and material properties.

The load-time function is applied as pressure time history perpendicular to the PWR containment cylindrical surface in a horizontal direction to impart the maximum force on the structure. For the free standing PWR steel containments, the force is initially applied to the concrete shield building. In case the concrete shield building fails on impact, the remaining pressure loading is applied to the steel containment. According to NEI Report 07-013 [60], the ductile failure strain limits for the rebar, prestressing steel, and steel plate are as follows:

TABLE 3.1
STRAIN LIMITS FOR PLATE, REINFORCEMENT, AND PRESTRESSING STEEL

Material	Strain Measure	Limiting Value
SA 516 Steel Plate	Membrane Principal Strain (Tensile)	0.050
	Local Ductile Tearing Effective Strain	0.070
304 Stainless Steel Plate	Membrane Principal Strain (Tensile)	0.067
	Local Ductile Tearing Effective Strain	0.137
Grade 60 Reinforcing Steel	Tensile Strain (Uniaxial)	0.050
Prestressing Steel (Ungrouted Tendons)	Tensile Strain (Uniaxial)	0.030
Prestressing Steel (Grouted Tendons)	Tensile Strain (Uniaxial)	0.020

The BWR containments are housed inside the reactor building structure. Therefore, a composite finite element model of the reactor building concrete, containment shield wall, and steel containment has to be developed. Usually, it requires analysis in several stages. First the load-time function is applied to the exterior reactor building wall. If the exterior wall fails, the remaining load is applied to the shield wall. In the event the shield wall fails, the load is then applied to the steel containment. In this analysis, the key decision is to determine when each structure is penetrated based on the failure criteria.

To check the containment structure for rigid components of the aircraft such as engine and landing gear, traditional missile impact formulas like the Modified NDRC (National Defense Research Committee) formula for penetration, and Design formula for perforation can be used. However, these formulas cannot be used for missile impact on a series of walls such as in case of the BWR reactor building. PENCURV computer code is a convenient and simple code that can be used to evaluate missile impact in such situations.

REFERENCES

[1] ASME BPV Code Section III, Div 2, Code for Concrete Containments, American Society of Mechanical Engineers, New York, 2010.

[2] ASCE SP-58, "Structural Analysis and Design of Nuclear Plant Facilities." American Society of Civil Engineers, New York, 1980.

[3] ACI 318-11, Building Code Requirements for Reinforced Concrete, American Concrete Institute, Farmington Hills, Michigan, 2011.

[4] Stevenson, J. D., "Advances in the Analysis and Design of Concrete Structures, Metal Containments and Liner Plate for Extreme Loads," SMiRT 10, 1989.

[5] Kohli, T., and Gurbuz, O., "Optimum Design of Reinforced Concrete for Nuclear Containments, Including Thermal Effects," Proceedings, Second ASCE Specialty, 1976.

[6] Topical Report BC-TOP-5A, Prestressed Concrete Nuclear Containment Structures, Bechtel Power Corporation, 1975.

[7] U.S. Nuclear Regulatory Commission, NUREG/CR-6906, Containment Integrity Research at Sandia National Laboratories, 2006.

[8] U.S. Nuclear Regulatory Commission, Overpressurization Test of a 1:4 Scale Prestressed Concrete Containment vessel Model (NUREG/CR-6810), Sandia National Laboratories, 2003.

[9] U.S. Nuclear Regulatory Commission, Standard Review Plan for the Review of Safety Analysis Reports for Nuclear Power Plants (NUREG-0800), 2007.

[10] Coronado, C., Malushte, S., Munshi, J., and Verma, N., (2009), "A Rational Seismic Design Approach for Reinforced Concrete Walls for Nuclear Power Plants," Transactions, SMiRT 20, Espoo, Finland, 6–11 August, 2009.

[11] Coronado, C., Reigles, D., Bae, S., and Munshi, J., (2011), "Finite Element Mesh Sensitivity Study Using ANSYS for Analysis and Design of Nuclear Concrete Structures," Transactions, SMiRT 21, New Delhi, India, 6-11 November, 2011.

[12] ACI 349-06, Code Requirements for Nuclear Safety Related Structures, American Concrete Institute, Farmington Hills, Michigan, 2006.
[13] Gurfinkel, G., "Thermal Effects in Walls of Nuclear Containments-Elastic and Inelastic Behavior," 1st SMIRT Conference, 1971.
[14] Oesterle, R. G., Report to ASME-ACI 359 Subgroup on design, Tangential Shear Code Provisions by Task Group on Shear, November 7, 1986.
[15] U.S. Nuclear Regulatory Commission, RG 1.35.1: Determining Prestressing Forces for Inspection of Prestressed Concrete Containment, 1990.
[16] Topical Report BC-TOP-8, Tendon End Anchor Reinforcement Test, Bechtel Power Corporation, September 1972.
[17] Wang, S. and Munshi, J., (2012), "Evaluation of Tendon Anchorage Zone Stresses in Prestressed Concrete Nuclear Containments using Finite Element Analysis," Proceedings of 20th International Conference on Nuclear Engineering and ASME 2012 Power Conference (ICONEPOWERR2012), Anaheim, California, USA, July 30–August 3, 2012.
[18] Ghosal, P. and Munshi, J., (2011), "Delaminations in Prestressed Concrete Containments," Transactions, SMiRT 21, New Delhi, India, 6–11 November, 2011.
[19] Johnson, T. E., Special Structures Group Technical Report, Concrete Containment Vessel Liner Plate Anchors and Steel Embedment Test Results , T. E. Johnson, P. C. Chang-Lo and B. P. Pfeiffer, Bechtel Power Corporation, July 1976.
[20] Topical Report BC-TOP-1, Containment Building Liner Plate Design Report, Bechtel Power Corporation, December 1972.
[21] Jovall, O. and Anderson, P. "Increase in steel liner strains due to concrete cracking — A reexamination of the tension liner strain allowable" Scanscot Technology AB, June, 2008.
[22] Jovall, O., Liner Anchors Interaction of Tensile and Shear Forces, Scanscot Technology AB, March, 2008.
[23] Hessheimer, M. F. and Dameron, R. A., "Containment Integrity Research at Sandia National Laboratories," NUREG/CR-6906, 2006.
[24] Dameron, R. A., Rashid, Y. R., and Sullaway, M. F., "Pretest Prediction Analysis and Posttest Correlation of the Sizwell-B 1:10 Scale Prestressed Concrete Containment Model Test," NUREG/CR-5671, 1998.
[25] Dameron, R. A., Rashid, Y. R., and Tang, H. T., "Leak Area and Leakage Rate Prediction for Probabilistic Risk Assessment of Concrete Containments under Severe Core Conditions," Nuclear Engineering and Design, 1995.
[26] Sheikh, A. H., "A Simplified Approach for Predicting Containment Performance During a Severe Accident," Transactions of the 19th International Conference on Structure Mechanics in Reactor Technology, August 2007.
[27] U.S. Nuclear Regulatory Commission, NUREG/CR-7110, Vol. 2: State-of-the-Art Consequence Analyses Project, 2012.
[28] Munshi, J., Johnson, W., Adediran, A., and Seo, C. Y. (2011), "Design of Concrete Structures for Impact and Impulse-Design Issues," Transactions, SMiRT 21, New Delhi, India, 6–11 November, 2011.
[29] ASME BPV Code Section III, Div 1 Subsection NE, Code for Class MC Components, American Society of Mechanical Engineers, New York, 2010.
[30] U.S. Nuclear Regulatory Commission, RG 1.57: Design Limits and Loading Combinations for Metal Primary Reactor Containment System Components, 2007.
[31] Orr, R. S., Tunon-Sanjur, and L., Tinic, S., "Finite Element Modeling of the AP1000 Nuclear Island for Seismic Analyses at Generic Soil and Rock Sites," 18th SMIRT Conference, 2005.
[32] Tinic, S., and Orr, R., "Finite Element Modeling of AP1000 Nuclear Island," 17th SMIRT Conference, 2003.
[33] Tunon-Sanjur, L., Orr, R. S., Tinic, S., and Ruiz, D. R., "Finite Element Modeling of the AP1000 Nuclear Island for Seismic Analyses at Generic Soil and Rock Sites," Nuclear Engineering and Design, 237, pp. 1474–1485, 2007.

[34] Bushnell, D., "BOSOR5 — Program for buckling of elastic-plastic complex shell of a revolution including large deflections and creep," Computers and Structures, Vol. 6, pp. 221–239, 1976.
[35] Mokhtarian, K. and Reedy, R. F., "Chapter 9 of Companion Guide to the ASME Boiler and Pressure Vessel Code: Subsection NE — Class MC Components," edited by Rao, K. R., ASME Press, 2002.
[36] Miller, C. D., "Research Related to Buckling Design of Nuclear Containment," Nuclear Engineering and Design, 79 , pp. 217–227, 1984.
[37] U.S. NUCLEAR REGULATORY COMMISSION, NUREG/CP-0065, Miller, C. D., Grove, R. B., and Bennett, J. G., "Pressure Testing of Large Scale Torispherical Heads Subject to Knuckle Buckling," Transactions of the International Conference on Structure Mechanics in Reactor Technology, August 1985.
[38] U.S. NUCLEAR REGULATORY COMMISSION, NUREG/CR-4926, Bennett, J. G., "An Assessment of Loss-of-Containment Potential Because of Knuckle Buckling for 4:1 Steel Containment Heads," LA-10972-MS, April 1987.
[39] U.S. NUCLEAR REGULATORY COMMISSION, NUREG/CR-5118, SAND88-7155, "Leak and Structural Test of Personnel Airlock for LWR Containments Subjected to Pressures and Temperatures beyond Design Limits," May 1989.
[40] Edwards, N. W., "Proposed Acceptance Criteria for Severe Accident Evaluations of Steel Containment Vessels," Nuclear Engineering and Design, 120, pp. 105–113, 1990.
[41] ASME BPV Code Case N-284-2, "Metal Containment Shell Buckling Design Methods, Class MC Section III, Division 1," American Society of Mechanical Engineers, New York, 2010.
[42] ASME BPV Code Case N-284-3, "Metal Containment Shell Buckling Design Methods, Class MC, TC, and SC Section III, Division 1," American Society of Mechanical Engineers, New York, 2011.
[43] U.S. NUCLEAR REGULATORY COMMISSION, SECY-93-087, "Policy, Technical, and Licensing Issues Pertaining to Evolutionary and Advanced Light-Water Reactor (ALWR) Designs," 1993.
[44] U.S. NUCLEAR REGULATORY COMMISSION, NUREG/CR-5096, SAND88-7016, "Evaluation of Seals for Mechanical Penetrations of Containment Buildings," 1988.
[45] U.S. NUCLEAR REGULATORY COMMISSION, NUREG/CR-5334, SAND89-0327, "Severe Accident Testing of Electrical Penetration Assemblies," November 1989.
[46] U.S. NUCLEAR REGULATORY COMMISSION, NUREG/CR-5561, SAND90-7020, "Analysis of Bellows Expansion Joints in the Sequoyah Containment," December 1991.
[47] U.S. NUCLEAR REGULATORY COMMISSION, NUREG/CR-6154, SAND94-1711, "Experimental Results from Containment Piping Bellows Subjected to Severe Accident Conditions," Vol. 1, September 1994.
[48] Barber, R. B., "Steel-Rod/Concrete Slab Impact Test (Experimental Simulation)," Bechtel Corp., October 1973.
[49] Rotz, J. V., "Results of Missile Impact Tests on Reinforced Concrete Panels," Vol. 1A, pp. 720–738, Second Specialty Conference on Structural Design of Nuclear Power Plant Facilities, New Orleans, LA, December 1975.
[50] Rotz, J. V., "Evaluation of Tornado Missile Impact Effects on Structures," Symposium on Tornadoes: Assessment of Knowledge and Implications for Man, Texas Tech Univ., Lubbock, TX, June, 1976.
[51] Stephenson, A. E., "Full Scale Tornado Missile Impact Tests," EPRI NP440, Sandia Laboratories, Tonopa, NV, prepared for Electric Power Inst., Palo Alto, CA, July 1977.
[52] Vassallo, F. A., "Missile Impact Testing of Reinforced Concrete Panels," prepared for Bechtel Power Corp. by Calspan Corp., January 1975.
[53] Emori, R. I., "Analytical Approach to Automobile Collisions," Paper No. 680016, Automotive Engineering Congress, Detroit, MI, 1968.
[54] Gwaltney, R. C., "Missile Generation and Protection in Light-Water-Cooled Power Reactor Plants," ORNL NSIC-22, Oak Ridge National Laboratory, Oak Ridge, TN, for the USAEC, September 1968.
[55] Norris, C. H., et al., Structural Design for Dynamic Loads, McGraw-Hill Book Co., Inc., New York, NY, 1959.

[56] Biggs, J. M., Introduction to Structural Dynamics, McGraw-Hill Book Co., Inc., New York, NY, 1964.

[57] Wang, S., Akinci, N. O., Li, J. and Johnson, W. H., "Design of Nuclear Safety-Related Steel Ducts/ Pipes for Tornado Missile Impact," Proceedings of American Society of Mechanical Engineers (ASME) 2011 Power Conference, Denver, CO, USA, July 2011.

[58] Riera, J. D., "A Critical Appraisal of Nuclear Power Plant Safety Against Accidental Aircraft Impact," Nuclear Engineering and Design, Volume 57, pp. 193–206, 1980.

[59] Computational Simulation of WS 1 and WS 2, "Water Slug Impact Tests: Validation of Aircraft Impact Analysis Methodology" (Official Use Only), Final Report by ABS Consulting and Anatech, Electric Power Research Institute, Palo Alto, CA, March 2004.

[60] NEI 07-13 Rev. 7, "Methodology for Performing Aircraft Impact Assessments for New Plant Designs." The Nuclear Energy Institute, May, 2009.

CHAPTER

4

INSERVICE INSPECTIONS AND LEAK RATE TESTING OF CONTAINMENTS

Hansraj Ashar

Containment reliability can be assured by periodic inspections and performance testing of containment components. This chapter describes the inspections and testing to be performed to assure that the containment will be capable of performing its intended function under the postulated loadings. The chapter also describes the deterministic inspection criteria and quasi risk-informed criteria so the uncertainties related to the use of risk-informed criteria are adequately addressed.

The chapter describes, in detail, the deterministic criteria for conducting containment inspections to verify the soundness of the containment structure, as well as the methodologies used to perform the leakage tests of the containment structures, containment access penetrations, and other systems (i.e., feed water pipes, main steamline pipes) penetrating the containment structure. The chapter also describes the use of risk-informed methodology used to determine the frequency of leak rate testing of containment structures and that of penetrations penetrating through the containment pressure-retaining components.

Keywords: Inservice Inspection (ISI), Integrated Leak Rate Testing (ILRT), Type A Test, Type B Test, Type C Test, Deterministic Approach (DA), Probabilistic Approach (PA), Risk Informed Approach (RIA), Large Early Release Frequency (LERF)

4.1 INTRODUCTION

Chapter 2 describes the containment-related generic design criteria (GDC) and briefly summarizes methods of implementing these criteria. In this chapter, I am discussing the detailed implementation of GDC 16, 52, 53, and 54 related to containment inservice inspections and containment periodic leak rate tests.

As required by GDC 16, all 104 operating reactors in the United States are provided with containments and the associated systems to serve as essentially leak-tight barriers against the uncontrolled release of radioactivity to the environment. The containment configurations of these reactors are discussed in Chapter 1, and the information related to containment structures is provided in Appendix 1A of Chapter 1. As required by GDC 52, all these containments, and their auxiliary systems are designed so that periodic integrated leak rate testing (ILRT) can be conducted at containment design pressure. As required by GDC 53, all containments are designed to permit periodic inservice inspections (ISI) of all important areas, appropriate surveillance program, and periodic leak rate testing of penetrations having resilient seals and expansion bellows.

As required by GDC 54, the piping systems penetrating the containment structures have been provided with leak detection and isolation capability. The leak detection and isolation capabilities are required to have adequate redundancy, reliability, and performance capabilities. Additional prescriptive requirements for containment leak rate testing are provided in Option A of Appendix J of 10 CFR Part 50 and are discussed in this chapter.

Based on the requirement of Option A of Appendix J, the plant owners used to perform visual examination of containment structures and associated piping penetrating the structures prior to performing an ILRT of containments to detect potential leakage areas. However, there was no consistency in examination methods, acceptance criteria, and personnel qualification requirements.

4.2 PURPOSE OF PERIODIC ISI AND LEAK RATE TESTING

Recognizing the importance of containments, GDCs set up basic requirements for inspecting and testing of containments. However, the detailed requirements have to depend upon the actual experience with the containment structures and containment pressure boundary components. Containment is a vital engineered safety feature of a nuclear power plant. In normal operating conditions, the containment is subjected to operating and environmental stressors, such as ambient pressure fluctuations, temperature variations, earthquakes, ice, and windstorms. In some containment designs, the principal leak-tight barrier is surrounded by another structure, such as a shield wall or a shield building. These structures protect the containment from some of the external influences, and in some cases, collect leakage from the primary containment for processing prior to release to the atmosphere. The mechanical stresses and strains generated by transients under normal conditions and the effects of high-probability ($>10^{-2}$) external influences are a small fraction of the limiting conditions for which the containment is designed. However, the fatigue life of the containment can be affected by the significant number of cycles of such low-stress transients. The containment is also subjected to various types of internal degradation (aging degradation), depending on the inherent characteristics of the materials, the fabrication processes, and the construction methods. The rate and extent of the degradation are influenced by the sustained environmental conditions such as temperature, humidity, water leakage, and borated water spills.

Thus, the reliability (or the success probability) of a containment to perform its intended function under design basis conditions as well as under higher loads due to severe accidents and beyond design basis earthquakes is influenced by the containment's inherent capability and the various stresses and degradation mechanisms that act on it. The basic concept in ensuring the reliability of the containment is to track the degradation of the containment components through periodic inspections and check the leak-tight integrity of the containment pressure-retaining components through periodic leak rate testing.

4.3 DETERMINISTIC APPROACH

As implemented in the United States, the deterministic approach (as differentiated from "Performance Based Approach") consists of conducting (1) periodic inspection of the containment pressure-boundary components at certain fixed interval, and (2) leak rate testing of the containment components at certain fixed interval. The basic features of the inspection and leak rate testing are discussed in the following paragraphs.

4.3.1 Containment Inservice Inspection

Until 1996, there was no formal program for inspecting containments, except for the inspection of prestressing tendons in prestressed concrete containments. For prestressed concrete containments with unbonded tendons (where the prestressing elements and associated anchorage components are protected from corrosion by a specially formulated grease), the monitoring of the prestressing system was performed by the implementation of Regulatory Guide (RG) 1.35 [1]. However, in the early 1990s, the NRC learned of a number of incidents of significant degradation of the prestressing tendon system, steel liner of concrete containment structures, and steel shells of steel containments. The NRC documented the major findings in NUREG-1522 [2]. The findings summarized in Reference 2 were identified by containment tendon inspections, or through ad hoc observations by the plant personnel, NRC inspectors, or, to a lesser degree, by inspections required before the leak rate testing of the containments pursuant to 10 CFR Part 50, Appendix J. Even when degradation was found, there was no formal procedure for documenting the degradation and implementing corrective actions.

In August 1996, the NRC amended its regulation 10 CFR 50.55a [3] to incorporate by reference the 1992 Edition and the 1992 Addenda of Subsections IWE (1992 thru 2010) [4] and IWL (1992 thru 2010) [5] of Section XI of the American Society of Mechanical Engineers (ASME) Boiler and Pressure Vessel (B & PV) Code (the Code) for inspection of containment structures in light-water-cooled nuclear power plants. The regulation required certain additions and modifications to the requirements of the Code. Subsection IWE provides the requirements for inservice inspection, repair, and replacement of Class MC pressure-retaining components (i.e., steel containments) and their integral attachments, and metallic shell and penetration liners of Class CC pressure retaining components (i.e., concrete containments) and their integral attachments. In conjunction with the additional requirements in Ref. [4], a general visual examination of the accessible portion of the entire containment is required to be performed three times in 10 years. Subsection IWL provides the requirements for pre-service examination, inservice inspection, and repair of the reinforced concrete and post-tensioning systems of Class CC components. In conjunction with the additional requirements in Ref. [5], a general visual examination of the accessible portion of the entire containment is required to be performed two times in 10 years.

For the readers not familiar with the ASME Code, it is useful to explain the way the relevant portions of the ASME Code has been organized and applied to the inservice inspections of steel and concrete containments.

4.3.1.1 ASME Code. ASME Code is organized in 12 sections, with sections relevant to nuclear power plants (NPP) are Section II (Section on Materials), Section III (Section on Nuclear Power), and Section XI (Section on Inservice Inspection of NPP structures, systems, and components). Some other sections, such as Section V (Non-Destructive Examination) and Section VIII (Pressure Vessels), have been used for older NPPs, when the nuclear sections were not available.

Each section is divided in subsections. Section XI (ISI) is divided in the following subsections:

- IWA — General Requirements
- IWB — Class 1 Components
- IWC — Class 2 Components
- IWD — Class 3 Components
- IWE — Class MC and CC (metal containments and liner of CC Containments)
- IWF — Classes 1, 2, 3, and MC Component Supports
- IWG — Core Internal Structures
- IWL — Class CC — Concrete and Post-tensioning

4.3.1.2 Requirements of Subsection IWE. Subsection IWE consists of the following Articles:

- IWE-1000: Scope & Responsibility
- IWE-2000: Examination & Inspections
- IWE-3000: Acceptance Standards
- IWE-4000: Repair Procedures (Delineated to IWA-4000 since 1998 Edition)
- IWE-5000: System Pressure Test
- IWE-7000: Replacements (merged with IWA-4000)

The essential portions of these articles are summarized in the following paragraphs:

The reader should note the following ASME Section XI definitions provided in Article IWA-7000, for "Inservice Examination," and "Inservice Inspection:"

Inservice Examination: the process of visual, surface, or volumetric examination performed in accordance with the rules and requirements of Section XI.

Inservice Inspection: methods and actions for assuring the structural and pressure-retaining integrity of safety-related nuclear power plant components in accordance with the rules of Section XI.

Subsection IWE provides requirements for inservice inspection of Class MC (ASME Subsection NE of Section III, Division 1) pressure retaining components and their integral attachments, and of metallic shell and penetration liners of Class CC (ASME Section III, Division 2) pressure retaining components and their integral attachments in light-water cooled plants.

Article IWE-1000 addresses the components to be examined, components exempted from examination, required accessibility for examination, inaccessible areas requiring examinations, and surface areas requiring augmented examinations. Article IWE-2000 addresses requirements for pre-service examinations, visual examination personnel qualifications, required qualifications for "Responsible Individual," inspection schedule, successive examinations. Table IWE-2500 provides the detailed examination requirements for various situations encountered in practice. For example, the required examination of wetted or submerged areas encountered in suppression chambers (or sometimes termed as wet well areas) of BWR Mk I, Mk II, and Mk III containments. Please see Chapter 1 of this book for description of theses containments. This Article also incorporates the visual examination requirements of IWA-2210. The visual examinations, method of performing these examinations, and qualifying personnel for these examinations, are described in IWA-2210. The Code defines three types of visual examination methods: termed as VT-1, VT-2, and VT-3 examinations.

VT-1 examination is conducted to detect discontinuities and imperfections on the surface of components, including such conditions as cracks, wear, corrosion, or erosion.

VT-2 examination is conducted to detect evidence of leakage from pressure retaining components, as required during the conduct of system pressure test.

VT-3 examination is conducted to determine the general mechanical and structural condition of components and their supports by verifying parameters such as clearances, settings, and physical displacements and to detect discontinuities and imperfections, such as loss of integrity at bolted or welded connections, loose or missing parts, debris, corrosion, wear, or erosion. VT-3 includes examination for conditions that could affect operability or functional adequacy of constant load and spring-type supports.

If it is necessary to identify the magnitude of a crack or a flaw identified during visual examinations, or to identify any degradation in inaccessible areas, IWA-2220 and IWA-2230 requires the use of surface and/or volumetric examinations. Detailed procedures for conducting these examinations, as well as qualifying the examination personnel, are provided in IWA-2200 and IWA-2300.

Article IWE-3000 describes acceptance standards when the examinations have shown degraded conditions. The article describes three methods of accepting a degraded component: (1) Accept, because, the degradation is within the acceptable limits, and the component could perform its intended function(s); (2) Accept, when appropriate repairs or replacement of the degraded area of the component are made so that its functionality is retained; and (3) Accept, if the engineering evaluation indicate that the component could perform its intended function. In the later Editions (i.e., 2001 thru 2010) of SS IWE, Article IWE-4000 on "Repair and Replacement," has been omitted from this SS, and the rules established in Article IWA-4000 (Repair and Replacement) are applicable. IWA-4000 of SS IWA (General Requirements) is also applicable to SS IWB (Class 1 Components), SS IWC (Class 2 Components), and SS IWD (Class 3 Components), unless specifically called out otherwise in respective subsections. Article IWE-5000 requires system pressure tests to verify integrity of pressure retaining components after repair or replacement of the components.

As Subsection IWE has been incorporated by reference in regulation 10 CFR 50.55a, NRC expanded or modified certain requirements of the subsection in the regulation. They are discussed in the following paragraphs.

(A) For steel shell or steel liner of concrete containment, the licensee is required to evaluate the acceptability of inaccessible areas when conditions exist in accessible areas that could indicate the presence of or result in degradation to such inaccessible areas. For each inaccessible area identified, the licensee is required to provide the following in the ISI Summary Report as required by Article IWA-6000 (Reporting Requirements). The details of this item is not discussed here, as the requirement has been incorporated in 2001 and later Edition of SS IWE.

(B) When performing remotely the visual examinations required by Subsection IWE, the maximum direct examination distance specified in Table IWA-2210-1 may be extended and the minimum illumination requirements specified in Table IWA-2210-1 may be decreased provided that the conditions or indications for which the visual examination is performed can be detected at the chosen distance and illumination. This NRC position is a relaxation of the code requirement, evolved as a result of recognizing the large size of the containment structure.

(C) The examinations specified in Examination Category E-B, Pressure Retaining Welds, and Examination Category E-F, Pressure Retaining Dissimilar Metal Welds, is optional. In the 1998 Edition of SS IWE of the Code, these requirements were deleted, as experiences with the steel containments indicated no concern for metal welds. However, the requirement for examination of dissimilar metal welds was added in the guidelines for license renewal. Please see Chapter 5 of the book for more information on license renewal guidelines.

(D) The following examination is required to be used as an alternative to the requirements of successive inspections of IWE-2430.

   (1) If the examinations reveal flaws or areas of degradation exceeding the acceptance standards of Table IWE-3410-1, an evaluation is required to be performed to determine whether additional component examinations are required. For each flaw or area of degradation identified which exceeds acceptance standards, the licensee is required to provide the following in the ISI Summary Report required by IWA-6000:

      (*i*) a description of each flaw or area, including the extent of degradation, and the conditions that led to the degradation;
      (*ii*) the acceptability of each flaw or area, and the need for additional examinations to verify that similar degradation does not exist in similar components, and;
      (*iii*) a description of necessary corrective actions.

   (2) The number and type of additional examinations to ensure detection of similar degradation in similar components.

   This requirement has been incorporated in Section IWE-2420 of 2010 Edition to an extent. However, NRC keeps the above requirements as an alternative to the Code requirements.

(E) A general visual examination as required by Subsection IWE must be performed once each period. This requirement has been incorporated in the Code since the 2001 Edition, in Table IWE-2500-1 of SS IWE.

(F) The 1998 Edition of Subsection IWE deleted the requirements for VT-1 and VT-3, and substituted them with Owner defined general and detailed examinations. NRC did not accept the substitution, and in 2001 requirements of the Regulation. SS IWE was revised. In the 2004 Edition of Subsection IWE, the Code incorporated the original requirements for examinations, and for qualifications of the personnel.

(G) The VT-3 examination method must be used to conduct the examinations in Items E1.12 and E1.20 of Table IWE-2500-1, and the VT-1 examination method must be used to conduct the examination in Item E4.11 of Table IWE-2500-1. An examination of the pressure-retaining bolted connections in Item E1.11 of Table IWE-2500-1 using the VT-3 examination method must be conducted once each interval. The "owner-defined" visual examination provisions in IWE-2310(a) are not approved for use for VT-1 and VT-3 examinations. In the 2007 Edition of IWE, the Table was revised to incorporate the NRC position.

(H) Containment bolted connections that are disassembled during the scheduled performance of the examinations, as required in Item E1.11 of Table IWE-2500-1, must be examined using the VT-3 examination method. Flaws or degradation identified during the performance of a VT-3 examination must be examined in accordance with the VT-1 examination method. The criteria in the material specification or IWB-3517.1 must be used to evaluate containment bolting flaws or degradation. As an alternative to performing VT-3 examinations of containment bolted connections that are disassembled during the scheduled performance of Item E1.11, VT-3 examinations of containment bolted connections may be conducted whenever containment bolted connections are disassembled for any reason. A Table 2500-1, Examination Category E-G was added in the 2007 Edition of Subsection IWE to incorporate the NRC position.

(I) The ultrasonic examination acceptance standard specified in IWE-3511.3 for Class MC pressure-retaining components must also be applied to metallic liners of Class CC pressure-retaining components. This NRC addition to the SS IWE is incorporated in IWE-3511.3 in 2007 Edition of the Subsection.

4.3.1.3 Requirements of Subsection IWL. Subsection IWL consists of the following articles:

Article IWL-1000 Scope and Responsibility
Article IWL-2000 Examination and Inspection
Article IWL-3000 Acceptance Standards
Article IWL-4000 Repair and Replacement Activities
Article IWL-5000 System Pressure Tests

The essential portions of these articles are summarized in the following paragraphs:

The ASME Section XI definitions of "Inservice Examination," and "Inservice Inspection:" provided in Article IWA-7000 are also applicable to Subsection IWL.

Subsection IWL provides requirements for preservice examination, inservice inspection, and repair/replacement activities of the reinforced concrete and the post-tensioning systems of Class CC components. This distinction is referred to as concrete containment as defined in CC-1000, except that the inservice inspection of steel liner of the concrete containment is included in the scope of Subsection IWE.

Article IWL-1000 addresses the components to be examined, components exempted from examination, and required accessibility for examination. Article IWL-2000 addresses requirements for pre-service examinations, visual examination requirements, personnel qualification requirements, required qualifications for "Responsible Engineer," ISI schedule, and successive examinations. Table IWL-2500 provides the detailed examination requirements for concrete in the table entitled Examination Category L-A, and that for the unbonded post-tensioning system in the Table L-B. As different from other subsections of Section XI, this Article defines the requirements for "General," and "Detailed" visual examinations of concrete surfaces, and of the components of the post-tensioning system. This Article also provides detailed requirements for testing corrosion protection medium (CPM, also called sheathing filler grease, or simply "grease") of the unbonded post-tensioning system. Moreover, the Article requires the additional examinations of concrete and post-tensioning tendons when plant owners have to cut a hole in the concrete containment structures to facilitate replacement of a steam generator, or a reactor vessel head.

Article IWL-3000 provides the acceptance standards for pre-service and in-service examination results. In both cases, the Article provides acceptability based on (1) examination, (2) engineering evaluation of the condition, and/or (3) repair/replacement of the affected components. Detailed acceptance criteria have been provided for acceptability of post-tensioning system and the associated CPM. Article IWL-4000 provides the requirements for repair/replacement activities. As part of this article, the article adopts certain specific requirements of Article IWA-4000. Article IWL-5000 provides the requirements for pressure testing of concrete containment structures following R/R activities.

As Subsection IWL has been incorporated by reference in regulation 10 CFR 50.55a, NRC expanded or modified certain aspects of the Subsection in the regulation. They are discussed in the following paragraphs.

(A) Grease (CPM) caps that are accessible must be visually examined to detect grease leakage or grease cap deformations. Grease caps must be removed for this examination when there is evidence of grease cap deformation that indicates deterioration of anchorage hardware. This requirement was incorporated in the 1998 Edition of Subsection IWL. Readers should note that this requirement was added as a result of the deformation of the anchorages of several vertical tendons at Farley containments please see Reference 2 for a brief description of this event.

(B) When evaluation of consecutive surveillances of prestressing forces for the same tendon or tendons in a group indicates a trend of prestress loss such that the tendon force(s) would be less than the minimum design prestress requirements before the next inspection interval, an evaluation must be performed and reported in the Engineering Evaluation Report as prescribed in IWL-3300. This requirement was incorporated in the 2001 Edition of Subsection IWL.

(C) When the elongation corresponding to a specific load (adjusted for effective wires or strands) during retensioning of tendons differs by more than 10% from that recorded during the last measurement, an evaluation must be performed to determine whether the difference is related to wire failures or slip of wires in anchorage. A difference of more than 10% must be identified in the ISI Summary Report required by IWA-6000. This requirement was incorporated in the 1998 Edition of SS IWL.

(D) The licensee shall report the occurrences of the following conditions, in the ISI Summary Report required by IWA-6000:

 (1) the sampled sheathing filler grease contains chemically combined water exceeding 10% by weight or the presence of free water;
 (2) the absolute difference between the amount removed and the amount replaced exceeds 10% of the tendon net duct volume;
 (3) grease leakage is detected during general visual examination of the containment surface.

These requirements were incorporated in the 2001 Edition of SS IWL

(E) For Class CC applications, the licensee is required to evaluate the acceptability of inaccessible areas when conditions exist in accessible areas that could indicate the presence of or result in degradation to such inaccessible areas. For each inaccessible area identified, the licensee is required to provide the following in the ISI Summary Report required by IWA-6000:

 (1) a description of the type and estimated extent of degradation, and the conditions that led to the degradation;
 (2) an evaluation of each area, and the result of the evaluation; and
 (3) a description of necessary corrective actions.

 This requirement was incorporated in the Examination Category L-A of Table IWL-2500-1 of the 2004 Edition of SS IWL.

(F) Personnel that examine containment concrete surfaces and tendon hardware, wires, or strands must meet the qualification provisions in IWA-2300. The “owner-defined” personnel qualification provisions in IWL-2310(d) are not approved for use. The 2007 Edition of SS IWL incorporated this requirement by incorporating the personnel qualification methodology in IWL-2300.

(G) Corrosion protection material must be restored following concrete containment post-tensioning system repair and replacement activities in accordance with the quality assurance program requirements specified in IWA-1400. This requirement has not been incorporated in IWA-1400 of the 2010 Edition of the ASME Code.

4.3.2 Prescriptive Leak Rate Testing Requirements

10 CFR Part 50, Appendix J (latest version 2010) [7] entitled, "Primary Reactor Containment Leakage Testing for Water-Cooled Power Reactors," establishes the testing requirements for preoperational and periodic verification of the leak-tight integrity of the primary reactor containment, including systems and components which penetrate the containments of light water-cooled power reactors, and establishes the acceptance criteria for such tests. The purposes of the tests are to assure that (a) leakage through the primary reactor containment and systems and components penetrating primary containment does not exceed allowable leakage rate values as specified in the technical specifications, or associated bases, and (b) periodic testing of reactor containment penetrations and isolation valves is performed so that proper maintenance and repairs are done during the service life of the containment.

The reactor containment leakage test program includes the performance of integrated leak rate testing (ILRT), also known as Type A test, containment penetration leak rate testing — Type B tests, and containment isolation valve leak rate testing — Type C test. The Type B and Type C tests are also referred to as local leak rate tests (LLRTs). The current (1995 to 2010) Appendix J requirements provide two options for performing the tests. Option A — deterministic, and Option B — performance-based.

Option A (prescriptive) requires that after the preoperational leakage rate tests, three integrated leak rate tests (Type A tests) be performed at approximately equal intervals during each 10year service period. Option A requires Type B tests to be performed during reactor shutdown or refueling, but in no case at intervals greater than 2 years. For containments employing continuous leakage monitoring, Option A requires the Type B tests to be performed every other reactor shutdown for refueling or every 3 years, whichever is less. Air locks are required to be tested every 6 months. Option A requires Type C testing to be performed every refueling outage, or every 2 years, whichever is less.

The NRC considers the implementation of these deterministic requirements, that is, periodic inspections, (as described in Section 4.3.1) and periodic leak rate testing of containment components, will maintain the high reliability (in the qualitative sense) of the containment. The prescriptive testing procedures are described in the following paragraphs.

Some of the Appendix J definitions (most of them common to Option A and Option B) are provided below for ready reference:

- "Primary reactor containment" means the structure or vessel that encloses the components of the reactor coolant pressure boundary, as defined in Section 50.2(v) of 10 CFR Part 50, and serves as an essentially leak-tight barrier against the uncontrolled release of radioactivity to the environment.
- "Leakage rate" for test purposes is that leakage which occurs in a unit of time, stated as a percentage of weight of the original content of containment air at the leakage rate test pressure that escapes to the outside atmosphere during a 24-hour test period.
- "Overall integrated leakage rate" (ILRT) means that leakage rate established from a summation of leakage through all potential leakage paths including containment welds, valves, fittings, and components which penetrate containment.
- "Type A Tests" means tests intended to measure the primary reactor containment overall integrated leakage rate (1) after the containment construction has been completed and is ready for operation, and (2) at periodic intervals thereafter.

- "Type B Tests" means tests intended to detect local leaks and to measure leakage across each pressure-containing or leakage-limiting boundary for the following primary reactor containment penetrations:
    1. Containment penetrations whose design incorporates resilient seals, gaskets, or sealant compounds, piping penetrations fitted with expansion bellows, and electrical penetrations;
    2. Fitted with flexible metal seal assemblies;
    3. Air lock door seals, including door operating mechanism penetrations which are part of the containment pressure boundary;
    4. Doors with resilient seals or gaskets except for seal-welded doors;
    5. Components other than those listed above in 1, 2, or 3, which must meet the acceptance criteria for the combined leak rate as defined in Section III.B.3 of Option A of Appendix J.
- "Type C Tests" means tests intended to measure containment isolation valve leakage rates. The containment isolation valves included are those that:
    1. Provide a direct connection between the inside and outside atmospheres of the primary reactor containment under normal operation, such as purge and ventilation, vacuum relief, and instrument valves;
    2. are required to close automatically upon receipt of a containment isolation signal in response to controls intended to effect containment isolation;
    3. are required to operate intermittently under post-accident conditions; and
    4. are in mainsteam and feedwater piping and other systems which penetrate containment of direct-cycle boiling water power reactors.

Appendix J uses the following abbreviations in defining the test parameters:

(1) Pa (psig or kPa) means the calculated peak containment internal pressure related to the design basis accident as specified either in the technical specification or associated bases.

(2) La (percent/24 hours) means the maximum allowable leakage rate at pressure Pa as specified for preoperational tests in the technical specifications or associated bases, and as specified for periodic tests in the operating license or combined license, including the technical specifications in any referenced design certification applicable new reactors.

Note: For licensing of new reactors, NRC has accepted the process of "combined licenses," in 10 CFR Part 52. Under this regulation, once certain prerequisites are met (e.g., Early Site Permit), a licensee can apply for construction permit and operating license in a combined license application. Chapter 9 of this book provides more information on this subject.

(3) Ld (percent/24 hours) means the design leakage rate at pressure, Pa, as specified in the technical specifications or associated bases.

(4) Lam, Ltm (percent/24 hours) means the total measured containment leakage rates at pressure Pa, obtained from testing the containment with components and systems in the state as close as practical to that which would exist under design basis accident conditions (e.g., vented, drained, flooded, or pressurized).

(5) "Acceptance criteria" means the standard against which test results are to be compared for establishing the functional acceptability of the containment as a leakage limiting boundary.

The following paragraphs briefly describe the procedures for conducting the leakage rate testing and are based on the regulatory requirements described in Option A of Appendix J:

4.3.2.1 Type A Testing. The prescriptive requirements for performing Type A tests are described in detail in Section IIIA of Option A of Appendix J. It provides prerequisites for performing the tests, together with how to achieve isolation of containment before conducting the tests and the test methods to be utilized.

It also provides the acceptance criteria applicable to the periodic Type A tests. Readers should note that at present, that is, June of 2011, there may be one or two NPPs, which still uses Option A of Appendix J for periodic ILRT. Majority of licensees have changed their technical specifications so that they can perform the periodic leak rate testing using Option B (performance based approach) of Appendix J. NRC has accepted some deviations from Option A requirements as discussed in the following paragraphs.

- Before performing an ILRT, the licensee, or his contractor used to conduct visual examination of the containment structure as required by Section V.A of Option A of Appendix J. As pointed out in Section 4.3.1, since the incorporation of Subsections IWE and IWL of the ASME Section XI Code in 10 CFR 50.55a, NRC has permitted the use SS IWE and SS IWL for the visual examinations prior to Type A leak rate testing, in lieu of that in Section V.A.
- Closure of containment isolation valves for the Type A test is required to be accomplished by normal operation and without any preliminary exercising or adjustments (e.g., no tightening of valve after closure by valve motor). Repairs of non-operating or leaking valves shall be made as necessary. Information on any valve closure malfunction or valve leakage that require corrective action before the test, shall be included in the summary report required by Section V.B of Option A of the Appendix.
- Option A requirements for "test methods" have not been changed to reflect the newer industry standards developed based on experience and available database. Though Option A refers to industry Standard ANSI/ANS 56.8 (1987) [8], for a method of leak rate calculations, the NRC endorses the use of test methods developed in ANSI/ANS 56.8 (1994) [9] for performing Type A, Type B and Type C tests. Though this approval is related to Option B of Appendix J, the 1994 ANS standard is also applicable to Option A of Appendix J.
- Though Option A of Appendix J allows the use of reduced pressure periodic tests, for a number of reasons, such as reconciliation with the peak pressure tests, this practice has not been used since mid-1980.
- The requirements of Section IV of Option A, though is not applicable to Option B, unless its use is specifically included in the documents related to Option B.
- Type A test Acceptance Criterion: The actual leakage rate "Lm" obtained from performing Type A test must be less than 0.75 La. If two consecutive periodic Type A tests fail to meet the acceptance criterion, the test schedule applicable to subsequent Type A tests has to be reviewed and approved by NRC.

4.3.2.2 Type B Testing. Section III.B of Option A, provides detailed requirements for (1) examination methods, (2) measurement of the rate of pressure loss, when pressurized, and (3) leakage surveillance by means of permanently installed penetration pressurization system. The allowable leak rate for the total leak rates from the combined Type B and Type C tests is limited to 0.6 La. All periodic Type B tests are required to be performed at a pressure not less than Pa. The acceptance criteria for the combined leakage rate from Type B and Type C tests must be less than 0. 6 La, except as explained in the Type C acceptance criteria.

4.3.2.3 Type C Tests. Type C tests are required to be performed by local pressurization of isolation valves as follows:

The pressure shall be applied in the same direction as that when the valve would be required to perform its function, unless it can be determined that the results from the tests for the pressure applied in a different direction will provide equivalent or more conservative results. Each valve to be tested is required to be closed by normal operation without prior exercising (e.g., no tightening of valves after closure by valve motor). Valves pressurized with air or nitrogen are required to be tested at Pa. The valves pressurized with fluid are required to be tested at a pressure not less than 1.10 Pa.

Acceptance Criteria for Type C tests: The combined leakage rate from all Type B and Type C tests is required to be less than 0.6La. However, the leakage from the valves that are sealed with fluid from a seal system may be excluded from the combined leakage rate, provided (a) these valves have been demonstrated to have leakage rates that do not exceed those specified in the TS or associated bases, and (b) the installed isolation seal water system fluid inventory is sufficient to assure the sealing function for at least 30 days at a pressure of 1.10 Pa.

4.3.2.4 Other Requirements. Section IV of Option A discusses the additional requirements for pressure tests, such as, when the containment needs to be cut for replacing a steam generator or a reactor vessel head. This requirement is applicable when the licensee has to replace a small size penetration or drill a small hole to add a penetration. After the completion of a specific repair or replacement, the licensee has to determine, which leak rate or structural test would be the most appropriate for confirming the operability of the containment as a whole. The Option also discusses the methods of performing leakage rate testing of containment components, when the containment consists of multiple leakage barriers, or it is subatmospheric. Section V describes the requirements for inspection and summary reports.

4.4 PERFORMANCE BASED APPROACH (PBA)

Since the 1979 accident at Three Mile Island, the Nuclear Regulatory Commission (NRC) has been studying various severe accident scenarios and gradually incorporating the use of probabilistic risk assessment (PRA) in the regulatory process. In 1988, the NRC staff requested information on the assessment of severe accident vulnerabilities from the operators of nuclear reactors, which is documented in Generic Letter (GL) 88-20 [10]. By the mid-1990s, in conjunction with the responses to these requests and further research by the NRC and its contractors in the use of the PRA methodologies, the NRC felt confident that PRA could be used in the regulatory process. Neither the NRC staff, nor the Commission, nor its advisory committee (known as Advisory Committee on Reactor Safeguards — ACRS) preferred the regulatory activities based on pure probabilistic approach. Hence, the NRC approached some of the regulatory activities based on performance with some input from risk parameters. 10 CFR 50.65, the Maintenance Rule (discussed briefly in Chapter 2 of this book) has been characterized as performance based regulation. Thus, the amount of efforts the licensees have to expend for maintaining the safety-related structures systems and components (SSC) would be based on the plant specific, as well as the industry (i.e., the nuclear industry) wide experience.

4.4.1 Inservice Inspections and PBA

Some intuitive performance-based criteria for inspection of containment structures are built into the existing codes (i.e., Subsections IWE and IWL of Section XI of the ASME Code). For example, in IWE-2420(d): "When the reexaminations required by IWE-2420(b) reveal that the flaws, areas of degradation, or repairs remain essentially unchanged for the next inspection periods, these areas no longer require augmented examination." Such logic is found in all prescriptive codes. There is no quantitative basis for the criterion to conclude that the conditions observed during the next inspection periods would represent the benign nature of the degradation. However, the criterion is based on the intuitive judgment of the knowledgeable committee members. In many cases, this kind of consensus approach is appropriate in setting prescriptive requirements.

The Committee members (responsible for developing SS IWE and SS IWL) also spent some efforts to relax the required frequency of the periodic inspections in Subsections IWE and IWL, similar to the risk-informed approach taken for some of the Subsections IWC and IWD piping systems. However, after thorough discussion, the Committee decided not to proceed in that direction. Hence, the ISI requirements for containments are deterministic and prescriptive.

4.4.2 Leak Rate Testing and PBA

The U.S. nuclear industry is using risk-informed methodologies in reducing unnecessary regulatory burden on the operators of the nuclear reactors and in addressing various issues of concern to the industry. Specifically, the nuclear industry had expressed its concern that the deterministic prescriptive requirements of Option A of 10 CFR 50, Appendix J are too costly to the owners of the nuclear power plants, and that steps should be taken to reduce the burden on the industry. As a result, NRC began looking at the costs and benefits of implementing the Option A requirements of Appendix J.

Earlier NRC studies documented in NUREG-1273 (1988) [11] had indicated that the public risk associated with undetected containment leakage as a result of implementing Option A of Appendix J, is very small. These studies recommended that the Type A, Type B, and Type C tests required by Option A should be continued since they provide the assurance of the continued high availability of the containment to perform its intended function. In the early 1990s, in conjunction with the database collected by the Nuclear Energy Institute (NEI), NRC attempted to quantify the risk associated with increasing the time interval for performing periodic leak rate testing. This study is documented in NUREG-1493 (1995) [12].

NUREG-1493 analyzed the risk data presented in NUREG-1150 (1990) [13] for five plants: Surry, Peach Bottom, Sequoyah, Grand Gulf, and Zion. The NUREG-1150 risk data was made applicable to the focus of the report, i.e., to provide technical bases for revising Option A of Appendix J requirements. Additionally, NUREG-1493 analyzed the operating data from North Anna and Grand Gulf Plants. Based on the operating data, the study calculated the change in risk (in person-rem) to the public for the 15 alternatives considered in the study. For example, Alternative 4 maintains the current Appendix J acceptance criteria for Type B, and Type C tests, and reduces the Type A (ILRT) frequency from three per 10 years to one per 10 years. The study concluded that the cost saving to the industry would be around $500 millions, with an insignificant increase in the risk to the public.

Option B of Appendix J does not provide prescriptive requirements for scheduling the periodic ILRTs and LLRTs. It simply states: "Type A test must be conducted (1) after the containment system has been completed and is ready for operation and (2) at a periodic interval based on the historical performance of the overall containment system as a barrier to fission product releases to reduce the risk from reactor accidents." For Type B and Type C tests, Option B states: "Type B pneumatic tests to detect and measure local leakage rates across pressure retaining, leakage-limiting boundaries, and Type C pneumatic tests to measure containment isolation valve leakage rates, must be conducted (1) prior to initial criticality and (2) periodically thereafter at intervals based on the safety significance and historical performance of each boundary and isolation valve to ensure the integrity of the overall containment system as a barrier to fission product release to reduce the risk from reactor accidents."

NEI TR 94-01 (1995) [14], contains the industry guidelines for implementing the Option B requirements. The report briefly discusses the Appendix J, Option B requirements and current testing methodologies for Type A, Type B, and Type C testing, and provides guidelines for determining the performance-based tests intervals for the three types of tests. For performing Type A, Type B, and Type C tests, the report recommends the use of ANSI/ANS 56.8 (1994) [9]. The report principally relies on the risk assessment in NUREG-1493 for establishing the guidelines. In Sections 9 and 10, NEI TR 94-01 [14] provides guidelines for establishing test intervals for leak rate tests. The following is a brief description of the relevant guidelines for Type A, Type B, and Type C tests.

4.4.2.1 Type A Test Requirements. The licensees are required to perform preoperational Type A test prior to reactor initial operation, and subsequent periodic Type A tests has to be performed within 48 months of the prior successful Type A test. The periodic Type A test interval can be increased to 10 years based on an acceptable performance history. An acceptable performance history is defined as completion of two consecutive periodic Type A tests where the "calculated performance leakage rate" was less than 1.0 La (where La is the acceptable leakage rate expressed in percentage of weight of dry air in the containment) at the containment peak calculated pressure (Pa) derived for the design basis accident. The report provides guidance for calculating the performance leak rate. It also establishes an acceptable criterion of 0.75 La for as-left Type A test performed just prior to the start of reactor operation, but after all the required repairs and replacements are made. Thus, the acceptance criteria for assessing the results of Type A tests are essentially the same as those required for Option A of Appendix J.

Section 9.2.6 of the report provides guidelines for corrective actions, if the as-found (performed after the plant shutdown, prior to making any major fixing) Type A performance leakage rate is not acceptable. The report states, "Once the cause determination and corrective actions have been completed, acceptable performance should be reestablished by performing a Type A test within 48 months following the unsuccessful Type A test." Furthermore, the report recommends that if the unacceptable Type A performance leakage rate is due to the failure of Type B or Type C test components, the unsuccessful test need not be included in the determination of the Type A performance leakage rate and Type A test interval. The corrective actions should follow the Type B and Type C corrective action guidelines.

4.4.2.2 Type B Test Requirements. Consistent with the current practice in Option A of Appendix J, Type B test interval criteria are established separately for containment access penetrations, such as air locks, and other penetrations. For containment air locks, preoperational test is required to be performed prior to the preoperational Type A test, and subsequent periodic testing of air lock is required to be performed once every 24 months. The air lock components, such as equalizing valves, door seals, and resilient seals are required be tested once every 24 months. The report refers to ANSI/ANS 56.8-1994 [9] for testing procedures for air locks and their components. Sections 10.2.2.2 and 10.2.2.3 recommend generic procedures for repair and adjustment of air locks and for corrective actions.

For other penetrations, Type B tests are required to be performed prior to initial reactor operation. Subsequent periodic Type B tests are required to be performed at a frequency of at least once every 24 months, until acceptable performance is established as defined in Section 10.2.1.2 of the report. The NEI report allows up to 10 years between Type B tests, based on the performance of the individual penetration.

4.4.2.3 Type C Test Requirements. NEI 94-01 recommends that Type C tests be performed prior to initial reactor operation. Subsequent Type C tests shall be performed at a frequency of at least 24 months. Test intervals for Type C isolation valves may be increased based upon completion of two consecutive periodic as-found Type C tests, where the results of each Type C test are within a licensee's allowable administrative limits. The administrative leakage rate limits are set by the licensees for individual penetrations and valves. However, the total leakage rates from these components must not exceed 0.6 La. When the Type C test performance criteria are met, the report permits the Type C test interval to be as much as 10 years.

Regulatory Guide 1.163 (1995) [15] endorsed the use of NEI 94-01 and ANSI/ANS-56.8-1994 for implementing the performance-based option of Appendix J with certain exceptions, such as that the maximum Type C test interval is 5 years instead of the 10 years recommended in NEI TR 94-01. Readers should note that ANSI/ANS-56.8-1994 has been revised as ANSI/ANS-56.8-2002 to make the standard applicable to the performance based approach.

4.4.3 Risk Informed Approach (RIA)

Though some of the elements of PBA (discussed in Section 4.4.2) were based on the assessment of change in risk of selected plants, NEI TR 94-01 set the requirements for leakage rate testing of containment components with variations in the frequencies of Type A, Type B, and Type C testing based on the plant-specific performance data.

NRC and the licensees of operating plants have recognized that from the standpoint of time and resources, the integrated leak rate testing (ILRT) of containment is costly. The objective of making the ILRT intervals performance-based was to help licensees reduce the refueling outage time while maintaining an adequate level of safety. The objective can be accomplished if the plant-specific performances, as well as the industry-wide data, show insignificant increase in risk from increasing the interval between Type A tests. NUREG-1493 [12] showed an insignificant increase in risk by extending the Type A test time interval to as much as 20 years. To help licensees performing plant-specific risk assessments, Electric Power and Research Institute (EPRI) developed a methodology to assess the risk impact of revising the ILRT intervals. The methodology is described in EPRI TR-104285 (1994) [16]. The EPRI report uses the basic methodology for risk assessment from NUREG-1493, but systematizes the steps to help licensees integrate their plant-specific risk assessment results for the relevant risk analysis. NUREG-1493 utilized the NUREG-1150 population dose model to estimate the risk.

When the NUREG-1493 and EPRI reports were developed (1993–1994), the NRC had not developed explicit guidelines for changing the licensing basis (LB) using risk-informed decision making. In 1998, the NRC issued RG 1.174 (1998) [17], which provided qualitative, as well as quantitative guidelines for making licensing basis changes. The criteria are established based on the Commission's Final Policy Statement for Probabilistic Risk Assessment (PRA) [18], which establishes the safety goals for nuclear power plants as follows:

> "Through the Safety Goal Policy Statement, the (NRC) Commission propagated its philosophy that the risk from the operation of a nuclear power plant should be no more than 0.1% of the risk to which people are exposed from other sources. This statement of risk translates into objectives on individual risk of 2E-06/yr of a latent fatality and 5E-07/yr of an early fatality." The subsidiary numerical objectives to be used with appropriate consideration of uncertainties in making regulatory judgments are: (1) a core damage frequency (CDF) of 1E-04/reactor-year (RY) as for accident prevention and (2) a conditional containment failure probability (CCFP) of 0.1 for accident mitigation. In its policy statement, the Commission also states: "The use of PRA technology should be increased in all regulatory matters to the extent supported by the state-of-the-art in PRA methods and data, and in a manner that complements the NRC's deterministic approach and supports the NRC's traditional defense-in-depth philosophy."

Based on this philosophy, the NRC staff developed RG 1.174 [17]. The acceptance criteria to be used for making licensing basis changes are provided in Figs. 3 and 4 of RG 1.174. Figure 3 shows how a licensing change would affect the relationship between the reactor core damage frequency (CDF) and change in CDF (ΔCDF). Figure 4 shows how a specific licensing bases change would affect the relationship between large early release frequency (LERF) and ΔLERF. These criteria have been found sufficient for assessing the impact of the change in the currently established requirements for the safety of the plant.

In containment-related risk assessment, LERF and ΔLERF relationship is of vital importance, and is shown in Fig. 4 of RG 1.174, and has been reproduced above as Fig. 4.1. In addition to the criteria, the guide provides a number of qualitative pointers on PRA quality assurance and how to account for uncertainties. It also permits the NRC staff to base its safety assessment on the risk-informed criteria in combination with deterministic assessment.

Recognizing the low-risk significance of increasing the ILRT intervals, a number of licensees performed risk assessments and submitted requests for approval of one-time licensing basis changes to extend the interval for performing the ILRT to up to 15 years. The requests included risk assessments based on the techniques of EPRI report [16], and RG 1.174 [17] criteria. The NRC staff had reviewed and approved such requests on a case by case basis. The industry also developed generic methodologies to obtain approval for a group of similar plants to change the containment ILRT interval to 20 years. After a number of discussions on the subject, NEI requested NRC to consider ILRT interval of 15 years on a permanent basis in NEI 94-01, Revision 2A (2008) [19]. As of May 2012, the NRC has not endorsed NEI TR 94-01(2008) [19] revision in RG 1.163 revision. However, the NRC staff's Safety Evaluation Report (2008) [20] has accepted the revised NEI TR 94-01, Rev. 2A (2008) [19], and accompanied EPRI Report 1009325, Revision 2 [21], for use in applying for permanent 15 year ILRT interval.

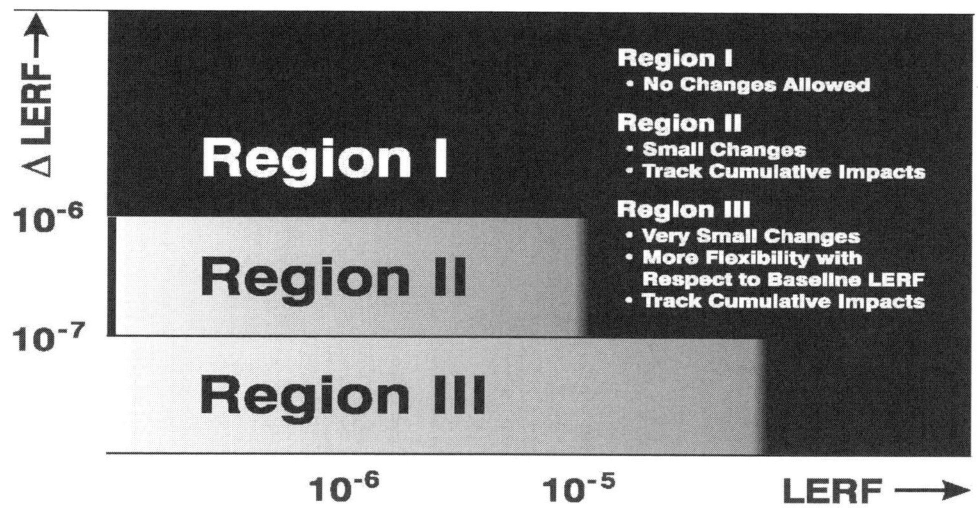

FIG. 4.1
ACCEPTANCE CRITERIA FOR LERF (SOURCE: REF. 17)

4.4.3.1 Discussion of RIA Issues. Incorporation of potential uncertainties is an important consideration in risk-informed decision making. Inservice inspections (ISIs) and leakage rate tests (LRTs) reduce the uncertainties related to the containment's ability to perform its intended function. When the ISIs and LRTs are performed in accordance with the prescriptive deterministic approach, they qualitatively assure high containment availability and reliability. However, when it can be shown that these deterministic approaches impose unnecessary burden on the industry resources, means have to be developed that would reduce such burden. The impact of reducing the frequency of performing the ISIs and LRTs needs to be addressed in relation to the increase in risk. A number of combinations of ISI frequencies, and LRT frequencies are possible. However, thus far, the NRC staff has received risk-informed applications for reducing the frequency of the ILRT, keeping the same performance-based LLRT frequencies and deterministic ISI frequencies. The following is a summary description of the factors involved in approving such applications.

The applications submitted by the nuclear power plant (NPP) licensees included a discussion of the risk consequences of extending the ILRT frequencies from three times in 10 years to once in 10 and 15 years. In assessing the risk impact of extended ILRT test intervals, the licensees assess the impact of change on LERF and the off-site consequences (person-rem per year), and the conditional containment failure probability (CCFP). Furthermore, the licensees stipulated that the containment structural integrity would be maintained between the ILRT time intervals through the ISI of containments performed in accordance with the requirements of 10 CFR 50.55a. During the in-depth reviews of the proposed licensee requests, the NRC staff pointed out that the containment ISIs and ILRTs supplement each other in ensuring the integrity of containments. Hence, the staff requested plant-specific information on containment ISIs to determine if extending the ILRT interval would adversely affect the containment integrity.

Over the last 25 years, a number of age-related degradation events have been identified as documented in NUREG-1522 [2] and ORNL/TM-2005/520 [22]. These degradations included the corrosion degradation of a steel shell of Mark I containments, and the steel liners of both Mark I and dry pressurized-water reactor containments. The information gathered in References 2 and 22, and additional information reported since then, convinced the NRC staff, that any attempt to reduce the frequency of ILRTs through risk-informed decision making must incorporate the potential for such degradation in the risk assessment. During the reviews of such applications, one of the generic questions raised by the NRC staff was related to the consequences of degradations that could occur in the uninspectable (e.g., embedded) areas of the containment. The NRC staff pointed out that such degradations could not be detected by the visual examination performed in accordance with Subsection IWE of Section XI of the ASME Boiler and Pressure Vessel Code. Accordingly, as part of the technical reviews of ILRT interval extension requests, the NRC staff sought information on how the potential leakage due to the age-related degradation mechanisms described above is factored into the risk assessment for the extension of the ILRT interval.

Neither NUREG-1493 [12], EPRI TR-104285 [16], nor NEI 94-01 [14] provides guidelines for incorporating the effects of degraded containments in licensees' risk assessments. The periodic containment ISIs cannot detect the existence of corrosion on the uninspectable side of containment. If such concealed corrosion has become widespread, an ILRT performed at peak design pressure could show unacceptable leakage through the containment. Such degradations at aging nuclear power plants were discovered at Oyster Creek in 1991, at Brunswick in 1999, at North Anna in 1999 and at D.C. Cook in 2000, and can potentially occur at other plants in the future. It is imperative to address the potential for such concealed degradations in the risk assessment related to the ILRT intervals.

A credible risk-informed analysis requires a credible database. The database should reflect the number of ILRT failures that have occurred in the past, with a stipulation that these failures would have gone undetected, if the ILRTs were not performed as required. However, the database cannot be directly used without some modifications for uncertainties and unidentified degradations of containment components. The readers should note that the databases discussed in Chapter 4 of NUREG 1493 and used for the statistical analysis were obtained from plants which were relatively new at that time (less than 15 years old), and the ILRT failure data was obtained when the tests were performed at higher frequencies (i.e., three times in 10 years). Unless licensees request a one-time extension to perform ILRT at more than 10 years, the ILRTs will be performed at 10 years. With plants experiencing environmental and aging degradations, the ILRTs for the plants tested in "as is" condition are likely to have higher failure rates than the ones analyzed in NUREG-1493. Depending upon the timing of performing the Type B tests, in relation to the ILRT test, the consequence of relaxing the test interval for Type B tests is likely to result in higher ILRT failure rates. Thus, to make a credible risk-informed decision on the frequency of containment ILRT, the licensees should consider the following factors:

1. The industry-wide database should be consistent and should include all operating plants. For example, Table Summary 1 of EPRI, report (Ref. 16) should be based on actual values from all plants rather than from a range of values from a few selected plants.
2. As discussed above, potential degradation on the uninspectable side of the containment, as well as the components which require ILRT to ensure their integrity (e.g., two-ply vent bellows in Mark 1 containments) should be factored in the risk analysis.
3. The plant-specific probabilistic risk assessment should be updated (as needed) to reflect the latest industry-wide database on the degradation of vital components.
4. Determination of large early release frequency (LERF), and adequate description of conditional containment failure probability (CCFP) of 0.1 for accident mitigation.

4.4.3.2 Industry Actions on ILRT Intervals. To get an approval of the extended ILRT interval of once in 15 years, NEI revised NEI 94-01 [14], together with EPRI TR-104285 (1994) [16]. Except for a few changes in references and clarification of some information, the basic recommendations of NEI 94-01, Rev. 2-A (2008) [19] are very similar to those of the earlier version. However, the new EPRI TR-1009325, Rev. 2, (2007) [21], incorporated a number of items that would guide the users (licensees) in developing the risk-related parameters required for the plant specific ILRT interval changes. Before discussing the basic features of the new EPRI report [21], it is important to briefly discuss the principles of relevant PRA parameters.

For NPPs, three levels of PRA have been identified. Level 1 deals with the analysis of internal processes that would affect calculation of CDF. Level 2 deals with the phenomena contributing to the containment vulnerabilities, and Level 3 analysis would provide the consequences of specific phenomena in terms of release of radionuclides that would affect public health and safety. Probabilistic risk assessments performed at levels 2 and 3 include an analysis of the containment. This analysis is important for differentiating among the consequences of various core-melt accident sequences and consists of two subtasks; (1) an identification of containment-failure modes, and (2) a prediction of the radionuclide inventory released to the environment for each accident sequence. They constitute the products of a level 2 PRA. They are also used in the subsequent tasks of more-extensive risk assessments.

A core-damage accident would induce a variety of physical processes in the reactor core, the pressure vessel, the reactor-coolant system. Computer codes have been developed to assist in the analysis of these processes. The results are insights into the phenomena associated with the accident sequence and a prediction of whether the containment fails.

A containment event tree (CET) is developed for each sequence of interest. If the containment is predicted to fail, the analysis predicts the time at which it will fail, where it will fail (i.e., whether radionuclides are released directly to the atmosphere through the containment building or to the ground through the basemat), and the energy associated with the release. Insights from this analysis could be used in the iterative process of constructing system event trees if accident phenomena affect system performance.

For each core-damage accident that is postulated to breach the containment, it is necessary to estimate the inventory of radionuclides that would be available for release to the environment. In this subtask, the analyst uses a computer model (1) to analyze the radionuclides released from the reactor fuel during the accident, and (2) to assess their transport and deposition inside the containment before containment failure. The EPRI report [21] has recognized these basic PRA features and systematically developed them to demonstrate the effect of ILRT interval on the risk parameter provided in RG 1.174 [17], that is, LERF. The following is a summary of the EPRI approach:

1. Collection of ILRT Failure Data

Data from ILRT tests has been collected at various times to support various applications. In summary, two NEI utility surveys, and an examination of recent ILRT results provided the ILRT data for 217 ILRT Type A tests that have been performed. Based on this data, the number of containment leakage events found during the performance of these tests is very small. The report emphasizes that no failures that would result in a large early release (approximately defined as a leakage rate of 35 La or higher) have been found. Leakage paths detectable only by Type A tests, and therefore influenced by the ILRT extension interval time have not been observed with magnitudes greater than 1.4 La. Appendix A of the report is a compilation of data from two NEI utility surveys, NUREG-1493, and other events discovered in reviewing other industry data (Licensee Event Reports [LERs] and other reportable events).

Note: 10 CFR 50.73, "Licensee event report system," requires the holder of an operating license under this part (Part 50) or a combined license under part 52, to submit a Licensee Event Report (LER) for any event of the type described in this paragraph within 60 days after the discovery of the event. Readers should refer to the regulation for precise definition of LER system.

After discussing the interpretation of ILRT leakage rates compiled in Appendix A, the report points out that there have been no events identified that could have resulted in a large early release as currently defined. Several ILRT events had unknown leakage rates. From the description of the events the report infers that the leakage was not large (for example, holes drilled in liner and penetration leakage). In any event, the limited ILRT data results in an inability to directly calculate an ILRT failure rate. However, the information that the data provide is valuable in an expert elicitation designed to estimate the probability of ILRT failure rates for a wide magnitude of leakage rates.

2. Expert Elicitation of Data

To allay the fears of misinterpretation of the data, EPRI decided to utilize the expertise and experience of the selected individuals familiar with the subject matter and were involved in this type of activities. Appendices B, C, D, and E of the report provide detailed description of the elicitation methodology. In short, Appendix B provides an overview of the expert elicitation process and its application to the solicitation of expert opinion for the ILRT Type A Testing Interval Optimization Project. The process is based on the "Recommendations for Probabilistic Seismic Hazard Analysis: Guidance on Uncertainty and Use of Experts" {NUREG/CR-6372 (1997)} [23] and "Branch Technical Position on the Use of Expert Elicitation in the High-Level Radioactive Waste Program" {NUREG-1563 (1996) [24]}.

Appendix C provides a description of the expert elicitation preparation process. Combined with the ILRT problem statement and the ILRT expert elicitation process, this appendix provides a full description of the expert elicitation inputs, process, and its application to the risk impact assessment of the ILRT interval optimization. Appendix D provides the results of the expert elicitation as well as the analysis of those results. Included are the changes made by the experts to the input form and processes. Appendix E presents a summary of the expert elicitation input. A total of eight tables are provided. The first four tables are associated with the large containment type, and the later four tables are associated with small containment type. For the purpose of this study, a large containment is defined as having free air volume of larger than 28,300 m^3 (~ 1,000,000 ft^3), and small containment is defined as having the free containment volume of less than 28,300 m^3 (~ 1,000,000 ft^3). Appendix F of the EPRI report provides detailed results of the statistical analysis of the expert elicitation. The report shows the results in 12 tables and three figures. Appendix G provides a tabulation of the risk studies performed by the licensees when applying for NRC approval of one time 15-year interval for performing ILRT.

Appendix H of the report contains a template for performing the risk impact assessment of the extended integrated leak rate testing intervals. The main purpose of the template is to illustrate the types of information that should be included in a plant-specific confirmation of risk impact associated with the extension of ILRT intervals. The report emphasizes that the template is one suggested approach for performing the assessment. Other approaches are not precluded. However, in applying the template the analyst should ensure that all relevant information is appropriately documented. In addition, the final assessment should comply with appropriate plant specific procedures for the documentation and control of similar types of assessments. The report recommends the following five steps for performing the assessment. The report assumes that the plants have performed the plant specific PRAs, as required by NRC's Generic Letter 88-20 (1988) [10].

Step 1 — Quantify the base-line risk in terms of frequency per reactor year for each of the eight accident classes as explained in Tables 5.1 to 5.5 of the report.
Step 2 — Develop plant-specific person-rem dose (population dose) per reactor year for each of the eight accident classes.
Step 3 — Evaluate risk impact of extending Type A test interval from 3 to 15 and 10 to 15 years.
Step 4 — Determine the change in risk in terms of LERF in accordance with RG 1.174.
Step 5 — Determine the impact on the CCFP.

RG 1.174 [18] does not address CCFP as part of the assessment process. However, the (NRC) Commission has established a subsidiary numerical objective for CCFP to be less than 0.1. Recognizing the importance of containment vulnerabilities for certain sequences, the EPRI report recommends the use of Step 5, in the assessment process. The report provides the following calculation methodology to determine CCFP and ΔCCFP.

$$\text{CCFP} = [1 - (\text{Class 1 frequency} + \text{Class 3a frequency})/\text{CDF}] \times 100\%$$
$$\text{CCFP}_3 = \text{X.XX\% (ILRT interval 3 times in 10 years)}$$
$$\text{CCFP}_{10} = \text{X.XX\% (ILRT interval once in 10 years)}$$
$$\text{CCFP}_{15} = \text{X.XX\% (ILRT interval once in 15 years)}$$
$$\Delta\text{CCFP} = \text{CCFP}_{15} - \text{CCFP}_3 = \text{X.XX\%}$$
$$\Delta\text{CCFP} = \text{CCFP}_{15} - \text{CCFP}_{10} = \text{X.XX\%}$$

The NRC has accepted ΔCCFP up to 1.1 percentage point for the one-time ILRT interval requests for extension to 15 years. In Chapter 2 of the report, EPRI recommends, an increase in CCFP of ≤1.5 percentage point. During the reviews of application for one time ILRT interval to 15 years, the NRC had found that the LERF parameter is not sensitive to the ILRT interval. Hence, the evaluation of CCFP parameter is essential.

In Table 4.4-1 of the report, EPRI provides a six step process for considering the effects of non-detected liner corrosion, in the basemat, cylinder and dome of the containment.

4.5 MISCELLANEOUS REMARKS

A. When the judgment regarding the containment reliability is based on risk-informed assessment, all factors that contribute to its structural and leak-tight integrity have to be considered in the risk assessment. It is feasible to develop a probabilistic model which would consider frequencies of performing inspections and leak rate testing in relation to the resources required to conduct such inspections and tests, keeping the conditional probability of containment failure to less than 0.1. Since risk-informed methodologies are used to analyze the response of the containment to severe accidents (see Chapters 6 and 7 of this book for more information), it is important to consider accident sequences that address the containment vulnerability to degradation which cannot be detected by visual examination methods. In the United States, reliance has been placed on combination of risk-informed assessment backed with deterministic analysis to assess the effects of containment degradation on containment reliability. The EPRI report [21] has explicitly considered the effects of corrosion on the containment event tree. The report has also addressed the concerns described in Section 4.4.3.1 of this chapter.

B. When decisions to make changes in the requirements are based on risk assessment, Section 2.3 (Element 3) of RG 1.174 states:

> "Careful consideration should be given to implementation and performance-monitoring strategies. The primary goal for this element is to ensure that no adverse safety degradation occurs because of the changes to the licensing basis. The NRC staff's principal concern is the possibility that the aggregate impact of changes that affect a large class of structures, systems, and components (SSCs) could lead to an unacceptable increase in the number of failures from unanticipated degradation, including possible increases in common cause mechanisms. Therefore, an implementation and monitoring plan should be developed to ensure that the engineering evaluation conducted to examine the impact of the proposed changes continues to reflect the actual reliability and availability of SSCs that have been evaluated."

C. The above RG process will ensure that the conclusions that have been drawn from the evaluation remain valid. Thus, in context of the containment reliability, a careful consideration should be given to implementation and performance monitoring strategies to ensure that no adverse consequence occurs because of the simultaneous relaxation of ILRT, LLRT frequencies. As of June 2011, except for a few licensees, all reactor licensees have adopted the performance based approach of one time extension (to 15 years) for performing Type A test. They are likely to adopt the 15-year Type A test interval, when fully endorsed by the NRC.

D. During containment inspections, the operating experience has shown that the amount of degradation could be outside the Code required acceptance criteria that would require risk significance determination. NUREG-1765 [25] provides guidelines for making such determination. The regulatory requirements for allowable leakage rates are discussed in Sections 4.3.2 and 4.4.2 above. Typical design values of La are 0.1 containment volume percent per day for PWRs, 0.5 volume percent per day for BWR Mk I and Mk II containments, and 0.2 volume percent per day for BWR Mk III containments. The studies [12, 26, 27] have indicated that the containment leak rate of 100 volume percent per day appears to constitute an appropriate threshold beyond which the release may become significant to LERF. Thus, a LERF significant leakage rate from containment would be a rate of greater than or equal to 1000 La for PWRs. 200 La for Mk I and Mk II BWRs, and 500 La for Mk III BWRs. The 100 volume percent per day leakage rate is approximately equal to a circular hole size in containments of 6.3 to 7.6 cm (2.5-3 inches) in diameter for large dry PWRs, 5 cm (2 inches) diameter for PWR Ice-condenser containments, 2.5 cm (1 inch) diameter for BWR Mk I and Mk II containments, and 6.3 cm (2.5 in.) for MK III BWRs.

Note: For more information about PWR and BWR containments, please see Chapter 1 of this book.

REFERENCES

[1] Regulatory Guide 1.35 (1990): "Inservice Inspection of Ungrouted Tendons in Prestressed Concrete Containments," U.S. NRC, Washington DC, 20555, July 1990.
[2] NUREG-1522 (1995): "Assessment of Inservice Conditions of Safety-Related Nuclear Power Plant Structures," Prepared by Ashar, H., Bagchi, G., U.S. Nuclear Regulatory Commission, Washington, DC, June 1995.
[3] 10 CFR 50.55a (2001): "Codes and Standards," U.S. Nuclear Regulatory Commission, Washington, D.C. 1996 through 2010 Revisions.
[4] ASME Section XI, Subsection IWE, (2010): "Inservice Inspection Requirements for Class MC Components, and Metallic Liners of Class CC Components of Light-Water Cooled Plants," ASME Boiler and Pressure Vessel Code, New York, N.Y.
[5] ASME Section XI, Subsection IWL, (2010): "Inservice Inspection Requirements for Class CC Components of Light-Water Cooled Plants," ASME Boiler and Pressure Vessel Code, New York, N.Y.
[6] ASME Section III, Division 2, (2010): "Code for Concrete Containments," ASME B & PV Code, New York, N.Y.
[7] 10 CFR Part 50, Appendix J (2010): "Primary Reactor Containment Leakage Testing for Water-Cooled Power Reactors," US NRC, Washington DC. 20555.
[8] ANSI/ANS 56.8 (1987): "Containment System Leakage Testing Requirements," Published by American Nuclear Society (ANS), La Grange Park, IL 60525, January 1987.
[9] ANSI/ANS 56.8 (1994): "Containment System Leakage Testing Requirements," Published by American Nuclear Society (ANS), La Grange Park, IL 60525, 1994.
[10] NRC Generic Letter (GL) 88-20 (1988): "Individual Plant Examination for Severe Accident Vulnerabilities," US NRC, Washington D.C., 20555, Nov. 1988.
[11] NUREG-1273 (1988): "Technical Findings and Regulatory Analysis for Generic Safety Issue II.E.4.3, Containment Integrity Check," Prepared by Serkiz, for U.S. Nuclear Regulatory Commission, Washington, D.C. 20555, June 1988.
[12] NUREG-1493 (1995): "Performance-Based Containment Leak-Test Program." Prepared by Dey, M., and Skoblar, L. etc., for U.S. Nuclear Regulatory Commission, Washington, D.C. 20555. July 1995.
[13] NUREG-1150 (1990): "Severe Accident Risks: An Assessment for Five U.S. Nuclear Power Plants," U.S. Nuclear Regulatory Commission, Washington, D.C. 20555, December 1990.
[14] NEI TR 94-01 (1995): "Industry Guideline for Implementing Performance-Based Option of 10 CFR Part 50, Appendix J," Nuclear Energy Institute, Washington, D.C., July 1995.
[15] Regulatory Guide 1.163 (1995): "Performance-Based Containment Leak Test Program," U.S. Nuclear Regulatory Commission, Washington, D.C. 20555, September 1995.
[16] EPRI TR-104285 (1994): "Risk Impact Assessment of Revised Containment Leak Rate Testing Interval," Electric Power Research Institute, Palo Alto, CA. September 1994.
[17] Regulatory Guide 1.174 (2011), "An Approach for Using Probabilistic Risk Assessment in Risk-Informed Decisions on Plant-Specific Changes to the Licensing Basis," U.S. Nuclear Regulatory Commission, Washington, D.C. 20555, May 2011.
[18] 60 FR 42622, USNRC (1995): "Use of Probabilistic Risk Assessment Methods in Nuclear Activities: Final Policy Statement," Federal Register, Vol. 60, p. 42622, August 16, 1995.
[19] NEI TR 94-01, Rev. 2A (2008): "Industry Guideline for Implementing Performance-Based Option of 10 CFR Part 50." Appendix J, October 2008.
[20] NRC's Final Safety Evaluation Report (2008): (1) NEI Topical Report 94-01, Revision 2A, "Industry Guideline for Implementing Performance-Based Option of 10 CFR Part 50. Appendix J," and (2) EPRI TR No. 1009325, "Risk Impact Assessment of Extended Leak Rate Testing Intervals, Revision 2." Published by U.S. NRC, Washington DC, June 2008.

[21] EPRI TR-1009325 (2007): "Risk Impact Assessment of Extended Leak Rate Testing Intervals," Electric Power Research Institute, Palo Alto, CA. August 2007.

[22] ORNL/TM-2005/520 (2005): "Inspection of Aged/Degraded Containments," Prepared by Naus, D., Oland, C., Ellingwood, B., Oakridge National Laboratories, for U.S. NRC, September 2005.

[23] NUREG/CR-6372 (1997): "Recommendations for Probabilistic Seismic Hazard Analysis: Guidance on Uncertainty and Use of Experts," U.S. NRC, Washington DC 20555. April 1997.

[24] NUREG-1563 (1996): "Branch Technical Position on the Use of Expert Elicitation in the High-Level Radioactive Waste Program," U.S. NRC, Washington DC, 20555. 1996.

[25] NUREG-1765 (2002): "Basis Document for Large Early Release Frequency (LERF) Significant Determination Process (SDP)," U.S. Nuclear Regulatory Commission, Washington DC 20555, 2002.

[26] NUREG/CR-4330 (1986), "Review of Light Water Reactor Regulatory Requirements," U.S. Nuclear Regulatory Commission, Washington DC 20555, 1986.

[27] NUREG/CR-6418 (1998): "Risk Importance of Containment and Related ESF System Performance Requirements," U.S. Nuclear Regulatory Commission, Washington DC 20555, 1998.

CHAPTER

5

LICENSE RENEWAL AND AGING MANAGEMENT FOR CONTINUED SERVICE

Dan Naus and Hansraj Ashar

As of August 2011, there were 104 commercial nuclear power reactors licensed to operate in 31 states in the United States. Initial operating licenses in the United States are granted for a period of 40 years. In order to help assure an adequate energy supply, the USNRC has established a timely license renewal process and clear requirements that are needed to ensure safe plant operation for an extended plant life. The principals of license renewal and the basic requirements that address license renewal are identified as well as additional sources of guidance that can be utilized as part of the license renewal process. Aging management program inspections and operating experience related to the concrete and steel containment structures are provided. Finally, several lessons learned are provided based on containment operating experience.

Keywords: Aging degradation, aging management program, aging management review, concrete containment, containment steel liner, GALL report, inservice inspection, integrated plant assessment, license renewal, operating experience, post-tensioning tendons, steel containment

5.1 INTRODUCTION

In the United States, the Atomic Energy Act (AEA) and regulations of the U.S. Nuclear Regulatory Commission (NRC) limit commercial power reactor licenses to a 40-year period and allow these licenses to be renewed for an additional period of 20 years, with no limit to these renewals. The original 40-year term for reactor licenses was based on economic and antitrust considerations, not on technical limitations. Due to this selected period, however, some of the structures and components may have been engineered on the basis of an expected 40-year service life. In order to ensure safe operation of nuclear power plants (NPPs), it is essential that the effects of age-related degradation of plant structures, as well as systems and components, be assessed and managed during both the current operating license period as well as subsequent license renewal periods [1].

5.2 LICENSE RENEWAL PROCESS, SAFETY PRINCIPLES, AND REGULATIONS

In December 1991, the NRC published 10 CFR Part 54 (License Renewal Rule) to establish the procedures, criteria, and standards governing NPP license renewal. In September 1994 the NRC proposed an amendment

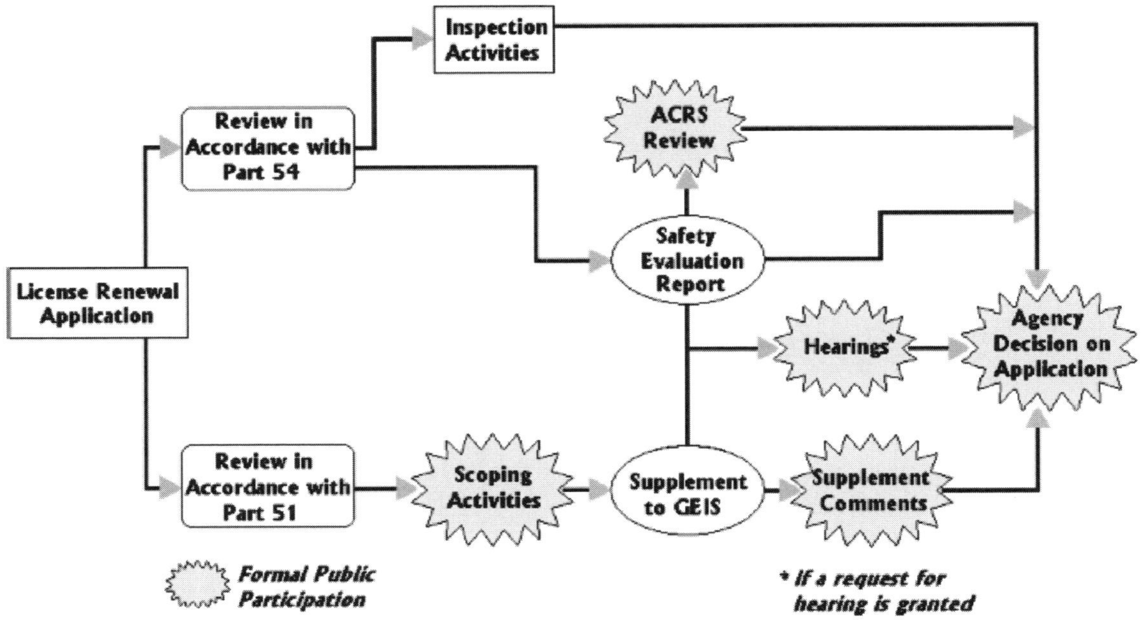

FIG. 5.1
SCHEMATIC OF LICENSE RENEWAL PROCESS
(Source: http://www.nrc.gov/reading-rm/doc-collections/fact-sheets/fs-reactor-license-renewal.html)

to the Rule, published in May 1995, that focuses on the effects of aging on long-lived passive structures and components and time-limited aging analyses (TLAAs) as defined in 10 CFR 54.21(a)1 and 54.3, respectively. The amendment also permitted greater reliance on the current licensing basis (CLB), the maintenance rule, and existing plant programs.

The License Renewal Rule process and application requirements for commercial power reactors are based on two key principles: (1) that the current regulatory process, continued into the extended period of operation, is adequate to ensure that the CLB of all currently operating plants provides an acceptable level of safety, with the possible exception of the detrimental effects of aging on certain systems, structures, and components (SSCs), and possibly a few other issues related to safety only during the period of extended operation; and (2) each plant's CLB is required to be maintained. The license renewal process proceeds along two tracks — one for safety issues (10 CFR Part 54) and one for environmental issues (10 CFR Part 51). Figure 5.1 provides a schematic of the license renewal process. The time from submittal of the License Renewal Application (LRA) to issuance of a renewed license is intended to be completed within a 22-month period unless it involves a hearing in which the process may require 30 months.

5.2.1 10 CFR Part 54 (Rule)

The scope of the Rule (10 CFR Part 54) includes: (1) safety-related SSCs that are relied upon to maintain integrity of the reactor coolant pressure boundary, ensure capability to shut down and maintain a safe shutdown condition, and prevent or mitigate offsite exposures comparable to 10 CFR Part 100 [2]; (2) non-safety-related SSCs whose failure could prevent safety-related functions as noted above; and (3) SSCs relied upon for compliance with regulations (i.e., fire protection, environmental qualification, pressurized thermal shock, anticipated transients without scram, and station blackout).

The LRA identifies reactor systems, structures, or components that would be affected by license renewal; demonstrates that it can manage the adverse effects of aging during the renewal period; and analyzes the environmental effects of extended reactor operation during the renewal term [3]. Applicants wishing to submit a LRA are responsible for preparing a plant-specific LRA that includes both general and technical information. The general information is similar to that provided with the initial plant operating license application. Technical information includes an Integrated Plant Assessment (IPA) [10 CFR Part 54.21(a)(1)], TLAAs (10 CFR Part 54.3), a supplement to the Final Safety Analysis Report (FSAR), and Technical Specification Changes (10 CFR Part 54.22). Readers should refer to Section 2.4.2 (Chapter 2) of this book for more information on FSAR and Technical Specification requirements.

The IPA identifies and lists structures and components subject to an aging management review (AMR) that perform intended functions without moving parts or without change in configuration or properties (passive) or that are not subject to replacement based on a qualified life or specified period (long-lived). For some passive structures and components within the scope of license renewal, no additional action may be required where an applicant can demonstrate that the existing programs provide adequate aging management throughout the period of extended operation. The containments and seismic Category I structures are identified as components subject to an AMR. Intended functions are those that the in-scope SSCs must be shown to fulfill that would form the basis for including the SSCs within the scope of the Rule. Methods used to identify SSCs subject to an AMR are to be identified. Finally, the applicant must demonstrate that the effects of aging will be adequately managed so that their intended function(s) will be maintained consistent with the CLB for the period of extended operation. Each year following submittal of the LRA and at least three months before the scheduled completion of the NRC review, an amendment to the LRA must be submitted that identifies any changes to the CLB that materially affects the contents of the LRA including the FSAR supplement.

Time Limited Aging Analyses are calculations or analyses that involve SSCs within the scope of the Rule that consider the effects of aging and involve assumptions based on the original 40-year operating term. For license renewal, TLAAs must be: (1) verified to bound the renewal period, (2) reanalyzed (recalculated) to determine if it will bound the renewal period, or (3) the applicant must show that the aging effects encompassed by the calculation will be managed. TLAAs are to be contained or incorporated by reference in the CLB.

As part of the LRA, the applicant is required to provide a supplement to the FSAR that provides a commitment to implement and summary descriptions of the required programs and activities for managing the effects of aging. The FSAR supplement is also required to provide evaluation of TLAAs for the period of extended operation. Technical specification changes or additions, with justification, necessary to manage the effects of aging during the period of extended operation are also to be included as part of the application.

5.2.2 10 CFR Part 51

In addition to its mission of protecting public health and safety under the AEA, the NRC is charged with protection of the environment in the use of nuclear materials. In 1996, the environmental regulations in 10 CFR Part 51 were revised to improve regulatory efficiency in environmental review for license renewal and codify the findings documented in the Generic Environmental Impact Statement (GEIS) for License Renewal [4]. The GEIS examines the possible environmental impacts that could occur as a result of renewing any commercial NPP license and, to the extent possible, establishes the bounds and significance of these potential impacts.

5.3 GUIDANCE DOCUMENTS

The NRC staff and industry have developed a number of guidance documents related to license renewal that are applicable to the containment structures.

5.3.1 NRC Guidance Documents

In addition to the Rule (10 CFR Part 54), the basic documents, that provide the regulatory requirements and regulatory framework for ensuring continued plant safety for license renewal include, Regulatory Guide

1.188, Standard Format and Content for Applications to Renew Nuclear Power Plant Operating Licenses [5], the Generic Aging Lessons Learned (GALL) Report [6], and the Standard Review Plan (SRP) for Review of License Renewal Applications for Nuclear Power Plants [7]. These guidance documents are living documents that are periodically updated. Additional sources of guidance related to license renewal include: inspection manual chapters, inspection procedures, information notices, license renewal interim staff guidance, regulatory guides, office instructions, nuclear plant aging research reports, and technical reports in NUREG series. Provided below are brief descriptions of selected basic and guidance documents related to license renewal.

5.3.1.1 Regulatory Guide 1.188, Revision 1. Regulatory Guide 1.188 [5] provides guidance to applicants on the format and content of a LRA to ensure uniformity of the format and level of technical content, and endorses NEI Guideline 95-10, Rev. 6 [8] for implementing the Rule. Conformance with the regulatory guide is not required, but its use facilitates both the applicant's preparation of a LRA and the NRC staff's timely and consistent review. Other formats are considered acceptable if they provide an adequate basis for approval of the application for license renewal.

5.3.1.2 Generic Aging Lessons Learned (GALL) Report. The GALL Report lists generic AMRs of SSCs that may be in the scope of LRAs and identifies aging management programs (AMPs) that are determined to be acceptable to manage aging effects of SSCs in scope of license renewal as required by 10 CFR Part 54. The GALL Report provides a technical basis for the SRP for License Renewal and contains the NRC staff's generic evaluation of the existing plant programs and documents the technical bases for determining where existing programs are adequate without modification and where existing programs should be augmented for the extended period of operation. Each structure and/or component is identified as well as its material(s) of construction, environment, aging effects/mechanisms, acceptable programs to manage the effects of aging, and if further evaluation is required. The adequacy of the generic AMPs in managing certain aging effects for particular structures and components is based on a review of ten program elements of an AMP for license renewal [7]: scope of program, preventative actions, parameters monitored or inspected, detection of aging effects, monitoring and trending, acceptance criteria, corrective actions, confirmation processes, administrative controls, and operating experience.

The GALL Report contains 11 chapters and an appendix [6]. Chapter I addresses application of the American Society of Mechanical Engineers (ASME) Code to license renewal. Selected sections of the ASME Code are endorsed by the NRC as incorporated in 10 CFR Part 50.55a. The NRC periodically amends 10 CFR Part 50.55a and issues Federal Register Notices about this rule in order to endorse, by reference, newer editions and ASME Code Addenda subject to the modifications and limitations identified in 10 CFR Part 50.55a. Some of the AMPs referenced in the GALL Report are based entirely or in part on compliance with the requirements of ASME Code Section XI, "Rules for Inservice Inspection of Nuclear Power Plant Components." The staff has determined that the referenced ASME Section XI programs or requirements provide an acceptable basis for managing the effects of aging during the period of extended operation for these AMPs, except where noted and augmented in the GALL report (e.g., Section XI Subsections IWE," Requirements for Class MC and Metallic Liners of Class MC Components of Light-Water Cooled Plants," and IWL, "Requirements for Class CC Concrete Components of Light-Water Cooled Plants"). The licensee is required to amend its CLB by updating its ASME Section XI edition and addenda of record to the most recently endorsed edition and addenda referenced in 10 CFR Part 50.55a 1 year prior to entering the next 10-year internal inservice inspection for its unit. Chapters II through VIII contain summary descriptions and tabulations of evaluation of AMPs for a large number of structures and components in major plant systems found in light-water reactor (LWR) NPPs. Chapter IX contains definitions of a selection of standard terms used within the GALL Report. Chapter X contains the TLAA evaluation of AMPs under 10 CFR Part 54.21(c)(1)(iii). Chapter XI contains AMPs for the structures and mechanical and electrical components. The Appendix to the GALL Report addresses quality assurance for the AMPs.

Chapter II of the GALL report addresses Pressurized Water Reactor (PWR) and Boiling Water Reactor (BWR) containment structures. Safety related (e.g., BWR reactor building and PWR shield building) and other structures (e.g., containment internal structures and water-control structures) are addressed in Chapter III of the GALL Report. Pressurized Water Reactor containments are subdivided into concrete containments (reinforced and prestressed), steel components, and common components. Boiling Water Reactor containments are subdivided into Mark I containments (steel and concrete), Mark II containments (steel and concrete), Mark III containments (steel and concrete), and common components. Information provided in Chapter II, as well as Chapters III through VIII, is in a tabular format.

AMPs related to NPP containment structures are identified in Chapter XI of the GALL Report. This chapter provides a description of each of these programs as well as the evaluation and technical basis related to a review of the ten program elements of an AMP for license renewal identified in the SRP-LR [7]. AMPs related to the concrete containment include: ASME Section XI, Subsection IWL (GALL AMP XI.S2) and Structures Monitoring (GALL XI.S6). AMPs related to the post-tensioning system include: ASME Section XI, Subsection IWL (GALL AMP XI.S2) and Concrete Containment Tendon Prestress Time-Limited Aging Analysis (GALL TLAA X.S1). AMPs related to the steel containments and liners of reinforced concrete containments include: ASME Section XI, Subsection IWE (GALL AMP XI.S1), 10 CFR Part 50, Appendix J (GALL AMP XI.S4), and Containment Liner Plate, Metal Containments, and Penetrations Fatigue Analysis Time-Limited Aging Analysis. With the exception of the Structures Monitoring Program and the TLAA associated with the post-tensioning system and the containment metal components, the AMPs are discussed in Chapter 4 of the GALL report.

The Structures Monitoring AMP (XI.S6) includes all structures, structural components, component supports, and structural commodities in the scope of license renewal that are not covered by other AMPs such as noted above. The structures monitoring program consists of periodic visual inspections by personnel qualified to monitor structures and components for applicable aging effects such as noted in American Concrete Institute Standards (ACI) 349.3R [9], ACI 201.1R [10], and American National Standards Institute/American Society of Civil Engineers Standard (ANSI/ASCE) 11 [11]. In general, all structures are monitored on a frequency not to exceed 5 years, but some structures subject to benign environmental conditions, may be monitored at an interval exceeding 5 years. Identified aging effects are evaluated by qualified personnel using criteria derived from industry codes and standards contained in the plant CLB, including ACI 349.3R, ACI 318 [12], ANSI/ASCE 11, and the American Institute of Steel Construction (AISC) specifications, as applicable. Acceptance criteria are selected for each structure/aging effect to ensure that the need for corrective actions is identified before loss of intended functions. The criteria are derived from design bases codes and standards that include ACI 349.3R, ACI 318, ANSI/ASCE 11, or relevant AISC specifications, as applicable, and consider industry and plant operating experience. The structures monitoring program also addresses detection of aging effects for inaccessible below-grade concrete structural elements. Groundwater is sampled at a frequency not to exceed 5 years and for plants with non-aggressive groundwater/soil (pH > 5.5, chlorides < 500 ppm, and sulfates < 1500 ppm), the program recommends: (1) evaluating the acceptability of inaccessible areas when conditions exist in accessible areas that could indicate the presence of, or result in, degradation to such inaccessible areas; and (2) examining representative samples of exposed portions of the below grade concrete, when excavated for any reason. A plant-specific AMP, accounting for the extent of degradation experienced, should be implemented to manage aging during the period of extended operation where aggressive groundwater is present or concrete structural elements have experienced degradation.

The Concrete Containment Tendon Prestress TLAA AMP (X.S1) provides reasonable assurance of the adequacy of prestressing forces in prestressed concrete containment tendons during the period of extended operation under 10 CFR Part 54.21(c)(1)(iii). The program consists of an assessment of inspections performed in accordance with the requirements of Section XI, Subsection IWL of the ASME Code, as supplemented by the requirements of 10 CFR Part 50.55a(b)(2)(viii). The assessment related to the adequacy of the prestressing force establishes: (1) acceptance criteria in accordance with NRC Regulatory Guide 1.35.1 [13], and (2) trend lines based on the guidance provided in NRC Information Notice 99-10 [14]. The estimated and measured prestressing forces are plotted against time, and the predicted lower limit, minimum required value, and trending lines are developed for the period of extended operation. Regulatory Guide 1.35.1 provides guidance for calculating the predicted lower limit and minimum required value. The trend line represents the trend of prestressing forces based on the actual measured forces. The prestressing force is acceptable when the trend line is above the minimum required value throughout the period of extended operation. If acceptance criteria are not met, then either systematic retensioning of tendons or a reanalysis of the containment is warranted to ensure design adequacy of the containment. Appendix 5A of this chapter provides additional information related to monitoring and trending the prestressing forces in NPP post-tensioned concrete containments.

If a plant's code of record (ASME Section III, Division 1 or Division 2) requires a fatigue analysis of the liner plate, then the analysis may be a TLAA and must be evaluated in accordance with 10 CFR Part 54.21(c)(1) to ensure that the effects of aging on the intended functions are adequately managed for the period of extended operation. The ASME Section III fatigue analysis requires the calculation of a cumulative usage factor based on the fatigue properties of the materials and the expected fatigue service of the component. The ASME Code limits the cumulative usage factor to a value less than or equal to one for acceptable design fatigue. Three acceptance criteria for fatigue of containment liner plates are available: existing calculations remain valid because the number of assumed cyclic loads will not be exceeded during the period of extended operation; the cumulative usage factor calculations are re-evaluated based on an increased number of assumed cyclic loads to cover the period of extended operation and the new cumulative usage factor does not exceed one; and a AMP provided by the applicant demonstrates that the effects of aging on the component's intended function(s) will be adequately managed during the period of extended operation.

5.3.1.3 Standard Review Plan for License Renewal (SRP-LR). The SRP-LR provides technical guidance for NRC staff reviewers and documents acceptance criteria and review procedures that are acceptable to the NRC staff [7]. The SRP-LR has the purpose of assuring the quality and uniformity of NRC staff reviews and presenting a well-defined base from which to evaluate applicant programs and activities for the period of extended operation. The SRP-LR is also intended to make regulatory information widely available to enhance communication with interested members of the public and the nuclear power industry and to improve their understanding of staff reviews. This report incorporates by reference the GALL Report and Regulatory Guide 1.188.

Each SRP-LR section is organized into subsections, generally consistent with "Standard Review Plan for Review of Safety Analysis Reports for Nuclear Plants [15]." Each of the individual SRP-LR sections address: (1) who performs the review, (2) the matters that are reviewed, (3) the basis for the review, (4) the way the review is accomplished, and (5) the conclusions that are drawn. Chapter 3 of the SRP-LR addresses how the AMRs and AMPs are reviewed.

Section 3.5 of the SRP-LR addresses the AMR and the associated AMPs for containments, structures, and component supports. Section 3.5.2 of the SRP-LR provides acceptance criteria describing methods to determine whether the applicant has met the requirements of 10 CFR Part 54.21. The AMR addresses three areas: (1) AMR results consistent with the GALL Report, (2) AMR results for which further evaluation is recommended in the GALL Report, and (3) AMR results that are not consistent with or not addressed in the GALL Report. For PWR and BWR containments the SRP-LR identifies a number of areas for which further evaluation is required: (1) cracking and distortion due to increased stress levels from settlement, reduction in foundation strength, and cracking due to differential settlement and erosion of porous concrete subfoundations; (2) reduction in concrete strength and modulus of elasticity due to elevated temperatures; (3) loss of material due to general, pitting, and crevice corrosion; (4) loss of prestress due to relaxation, shrinkage, creep, and elevated temperature; (5) cumulative fatigue damage; (6) cracking due to stress corrosion cracking; (7) loss of material (scaling, spalling) and cracking due to freeze–thaw cycles; (8) cracking due to expansion from reaction with aggregates; and (9) increase in porosity and permeability due to leaching of calcium hydroxide and carbonation. The AMR consists of identifying the material, environment, aging effects, and the AMP(s) credited for managing the aging effects.

Review of the AMPs requires assessment of the ten program elements as defined in the SRP-LR to verify their technical adequacy. The staff verifies that the applicant's programs are consistent with those described in the GALL Report and/or with plant conditions and operating experience during the performance of an AMP audit and review that is conducted at the applicant's facility to evaluate the AMPs that the applicant claims to be consistent with the GALL Report.

5.3.1.4 Nuclear Plant Aging Research (NPAR) Reports. A comprehensive Nuclear Plant Aging Research (NPAR) Program was implemented by the NRC in 1985 to identify and resolve technical safety issues related to the aging of SSCs in operating NPPs, both during the initial operating license period as well as periods of license renewal, that have important safety functions [16, 17]. The NPAR Program was directed at gaining knowledge and understanding of the degradation processes within NPPs with an emphasis placed on identifying and characterizing the mechanisms of material and component degradation during service and using research results in the regulatory process. The research included evaluating methods of inspection, surveillance, condition monitoring, and maintenance as a means of managing aging effects that may impact safe plant operation. Goals of the program were to: (1) identify and characterize aging effects that, if unchecked, could cause degradation of SSCs and thereby impact plant safety; (2) identify methods of inspection, surveillance, and monitoring, and evaluate the residual lifetime of SSCs that will ensure the timely detection of significant aging effects before loss of safety function; and (3) evaluate the effectiveness of storage, maintenance, repair, and replacement practices in mitigating the rate and extent of degradation caused by aging. During the conduct of the NPAR Program over 80 technical reports were prepared. Of these reports, one addressed concrete structures [18] and one addressed metal containments [19].

5.3.1.5 Technical Reports in NUREG Series (NUREGs). NUREG Series Publications are reports or brochures on regulatory decisions, results of research, results of incident investigations, and other technical and administrative information. The publications can be in the form of: publications prepared by NRC staff (NUREG-nnnn) or contractors (NUREG/CR-nnnn), brochures prepared by NRC staff (NUREG/BR-nnnn), conference proceedings prepared by NRC staff or contractors (NUREG/CP-nnnn), and publications resulting from international agreements (NUREG/IA-nnnn). NUREGs prepared by the NRC staff are available (http://nrc.gov/reading-rm/doc-collections/nuregs/staff/) as well as those prepared by contractors (http://nrc.gov/reading-rm/doc-collections/nuregs/contract/). Two NUREG publications, in addition to the two reports discussed in the previous section prepared under the NPAR Program, provide information related to NPP containment structures and license renewal [20, 21].

5.3.2 Industry Guidance Documents

Several industry documents have been prepared pertinent to license renewal that are related to NPP containment structures.

5.3.2.1 NUMARC Reports. The U.S. nuclear power industry, through coordination by the NUMARC, and sponsorship by the DOE and the EPRI, has evaluated age-related degradation effects for a number of major plant SSCs in the license renewal technical industry reports (IRs). In 1990, NUMARC, now the Nuclear Energy Institute (NEI), submitted for NRC review, ten IRs dealing with specific structures and components of NPPs and one IR addressing a screening methodology for performing an IPA under 10 CFR Part 54 [22]. Of these reports, one addressed PWR containments [23], one addressed BWR containments [24], and one addressed Class I Structures [25]. No safety evaluations were developed for review of these IRs; however, a summary of the technical information and NUMARC/NRC agreements resulting from a review of nine of these reports was developed [26].

5.3.2.2 NEI 95-10. The industry, through the NEI, has developed a guidance document providing an acceptable approach for implementing the requirements of 10 CFR Part 54 and how to perform a license renewal assessment for a plant and what information to submit in a LRA [8]. The guidance provided in this document is founded on industry experience in implementing the Rule. As such, the document addresses: identification of SSCs within the scope of license renewal; identification of the intended functions of SSCs within the scope of license renewal; identification of the structures and components subject to AMR and their intended functions; assurance that the effects of aging are managed; application of new programs and inspections for license renewal; identification and resolution of TLAAs; identification and evaluation of exemptions containing TLAAs; and identification of a standard format and content of a LRA.

5.4 LICENSE RENEWAL INSPECTIONS

The fundamental task of the license renewal inspection program is to ensure that there is reasonable assurance that the effects of aging will be managed consistent with the CLB during the period of extended operation. The inspection program objectives are to [27]: (1) provide a basis for recommending issuance or denial of a renewed license, (2) identify weaknesses within the applicant's overall license renewal program or an individual AMP that fail to provide reasonable assurance that the applicable aging effects will be adequately managed during the period of extended operation, and (3) determine the status of compliance with 10 CFR Part 54 and other areas related to maintaining and operating the plant such that the continued operation beyond the current licensing term will not be inimical to the public health and safety. License renewal site inspections include scoping and screening inspection and AMP inspections. In addition, post-renewal site inspections are conducted.

Guidance is available for conduct of the license renewal site inspections [28]. The site inspections are assessments of an applicant's implementation of and compliance with 10 CFR Part 54 requirements. The site inspections are performed by a team inspection in the areas of scoping and screening activities, observation of the condition of plant equipment, and implementation of the AMPs and review of associated documentation.

The scoping portion of the inspection verifies on a sampling basis through on-site review and a walk down of selected areas of the plant that nonsafety-related SSCs whose failure could prevent safety-related SSCs from accomplishing a safety-related function are correctly included within the scope of license renewal. The inspections verify that there is reasonable assurance that the applicant has adequately documented all the identified passive and long-lived SSCs requiring an AMR (The AMR was addressed in Section 5.3.1.3).

The AMP review part of the license renewal inspection program is intended to assess the implementation of the AMPs resulting from the applicant's license renewal program and may be performed in conjunction with the scoping and screening inspection. The AMP inspections verify that there is reasonable assurance that the applicant has adequately addressed all the identified passive and long-lived SSCs identified during the AMR, and, through review of supporting documentation and a walk down of selected systems, that the effects of aging can be adequately managed in the period of extended operation. Activities to be performed under the AMP license renewal inspections include [28]: (1) determination from the LRA the AMPs that will be credited with preventing applicable aging effects for selected SSCs and verification that the AMPs will ensure that the aging effects will be managed so that there is reasonable assurance that the intended function(s) will be maintained consistent with the CLB throughout the period of extended operation; (2) review of the descriptions of the AMPs from the LRA, updated FSAR, plant procedures, and related engineering support documentation, supplemented by interviews with on-site engineering staff responsible for implementation of the AMPs to assess their knowledge and involvement; (3) verification that the applicant evaluated site-specific information such as surveillance test results, preventative maintenance records, corrective maintenance records, equipment history files, and inservice test and inspection results, and that industry operational experience was evaluated in determining aging effects; (4) performance of walk downs of selected in-scope SSCs to verify that any observable aging effects were identified and conduct of a containment inspection if there is evidence that some aging effects are not adequately addressed in the application or there is an open item generated either by an inspection or the application review that is related to an area inaccessible during the regularly scheduled inspection; (5) discussion with the applicant of observed aging effects not addressed by the LRA; and resulting AMPs; (6) selection of a system to perform a review in order to determine if the applicant properly accounted for all possible environmental aging effects on that system in the LRA: and (7) review of results for previously existing AMPs, results of past tests and inspections, and verification that the proposed and existing programs adequately demonstrate adequate scope and methodology to detect, monitor, trend, and correct age-related degradation through performance and/or condition monitoring, technical specification surveillance, and other aging management activities.

If during the safety review process additional information is required of the applicant, this is addressed through a Request for Additional Information (RAI) prepared by the NRC staff or their contractors. Results of the safety review process are provided in the Safety Evaluation Report (SER) that provides the technical and legal basis for the NRC's disposition of a LRA by documenting the bases for the NRC staff's conclusion and providing sufficient information to explain the NRC staff rationale to someone unfamiliar with the licensee's application.

The post-approval site inspection is part of the license renewal inspection program and was implemented to ensure compliance and assess performance in relation to 10 CFR Part 54. The post-renewal inspections verify on a sampling basis that the licensee [29]: (1) has completed the necessary actions to comply with the license conditions that are a part of the renewed operating license and has implemented the AMPs and TLAAs included in the NRC staff's license renewal SER and (2) followed the guidance in NEI 99-04 [30] for the license renewal commitment change process, including the elimination of commitments, and properly evaluated, and reported, when necessary, changes to the license renewal commitments listed in the Updated FSAR (UFSAR) in accordance with 10 CFR Part 50.59. This inspection also verifies that the UFSAR supplement describes the AMPs and TLAAs approved by NRC in the SER issued with the renewed license.

5.5 OPERATING EXPERIENCE

As part of the license renewal process, the NRC conducts an AMR that includes an assessment of the applicant's proposed programs to manage aging of structures and components. One of the elements associated with an assessment of an AMP addresses operating experience, and for the concrete structures this includes a walk-down of areas of interest (e.g., buildings, spent fuel pool, supports, masonry walls, and water-control structures). Occurrences of age-related degradation have been observed related to the NPP concrete structures and containments and steel containments (see Appendix 5B).

Some general comments are provided below relative to operating experience of the concrete and steel containments.

From a safety perspective, the containment is one of the most important structures in a nuclear power plant, because it provides the final barrier to the release of radioactive fission products to the environment in the unlikely event of an accident. Due to potential difficulty and cost of replacement of a nuclear power plant containment, if significant degradation were to occur, it has the potential to be life-limiting for the plant.

In general, the performance of the nuclear power plant containments has been good; however, there have been degradation occurrences that if not addressed could challenge the capacity of the containment (e.g., corrosion of metallic pressure boundary).

As a result of the maturity of the nuclear power plants in the United States, degradation due to initial construction, materials selection, or design problems has been addressed; however, degradation occurrences related to environmental effects are likely to increase.

Utilization of input derived from operating experience is of great importance relative to identification of age-related degradation and its management during the term of a renewed operating license.

Primary factors that have lead to degradation of the concrete containments have been corrosion of embedded steel reinforcement, post-tension anchor head failure, wire breaks, Also, there are instances of groundwater intrusion associated with concrete cracking into the containment building and tendon galleries with associated leaching of the concrete, larger than anticipated loss of tendon prestressing forces, and concrete cracking and spalling due to freeze–thaw cycles.

The primary degradation factor that has produced degradation in steel containments and metal liners of reinforced concrete containments has been corrosion due to the presence of water that in some cases can be in the form of boric acid. The presence of water in inaccessible areas leading to degradation of coatings and corrosion of steel containments and liners of reinforced concrete containments, and the extensive coating degradation and pitting corrosion of the torus steel shell of some BWR Mark I plants is an aging concern.

A properly established in-service inspection program and timely maintenance are keys to ensuring that nuclear power plant safety-related containment structures will continue to meet their functional and performance requirements.

Programs in the GALL Report such as 10 CFR Part 50, Appendix J and ASME Section XI Subsections IWL and IWE for conduct of condition assessments of the containments have been effective in managing aging; however, these programs may need to be revised to address unanticipated challenges that could impact the containment (e.g., development of criteria to address the presence of alkali-silica reactions in the concrete).

The license renewal process for initial renewal of operating licenses is in place and has been effective; however, potential issues related to subsequent license renewals (e.g., unanticipated aging factors) need to be identified and addressed relative to the need for changes in the regulatory process and adequacy of current aging management programs.

REFERENCES

[1] Auluck, R., 2011, “Containment Degradation Issues at U.S. Nuclear Power Plants — Lessons Learned,” Paper ID# 614, Transactions of SMiRT 21, Division 8, New Delhi, India, 6–11 November.

[2] Code of Federal Regulations 10 CFR Part 100, 2012, “Reactor Site Criteria,” Office of Federal Register, Washington, DC.

[3] Code of Federal Regulations 10 CFR Part 54, 2012, “Requirements for Renewal of Operating Licenses for Nuclear Power Plants,” Office of Federal Register, Washington, DC.

[4] “Generic Environmental Impact Statement for License Renewal of Nuclear Plants,” NUREG-1437, U.S. Nuclear Regulatory Commission, Washington, May 1996.

[5] “Standard Format and Content for Application to Renew Nuclear Power Plant Operating Licenses,” Regulatory Guide 1.188, U.S. Nuclear Regulatory Commission, Washington, D.C., September 2005.

[6] “Generic Aging Lessons Learned (GALL) Report,” NUREG-1801, Rev. 2, U.S. Nuclear Regulatory Commission, Washington, D.C., December 2010.

[7] “Standard Review Plan for License Renewal Applications for Nuclear Power Plants,” NUREG-1800, Rev. 2, U.S. Nuclear Regulatory Commission, Washington, D.C., December 2010.

[8] “Industry Guidelines for Implementing the Requirements of 10 CFR Part 54 — The License Renewal Rule,” NEI 95-10, Rev. 6, Nuclear Energy Institute, Washington, D.C., June 2005.

[9] “Evaluation of Existing Nuclear Safety-Related Concrete Structures,” ACI 349.3R, American Concrete Institute, Farmington Hills, Michigan, 2002.

[10] “Guide for Making a Condition Survey of Concrete in Service,” ACI 201.1R, American Concrete Institute, Farmington Hills, Michigan, 1992.

[11] “Guideline for Structural Condition Assessment of Existing Buildings,” ANSI/ASCE 11-99, American National Standards Institute/American Society of Civil Engineers, Washington, D.C., 1999.

[12] “Building Code Requirements for Reinforced Concrete and Commentary,” ACI 318, American Concrete Institute, Farmington Hills, Michigan, 2011.

[13] “Determining Prestressing Forces for Inspection of Prestressed Concrete Containments,” Regulatory Guide 1.35.1, U.S. Nuclear Regulatory Commission, Washington, D.C., July 1990.

[14] "Degradation of Prestressing Tendon Systems in Prestressed Concrete Containments," Information Notice 99-10, U.S. Nuclear Regulatory Commission, Washington, D.C., April 1999.

[15] "Standard Review Plan for the Review of Safety Analysis Reports for Nuclear Power Plants: LWR Edition," NUREG-0800, U.S. Nuclear Regulatory Commission, Washington, D.C., March 2007.

[16] "Nuclear Plant Aging Research (NPAR) Program Plan — Status and Accomplishments," NUREG-1144, Rev. 2, U.S. Nuclear Regulatory Commission, Washington, D.C., June 1991.

[17] Vora, J. P., "Potential Application of Results of NRC Aging Research to License Renewal," *Nuclear Engineering and Design* **118**, pp. 343–354, North-Holland Publishers, 1990.

[18] Naus, D. J., "Concrete Component Aging and Its Significance Relative to Life Extension of Nuclear Power Plants," NUREG/CR-4652, U.S. Nuclear Regulatory Commission, Washington, D.C., September 1986.

[19] Shah, V. N., Smith, S. K., and Sinha, U. P., "Insights for Aging Management of Light Water Reactor Components," NUREG/CR-5314, Vol. 5, U.S. Nuclear Regulatory Commission, Washington, D.C., March 1994.

[20] "Aging Management of Nuclear Power Plant Containments for License Renewal," NUREG-1611, U.S. Nuclear Regulatory Commission, Washington, D.C., September 1997.

[21] Naus, D. J., Oland, C. B., and Ellingwood, B. R., "Report on Aging of Nuclear Power Plant Reinforced Concrete," NUREG/CR-6424, U.S. Nuclear Regulatory Commission, Washington, D.C., March 1996.

[22] Regan, C., "Technical Information from Industry Reports Addressing License Renewal," Paper D01/4, Transactions of the 14th International Conference on Structural Mechanics in Reactor Technology, Lyon, France, August 17–22, 1997.

[23] "Pressurized Water Reactor Containment Structures License Renewal Industry Report," NUMARC Report Number 90-01, Rev. 1, Nuclear Management and Resources Council, Washington, D.C., December 1991.

[24] "Boiling Water Reactor Containments License Renewal Industry Report," NUMARC Report Number 90-10, Rev. 1, Nuclear Management and Resources Council, Washington, D.C., December 1991.

[25] "Class I Structures License Renewal Industry Report," NUMARC Report Number 90-06, Rev. 1, Nuclear Management and Resources Council, Washington, D.C., December 1991.

[26] "Summary of Technical Information and Agreements from Nuclear Management and Resources Council Industry Reports Addressing License Renewal," NUREG-1557, U.S. Nuclear Regulatory Commission, Washington, D.C., October 1996.

[27] "Policy and Guidance for License Renewal Inspections Programs," IMC 2516, U.S. Nuclear Regulatory Commission, Washington, D.C., February 18, 2005.

[28] "License Renewal Inspection," IP 71002, U.S. Nuclear Regulatory Commission: Washington, D.C., November 23, 2011.

[29] "Post-Approval Site Inspection for License Renewal," IP 71003, U.S. Nuclear Regulatory Commission: Washington, D.C., October 31, 2008.

[30] "Guidelines for Managing NRC Commitment Changes," NEI 99-04, Nuclear Energy Institute, Washington, D.C., July 1999.

Appendix 5A
Monitoring and Trending of Prestressing Forces in Prestressed Concrete Containments

This paper illustrates a statistically valid procedure for developing a trend line for future projections of prestressing forces in a prestressed concrete containment with unbonded tendons.

The United States has 38 prestressed concrete containments (PCCs). The prestressing tendons in the containments are arranged so that the pre-compression imparted to the concrete is adequate to withstand the tensile stresses produced by the internal pressure during a postulated design-basis accident (DBA) without significant cracking of the containment structure. The installation process and the time-dependent characteristics of the containment concrete and the prestressing steel affect prestressing forces in the tendons after their installation.

The design process estimates and accounts for these effects in establishing the minimum prestressing forces required to withstand the anticipated DBA stresses up to the licensed life of the containments. In the United States, applicants for the plants with PCCs must perform a time-limited aging analysis (TLAA) of the prestressing forces. The TLAA must demonstrate that the prestressing forces at the end of the extended period of operation will be adequate to serve their intended functions. Because the prestressing tendons in the containments of U.S. nuclear power plants are unbonded, the prestressing forces in the containments can be measured and the prestressing tendons retensioned, if necessary.

The prestressing forces in prestressing tendons are monitored as stipulated in Subsection IWL of Section XI of the American Society of Mechanical Engineers (ASME) Boiler and Pressure Vessel Code, as incorporated by reference in Title 10, Section 50.55a, of the *Code of Federal Regulations* (10 CFR 50.55a). As part of this monitoring, the tendon forces are measured and recorded. These measured prestressing forces are used to project the trend of the prestressing forces during the licensed life of the plant. The trending is used for detecting a need for retensioning the tendons, as well as for TLAA required for plant-life extension.

This paper discusses the factors involved in monitoring and trending the measured prestressing forces, as well as related operating experience.

Keywords: containment, prestressing force, trending, regression analysis, unbonded tendons

5A.1 INTRODUCTION

Nuclear power plants in the United States use prestressed concrete for a number of structures, such as containments, a reactor pressure vessel, missile shield members, reactor cavity walls, spent fuel pool girders, and structural supports for ice condenser floors. However, it is used principally in the construction of containment structures (or containments). Hence, this paper addresses the monitoring and trending of prestressed forces in prestressed concrete containments (PCCs).

The containment is a vital engineered safety feature of a nuclear power plant. It encloses the entire reactor and the reactor coolant system and serves as a final barrier against the release of radioactive fission products to the environment under various accident conditions. The containment design is based on pressure and temperature loadings associated with a loss-of-coolant accident (LOCA) resulting from a double-ended rupture of the largest pipe in the reactor coolant system. The containment is also designed to retain its integrity under low-probability (i.e., less than 10^{-4}) environmental loadings, such as those generated by an earthquake, tornado, hurricane, seiche, or tsunami. Additionally, the containment must provide biological shielding under normal and accident conditions and protect the internal equipment from external missiles, such as tornado- or turbine-generated missiles and aircraft impact (where postulated).

5A.2 CONSTRUCTION AND DESIGN FEATURES

With the introduction of concrete containment construction in the United States, the vertical cylinder with a shallow or hemispherical dome and a flat foundation mat became the preferred shape for containments. In the 1960s, the first PCCs were prestressed only in the vertical direction, with non-prestressed reinforcing in the circumferential (hoop) direction of the cylinder and in the dome (e.g., containments at the R.E. Ginna and the H.B. Robinson Unit 2 nuclear stations).

Fully prestressed concrete containments were first built in the late 1960s and typically consisted of a cylindrical wall, a shallow dome, a large ring girder at the intersection of the dome and wall, six buttresses, and a flat reinforced-concrete base slab. Hoop tendons anchored at two alternate buttresses and spanning 120 degrees (six buttresses total), as well as vertical tendons anchored at the top of the ring girder and at the bottom of the foundation mat in a specially constructed tendon gallery, served to prestress the wall. Three sets of tendons oriented at 120° to one another prestressed the dome. The dome tendons were anchored to the ring girder. Because of the number of tendons (more than 900), which were very labor intensive to fabricate, install, tension, and make resistant to corrosion, and because the plants were increasing in size, a second generation of fully PCC designs with three or four buttresses evolved. Although the use of three buttresses instead of six increased the length of hoop tendons and the friction force, the reduced number of buttresses and anchorages produced considerable cost savings. In addition, the capacity of the posttensioning tendons was approximately doubled, significantly reducing the total number of tendons. In the third generation of PCCs, a hemispherical dome replaced the shallow dome, the ring girder was eliminated, and inverted U-shaped tendons replaced the dome and vertical tendons. The inverted U-shaped tendons were divided into two sets of tendons oriented at 90 degrees to one another. Hoop tendons were still anchored to the three buttresses. Figure 5A.1 shows various configurations of prestressing tendons in U.S. PCCs.

5A.2.1 Prestressing Systems

The three major categories of posttensioning systems include wire, strand, and bar. The tendons are installed within preplaced ducts in the containment structure and are posttensioned from one or both ends after the concrete has achieved sufficient strength. Prestressing systems have evolved over the years primarily in terms of system capacity. Before the advent of PCCs, prestressing systems were relatively small, with an ultimate capacity of less than 4.45 megaNewtons (MN) (500 tons). The necessity to withstand high forces resulting

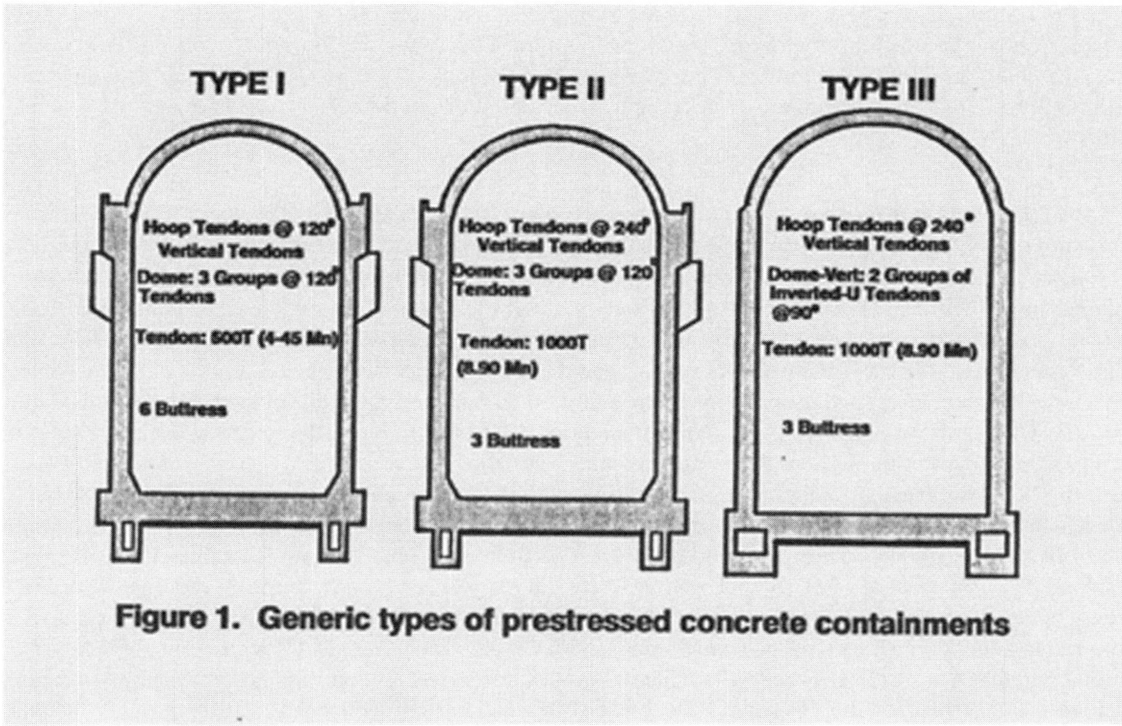

Figure 1. Generic types of prestressed concrete containments

FIG. 5A.1
GENERIC TYPES OF PRESTRESSED CONCRETE CONTAINMENTS

from a combination of increased volumes and pressures of the dry containments of the pressurized-water reactors led to the development of tendon systems with capacities ranging from 8.0 MN (900 tons) to 10.7 MN (1200 tons). This development permitted increased spacing of tendons, reduced congestion, and almost halved the number of tendons, tendon ducts, and anchorages. The large tendons were developed using groupings of multiwire or multistrand systems. A number of static and cyclic tests on prototype full-size tendons and detailed stress analyses of anchorage hardware verified the satisfactory performance of the large tendons. The American Society of Mechanical Engineers (ASME) provides current material and performance test requirements for the posttensioning systems [9]. In the United States, the 8.9-MN (1000-ton) systems approved for use include (1) the Birkenmaier Brandestioni RVS and Vogt (BBRV) wire system, (2) the Vorspann System Losinger (VSL) strand system, and (3) the Stress Steel S/H strand system. The BBRV and VSL are the most commonly used systems for PCCs.

The large-capacity BBRV tendon systems consist of 163, 169, 170, or 186 wires. The wire diameter is 6.35 mm (0.25 in.) except for the 163-wire tendon, where 7-mm (0.28-in.) wires are used. Each wire is anchored by a buttonhead that is cold-formed and bears on anchor heads at each end of the tendon. The large-capacity VSL tendon system consists of 55 seven-wire strands with a wire diameter of 12.7 mm (0.5 in.). A two-piece split cone wedge held tightly against the inner surface of the anchor head keeps it in place.

5A.2.2 Corrosion Inhibitors for Prestressing Tendons

Prestressing plays a vital role in ensuring the structural integrity of the PCC throughout its design life (normally 40 years). However, because the tendons and their anchorage hardware are fabricated from high-strength, high-hardness materials and are subjected to sustained high stresses, they are susceptible to stress-corrosion cracking and hydrogen embrittlement. Filling the tendon ducts with Portland cement grout (bonded tendons) or petrolatum-based microcrystalline waxes compounded using organic corrosion inhibitors (unbonded tendons) protect the tendons and their load-bearing hardware against corrosion influences.

5A.2.2.1 Portland Cement Grout. Portland cement has been used effectively as a deterrent to the corrosion of steel in prestressed concrete for 60 years and in reinforced-concrete construction for more than 100 years. Schupack reviewed the durability performance of posttensioned tendons in conventional civil engineering structures [1]. The review shows only a small incidence of corrosion (less than 1 in 100,000) related to ill-conceived detailing, poor construction practices, or the presence of contaminants. Wern et al. [2], Harstead et al. [3], and Schupack [4] supplement this performance history with reports of extensive tests to ensure the adequate penetration of grout through vertical bars and curved hoop and long vertical strand tendons. The regulators, however, were concerned about the inability to positively check the integrity of the posttensioning system throughout the service life of the structure. As a result of discussions and public meetings, the U.S. Nuclear Regulatory Commission (NRC) developed two regulatory guides (RGs), RG 1.107, "Qualification of Cement Grouting for Prestressing Tendons in Containment Structures," issued August 1977 [5], and RG 1.90, Revision 1, "Inservice Inspection of Prestressed Concrete Containment with Grouted Tendon," issued February 1977 [6]. Though the NRC intended these guides to ensure thorough scrutiny of grout material and installation and the periodic review of the status of PCCs with grouted tendons, these actions did not encourage the use of grouted tendons in PCCs. At present, two PCCs with grouted tendons exist in the United States — (1) at Three Mile Island Unit 2 (it has strand tendons and has not operated since the accident in 1979) and (2) the PCC at H.B. Robinson Unit 2 (it has bar tendons).

5A.2.2.2 Petrolatum-Based Grease. Although a more recent development, the use of organic petrolatum-based corrosion protection compounds (greases) has gained prominence in PCCs in the United States because of the relative ease with which the tendons can be inspected and tested. Additional advantages include the following:

- Encapsulation provides an approximately 50% reduction in the friction factor, thus permitting the use of longer tendons and fewer buttresses and anchorages.
- Tendons may be relaxed, retensioned, and replaced as required.
- A corrosion-protection coating applied in the shop (before shipment) permits the efficient scheduling of installation and tensioning of tendons and the installation of grease during the construction sequence.

The petrolatum-based filler greases have evolved over the years to make the products more suitable to the application of unbonded (greased) tendons in PCCs. Initially, the product consisted of a sheathing filler containing polar wetting agents, rust-prevention additives, microcrystalline waxes, and proprietary constituents formulated to displace water, self heal, and resist electrical conductivity. The next generation of materials added a plugging agent to raise the low-flow point of the products (approximately 39°C (100°F)) to keep them from seeking loose sheathing joints and flowing into hairline cracks in concrete. A subsequent refinement involved the incorporation of a light base number to provide alkalinity (3 mg KOH/g of product) for improved corrosion protection. According to Naus, the current generation of products has been formulated to (1) increase the viscosity without sacrificing pumpability, (2) raise the congealing point to 57–63°C (135–145°F), (3) increase the resistance to flow from sheathing joints, (4) increase water-displacing characteristics, and (5) raise the base number (35 mg KOH/g of product) to provide higher reserve alkalinity [7].

Although the prestressing forces can be monitored directly when the tendons are greased or unbonded, the prestressing forces in grouted or bonded tendons cannot be directly monitored.

5A.2.3 Design Considerations

Early light-water reactor plant PCCs were designed and constructed using the provisions of American Concrete Institute (ACI) 318 [8], supplemented by the specific loads and load combinations pertinent to their design, as stipulated by the NRC. Section III, Division 2, of the ASME Code delineates the current requirements for the design and construction of concrete reactor vessels and containments [9]. RG 1.136, Revision 2, "Materials, Construction and Testing of Concrete Containment," issued 1978 [10], and Section 3.8.1, "Concrete Containments," of NUREG-0800, "Standard Review Plan for the Review of Safety Analysis Reports," issued 1981 [11], delineate supplemental criteria or endorsement of the ASME Code.

The conservatively calculated internal pressure load (P) imposed on the containment during a LOCA event principally governs the design for the required prestressing forces (F) in a PCC. Prestressing forces required in each of the major directions (i.e., hoop and vertical in cylinders, and two-way or three-way in the dome) are estimated so as to neutralize the tensile forces generated by P, 1.25P, or 1.5P (depending upon the design requirements). For a preliminary design, the effective prestressing force in a tendon is the initial force (Fi) (approximately 70% of the guaranteed ultimate tensile strength (GUTS) of the tendon), minus 15% to 25% of Fi, to account for the initial and time-dependent losses. The final design incorporates the detailed calculations of the losses and predicted forces for the tendons.

5A.3 FACTORS CONTRIBUTING TO PRESTRESS LOSSES

The initial losses to be considered in predicting forces include the losses caused by (1) the slip of prestressing elements at the anchorages and (2) the elastic shortening of the structure. The time-dependent losses include those induced by (1) concrete shrinkage, (2) concrete creep, and (3) the relaxation of prestressing steel. In RG 1.35.1, "Determining Prestressing Forces for Inspection of Prestressed Concrete Containments," issued July 1990 [12], the NRC provides a detailed discussion of the initial prestressing losses, including guidelines for calculating the losses and the predicted forces that could be used to compare with the measured prestressing forces during the periodic inspection of the containment posttensioning tendons. The RG also recommends giving appropriate consideration to the effects of the potential degradation of prestressing elements and variation in sustained temperature. The following subsections discuss major contributors to the time-dependent losses.

5A.3.1 Shrinkage of Concrete

Given the construction schedule for a typical PCC, a substantial portion of the expected long-term shrinkage takes place before the structure is prestressed. Hanson et al. presents formulas for predicting the long-term shrinkage based on the assumption that the shrinkage approximately follows the laws of diffusion and supports the formulas by experimental investigation [13]. An appropriate extrapolation of these formulas, for the volume-to-surface ratio of the structure in excess of 60 cm (24 in.) and the contributing shrinkage that could occur 100 days after the average time of the construction of the structure, would yield a value of $100 \times 1 \times 10^{-6}$. This value is considered to be reasonable at a temperature of 21°C (70°F) and a relative humidity of 50%. However, this value must be modified to account for the significantly higher shrinkage in a low-humidity environment and the significantly lower shrinkage in a high-humidity environment.

5A.3.2 Creep of Concrete

The influence of concrete creep, consisting of basic creep and drying creep, is one of the most significant and variable factors in the computation of time-dependent losses in PCC structures. The exchange of moisture between the structure and its environment causes drying creep, or stress-induced shrinkage. Its characteristics are similar to those of shrinkage, except that they represent an additional moisture movement resulting from the stressed condition of a structure. The amount of drying creep depends mainly on the volume-to-surface ratio of the structure and the mean relative humidity of the environment. For PCC structures having a volume-to-surface ratio in excess of 60 cm (24 in.), the relative influence of drying creep (compared to basic creep) is negligible.

Four parameters significantly influence the magnitude of basic creep:

(1) Concrete mix — cement and aggregate type; proportion of cement, water, and aggregates; and the influence of admixtures

(2) Age at loading — the basic creep value as a function of the degree of hydration that has taken place at the time of loading

(3) Magnitude of the average sustained stress on the concrete

(4) Temperature

Investigations by Hansen support the assumption that basic creep varies linearly with the intensity of sustained stress, as long as the average stress in the concrete does not exceed 40% of the ultimate strength of the concrete [14]. The specific creep is thus defined as the ratio of basic creep to the average stress intensity.

Appendix A to RG 1.35.1 provides an acceptable method for determining basic creep at various times for a given concrete mix as a function of the age at loading [12]. This method is based on concepts and equations derived by Hansen from a rheological model representing concrete creep. Most investigators agree that no one formula applies generally for determining the long-term creep for various concrete mixes. Hence, RG 1.35.1 recommends a method of predicting the long-term basic creep from the results of short-term creep tests.

5A.3.3 Relaxation of Prestressing Steel

The stress relaxation properties of prestressing steel vary with its chemical composition and thermal and mechanical treatment. Manufacturers should have the ability to provide data on the long-term loss in prestressing steel stress from pure relaxation. Section CC-2424 of the ASME Code requires a minimum of three 1000-hour relaxation tests for the prestressing steel proposed for use. Each of the three tests should have a sufficient number of data points to extrapolate the 1000-hour pure relaxation data to the planned useful life of the structure.

5A.3.4 Losses Caused by Degradation of Prestressing Elements

In the United States, utilities make allowance for both overall and localized wire breakage. Such an allowance in the design of the containment would allow the breakage of a few wires during construction without the need to replace these wires. For a tendon with a few broken wires, care should be taken not to overstress intact wires to bring the tendon force to a prescribed value. Instead, the tendon should be extended to the same strain level as other similar tendons (without broken wires). The procedure will leave the tendon at a prestress level lower than the prescribed level (generally 70% of the GUTS). This practice is acceptable provided that the design includes an allowance for the breakage of wires.

5A.3.5 Effects of Temperature

The effect of differences between the average temperature of the structure during installation and during inspections has particular importance when comparing prestress forces. Localized hot spots and temperature variations along the length of a tendon can cause variations in the force along its length. The differences between the coefficients of expansion or contraction of concrete and steel can also cause modifications of tendon forces. These effects, as appropriate, should be considered in comparing the measured prestressing forces with the predicted forces. The United States experience indicates the distinct effect of elevated temperatures on the relaxation of prestressing steel. Ashar et al. provides a comprehensive report on the effects of temperature on the relaxation of prestressing steel [15].

5A.4 MONITORING PRESTRESSING FORCES

5A.4.1 Grouted or Bonded Tendons

Because grouted tendons cannot be inspected directly, alternative methods for assessing the integrity of the PCCs with grouted tendons were investigated in the early 1970s, when some utilities in the United States were exploring the use of grouted tendons for their PCCs. After a number of public meetings and discussions among knowledgeable professionals, the NRC promulgated the final provisions for inspecting the PCCs with grouted tendons in RG 1.90, Revision 1. This guide provides two distinct alternatives. The first requires the monitoring of the prestress level in the PCC with strategically located instrumentation (e.g., strain gauges, stress meters, load indicators). The second alternative requires the monitoring of PCC deformations at critical locations under prescribed pressures. Both alternatives include common inspection provisions for monitoring prestressing forces in strategically located ungrouted tendons and visually examining tendon-anchorage areas and structurally critical areas. The designers of PCCs in the United States opted for greased or unbonded tendons. However, Marneffe et al. reports the successful use of RG 1.90, Revision 1, in a number of PCCs in Belgium [16].

5A.4.2 Greased or Unbonded Tendons

Inservice inspection (ISI) programs for the earlier PCCs (i.e., those licensed before 1973) with greased tendons were developed on a case-by-case basis. In general, the inspection consisted of (1) monitoring nine selected tendons for prestressing force, grease, and condition of anchor hardware and (2) visually examining the exterior of the containment. The inspections were to be performed more frequently in the early years (typically, the first 5 years) and less often during the later years. The practice of repeatedly inspecting the same tendons was a subject of considerable discussion in the early 1970s. A number of industry professionals and members of the regulatory staff felt that monitoring the same nine tendons out of approximately 1000 tendons would not provide adequate information about the condition of the entire tendon population. They also recognized that repeatedly tensioning and detensioning the same tendons during all inspections would increase the possibility of damaging these tendons, and they agreed that inspection of the tendons on a random but representative basis is necessary to ensure the effectiveness of the inspection program.

After a number of discussions with industry groups and considering the performance of prestressing systems in other civil engineering structures, the NRC staff decided that 21 tendons should be inspected during each of the first three inspections. If these inspections did not uncover any significant problems with the post-tensioning system, the subsequent inspections could be performed using a reduced sample size. The first effective version (i.e., Revision 1) of RG 1.35, "Inservice Inspection of Ungrouted Tendons in Prestressed Concrete Containments," issued 1974 [17], prescribed a scheme calling for the inspection of 21 tendons (10 hoop, 5 vertical, and 6 dome tendons) during the first three inspections and nine tendons (three from each group) during subsequent inspections. Additionally, the RG provided for (1) inspecting tendon-anchorage areas, (2) checking for grease coverage in the selected tendons, (3) testing the material properties of the prestressing elements removed from the tendons, and (4) visually examining the exterior surfaces of the PCC. The NRC promulgated Revision 2 of the guide for the Type III prestressing tendon arrangement shown in Fig. 5A.1 [18].

The regulatory staff issued Revision 3 to RG 1.35 in July 1990 to update and clarify the guide on the basis of experience obtained during prior inspections [19]. Various interpretations by the utilities of the acceptability of measured prestressing forces in tendons concerned the regulatory staff enough to issue a companion guide, RG 1.35.1, together with Revision 3 of RG 1.35. Dougan [20] and Ashar et al. [21] provide additional information on the development of ISI provisions and an assessment of their effectiveness.

In August 1996, the NRC staff issued an amendment to its regulation for the use of codes and standards. The amended regulation endorses Subsection IWL, Section XI, of the ASME Code [22]. Subsection IWL, which covers the ISI of reinforced and PCCs, incorporates the provisions of RG 1.35, Revision 3, for the ISI of post-tensioning tendons.

5A.5 TRENDING PRESTRESSING FORCES

5A.5.1 Bonded Tendons

In the United States, only one plant uses bonded tendons. All of the vertical bar tendons at this plant are grouted. Thus, no database exists for trending the behavior of bonded tendons in the United States.

5A.5.2 Unbonded Tendons

As discussed above, RG 1.35.1 provides detailed guidance for constructing the upper and lower bounds of prestressing forces, based on the expected variation in the systemic time-dependent losses in prestressing forces. Figure 5A.2 provides the upper and lower bounds and the minimum required prestressing forces as functions of time. To accept the measured prestressing forces, they should exceed the minimum required prestressing force derived from the design requirement to counteract the tensile stresses in the containment shell induced by the postulated design-basis accident. Because the sample size of the prestressing tendons

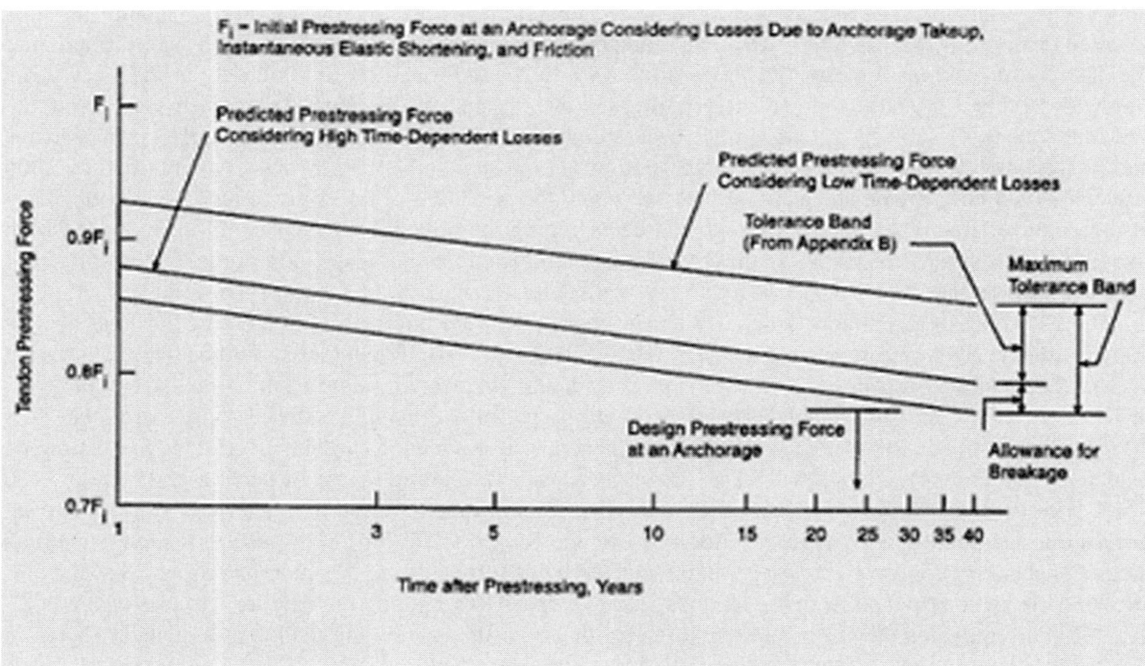

FIG. 5A.2
TOLERANCE BANDS OF ACCEPTABLE PRESTRESS

selected for force measurement could be as low as 2% of the tendon population, comparing the average of the measured tendon forces with the predicted prestressing force is not very useful because the process (1) does not provide confidence regarding the status of the other 98% of the tendons and (2) does not help determine the trend of prestressing forces for future projection. These deficiencies in the present process dictate the need to develop procedures that would enable the utilities to develop trend lines based on measured forces recorded during all prior inspections.

Figure 5A.2 assumes that the prestressing force is linearly related to the logarithm of time. This assumption makes sense because the concrete creep and relaxation of prestressing steel, the dominant sources of prestress loss, are similarly related to time. Based on an extensive database of 11 nuclear plants in the United Kingdom, Irving et al. established a similar relationship between the prestressing tendon force and time [23].

Based on the above assumption and using the least-square method for performing linear regression analysis,

$$\mathrm{F} = \mathrm{A} \log \mathrm{T} + \mathrm{B} \text{ or setting } Y = AX + B;$$

obtaining the sum of the squares of deviations from the true line,

$$\mathrm{S} = \sum_{i=1}^{n} (Y_i - B - AX_i)^2;$$

taking partial derivatives with respect to A and B,

$$\frac{\partial S}{\partial A} = -2\sum_{i=1}^{n} X_i(Y_i - B - AX_i); \text{ and}$$

$$\frac{\partial S}{\partial B} = -2\sum_{i=1}^{n} (Y_i - B - AX_i); \text{ and}$$

setting these equations to zero and solving for A and B would yield

$$A = \frac{\sum (X_i - \bar{X})(Y_i - \bar{Y})}{\sum (X_i - \bar{X})^2} \quad \text{and} \quad B = \bar{Y} - A\bar{X},$$

where;

$$\bar{X} = \sum X_i/n \quad \text{and} \quad \bar{Y} = \sum Y_i/n.$$

Substituting measured prestressing forces for Y_i and the logarithm of corresponding time as X_i provides the slope and intercept of a trend line. This is the simplest way to compute a trend line. Some applicants have refined the procedure by incorporating potential error ($\bar{e}$) in the measurement of forces.

5A.6 DISCUSSION

Based on the requirement for trending prestressing force values in Title 10, Section 50.55a (b) (2) (viii) (B), of the *Code of Federal Regulation* [24], Subsection IWL, Section XI, of the ASME Code was revised in 1998 to incorporate a similar requirement. The provision essentially mandates corrective actions if, during an inspection, the trending of prestressing forces based on the prior tendon force measurements indicates that the prestressing force value is likely to fall below the minimum required value. In addition, in the license renewal applications, the owners of plants with PCCs are required to perform a time-limited aging analysis of the tendon prestressing forces to provide assurance that, at the end of the license renewal period, the tendon forces will meet the licensing basis requirements. These two provisions required the owners of the plants to perform a trending analysis of prestressing forces. Figure 5A.3 shows the parameters used by one plant owner.

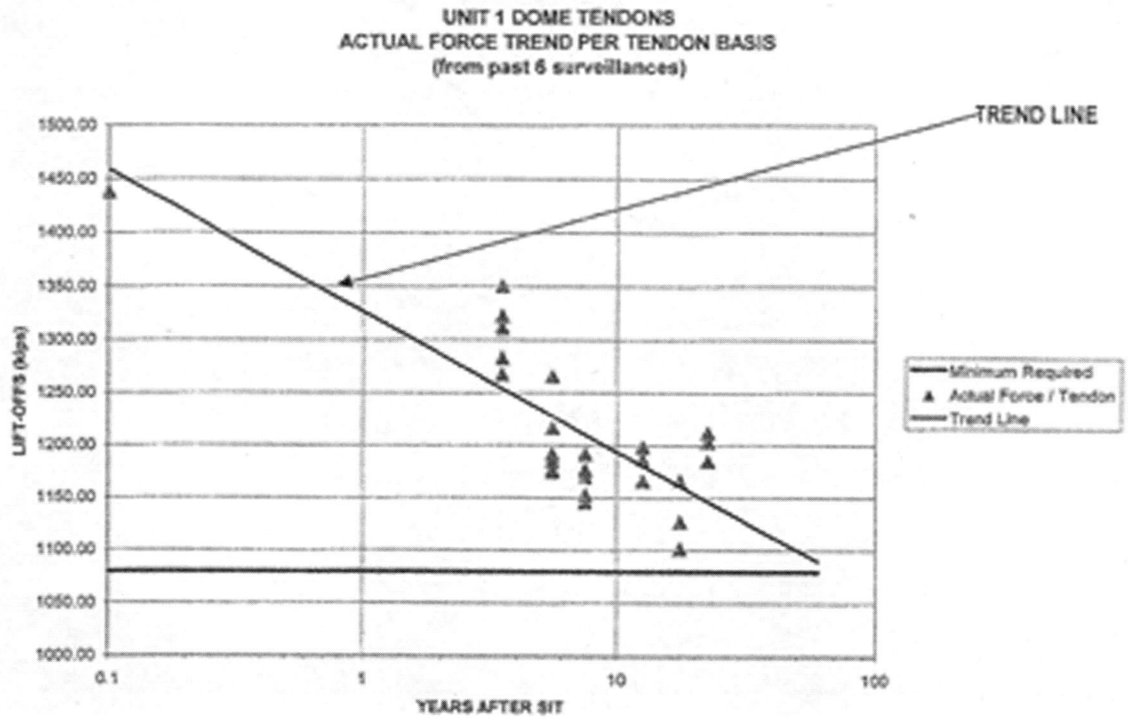

FIG. 5A.3
BASIC PARAMETERS THAT WERE USED BY ONE PLANT OWNER

Section 5A.5.2 above proposes the simple method of trending a number of prestressing force measurements taken at various times after the installation of tendons. The trend line will only reflect the prestressing force values at tendon anchorages. Trend lines do not represent the minimum prestressing forces along the tendons. The minimum required prestressing force values must be adjusted to account for friction to compare them with the values obtained from the trend lines.

To maintain consistency when comparing the projected prestressing forces from the trend lines with the minimum required prestressing forces, the tendons should be grouped so that they exhibit similar time-dependent characteristics. For Types I and II (see Figure 5A.1), the hoop (or horizontal), vertical, and dome tendons comprise the logical groups. For Type III tendons, the hoop and inverted-U tendons form the logical groups.

5A.7 CONCLUDING REMARKS

The monitoring and trending of prestressing tendon forces assure that the containment has adequate compression to counteract the tension produced by various loadings for which the containment is designed. The tension generated by the pressure associated with various postulated accident conditions represents the major loading for which the prestressing force in a prestressed concrete containment is provided. The trending of prestressing forces is essential for projecting the potential behavior of prestressing forces during plant-life extension. A statistically valid approach recommended for developing trend line is to perform the regression analysis of the measured prestressing forces.

REFERENCES

[1] Schupack, M. (1978), "A Survey of Durability Performance of Post-Tensioning Tendons," *Journal of American Concrete Institute*, 75(10):501–10, October, 1978.

[2] Wern, A., M..Schupack, W. Larsen (1971), "Prestressing System for H.B. Robinson Nuclear Power Plant," *Journal of the Power Division*, American Society of Civil Engineers (ASCE), Vol. 97, March 1971.

[3] Harstead, G., E. Kummerle, J. Archer, M. Porat (1971), "Testing of Large Curved Prestressing Tendons," *Journal of the Power Division*, ASCE, Vol. 97, March, 1971.

[4] Schupack, M. (1975), "Grouting Aid for Controlling the Separation of Water for Cement Grout for grouting Vertical Tendons in Nuclear Concrete Pressure Vessels," Conference Paper 151/75, "Experience in Design, Construction and Operation of Prestressed Concrete Pressure Vessels and Containments for Nuclear Reactors," Institution of Mechanical Engineers, London, England, September, 1975.

[5] U.S. Nuclear Regulatory Commission (1977), "Qualification of Cement Grouting for Prestressing Tendons in Containment Structures," RG 1.107, August, 1977.

[6] U.S. Nuclear Regulatory Commission (1977), "Inservice Inspection of Prestressed Concrete Containment with Grouted Tendon," RG 1.90, Revision 1, February, 1977.

[7] Naus, D. (1979), "An Evaluation of the Effectiveness of Selected Corrosion Inhibitors for Protecting Prestressing Steels in PCPVs," ORNL/TM-6479, Oak Ridge National Laboratory (ORNL), Oak Ridge, Tennessee, March, 1979.

[8] American Concrete Institute (1963), "Building Code Requirements for Reinforced Concrete," ACI Standard 318, Detroit, Michigan.

[9] American Society of Mechanical Engineers (2004), "Code for Concrete Reactor Vessels and Containments," Section III, Division 2, ACI Standard 359.

[10] U.S. Nuclear Regulatory Commission (1978), "Materials, Construction and Testing of Concrete Containment," RG 1.136, Revision 2.

[11] U.S. Nuclear Regulatory Commission (1981), "Standard Review Plan for the Review of Safety Analysis Reports," NUREG-0800.

[12] U.S. Nuclear Regulatory Commission (1990), "Determining Prestressing Forces for Inspection of Prestressed Concrete Containments," RG 1.35.1, July, 1990.

[13] Hanson, T., H. Mattock (1966), "Influence of Size and Shape of Member on Shrinkage and Creep of Concrete," *Journal of the American Concrete Institute*, Volume 63. February, 1966.

[14] Hansen, T. (1960), "Creep and Stress Relaxation of Concrete," Swedish Cement and Concrete Research Institute, Stockholm, Sweden.

[15] Ashar, H., J., Costello, H. Graves (1997), "Prestress Force Losses in Containments of U.S. Nuclear Power Plants," Proceedings of Joint WANO/OECD—NEA Workshop (August 25–26, 1997), "Prestress Loss in NPP Containments," Poitiers, France, August, 1997.

[16] Marneffe, L., P. Simoens, L. Imschoot, (1997), "Prestress Behaviour in Belgian NPP Containments," Proceedings of Joint WANO/OECD—NEA Workshop (August 25–26, 1997), "Prestress Loss in NPP Containments," Poitiers, France, August, 1997.

[17] U.S. Nuclear Regulatory Commission (1974), "Inservice Inspection of Ungrouted Tendons in Prestressed Concrete Containments," RG 1.35, Revision 1, May, 1974.

[18] U.S. Nuclear Regulatory Commission (1976), "Inservice Inspection of Ungrouted Tendons in Prestressed Concrete Containments," RG 1.35, Revision 2, January, 1976.

[19] U.S. Nuclear Regulatory Commission (1990), "Inservice Inspection of Ungrouted Tendons in Prestressed Concrete Containments," RG 1.35, Revision 3, July, 1990.

[20] Dougan, J. (1982), "Evaluation of In-Service Inspections of Greased Prestressing Tendons," NUREG/CR-2719 (ORNL/TM-8278), September, 1982.

[21] Ashar, H., D. Jeng (1990), "Effectiveness of In-Service Inspection Requirements of Prestressed Concrete Containments—U.S. Experience," Proceedings of the Second International Conference on Containment Design and Operation, Toronto, Canada, October, 1990.

[22] American Society of Mechanical Engineers (1992 and 1998), Boiler and Pressure Vessel Code, Subsection IWL, Section XI, "Requirements for Class CC Concrete Components of Light-Water Cooled Plants," New York, New York.

[23] Irving, J., M. Hinley, D. McClusky (1997), "The Measurement of Unbonded Tendon Loads in PCCV and Primary Containment Buildings," Proceedings of Joint WANO/OECD—NEA Workshop (August 25–26, 1997), "Prestress Loss in NPP Containments," Poitiers, France, August, 1997.

[24] U.S. Nuclear Regulatory Commission (1996), "Section 50.55a, Codes and Standards, of Title 10, Part 50 of the Code of Federal Regulation," August, 1996.

Appendix 5B Summary of Major Degradation in Containments

5B.1 INTRODUCTION

As noted in Chapter 5, the general performance of the containments has been very good. However, as the plants age the incidences of degradation will likely increase, primarily due to environmental effects. Provided below are examples of operating experience that has impacted the reinforced concrete containment and its liner, steel containments, and the torus of BWR MK I plants.

5B.2 REINFORCED CONCRETE CONTAINMENTS AND STEEL LINERS

Degradation of the post-tensioning system, reinforced concrete structures, and steel liner has been reported.

5B.2.1 Post-Tensioning System

Tendon wire corrosion — Calvert Cliffs Nuclear Power Plant, Units 1 and 2

During the 1997 tendon surveillance of the Calvert Cliffs Nuclear Power Plant Unit 1 containment corrosion was identified on some vertical tendons [1]. The root–cause analysis by the licensee concluded that the tendon wire failures and corrosion problems resulted from combinations of water and moist air intrusion into the end caps and inadequate initial grease coverage of wires in the area just under the stressing washer, thus creating a void. To address the deficiencies identified in the root cause analysis, short-term corrective actions were taken by the licensee during the 1997 inspection and the following years that included spraying hot grease under the stressing washer, re-orienting the shims to leave a gap between the shims to allow a vent path to help eliminate the void, re-greasing non-corroded tendons, and replacing the existing grease caps with a new redesigned grease cap at the upper bearing plate to prevent water intrusion. The licensee also performed additional actions and inspections related to development of a long-term corrective action plan that included: (1) degraded tendon replacement (criteria developed resulted in identification of 46 vertical tendons in each unit for replacement), (2) validation of tendon wire breakage statistical model through enhanced inspections, (3) use of radiography to examine the button head area for wire breaks, and (4) comparison of the minimum required versus the realistically predicted tendon forces at years 2034 and 2036 for Units 1 and 2, respectively, to demonstrate that they are adequate.

Larger than estimated loss of prestressing force — several units

Licensees at several plants have reported lower than predicted prestressing forces for vertical, hoop, and dome tendons [2]. In 1979–1980, R.E. Ginna Nuclear Power Plant reported lower that predicted prestressing forces for several of the tendons in its vertically-prestressed containment. A similar experience occurred at the Virgil C. Summer Nuclear Station when in February 1990 prestressing forces in the 115 vertical tendons were found lower than predicted. Appreciably lower prestressing forces were found in a number of randomly selected vertical, hoop, and dome tendons at Turkey Point Station Units 3 and 4 during performance of the 20th year tendon surveillance in November–December 1992. At Crystal River 3 Nuclear Power Plant several prestressing tendon surveillance inspections over the last 20 years have indicated that the lift-off forces in the hoop prestressing tendons were consistently found to be lower than the 95% predicted values. The cause of the low lift-off forces experienced at the above plants was attributed to high tendon wire relaxation as a result of elevated temperature effects.

Although tendons in containments of the above referenced plants have experienced larger than estimated losses of prestressing force, the tendons are ungrouted so that the typical solution has been to simply re-tension the tendons to the required prestressing force level.

5B.2.2 Concrete Containment Vessel

Containment cylinder delamination — Crystal River 3 Nuclear Power Plant

On September 26, 2009, the steam generator outage at Crystal River 3 Nuclear Power Plant began and included the creation of a construction opening in the containment wall by hydrodemolition to facilitate replacement of the steam generator [3]. The construction opening was rectangular in shape, 8.2 m by 7.62 m (27 ft. by 25 ft.) running from elevation 55.78 m to 64.10 m (183 ft. to 210 ft.). Following removal of the concrete to form the opening, plant personnel identified a delamination adjacent to the construction opening that was about 25.4 cm (10 in.) into the 1.07 m (42 in.) thick concrete containment wall in the plane of the horizontal tendon conduits (Fig. 5B.1). Impulse-response non-destructive testing (NDT) was performed to identify the extent of the delamination and concrete cores were obtained to validate the NDT results, provide samples to support the root cause analysis, and to provide access for boroscope examinations to look at the internal structure of the concrete (e.g., presence of cracking). Figure 5B.2 presents the results of this testing and examination and indicates that the delaminated area included most of the containment wall in Bay 3-4 extending from Buttress 3 to Buttress 4 horizontally and vertically from the top of the equipment opening to the ring beam where the dome tendons were anchored. The concrete cracking was apparently arrested by the increased density of steel reinforcement located in these areas. Non-destructive testing of accessible concrete surfaces in other bays of the containment wall indicated that the concrete was sound.

A root cause team was assembled that identified 75 failure modes that could potentially cause or contribute to the delamination. The failure modes were grouped into nine general areas: (1) containment design and analysis, (2) concrete construction, (3) use of concrete materials, (4) concrete shrinkage, (5) chemically- or environmentally-induced distress, (6) concrete-tendon-liner interactions, (7) concrete cutting, (8) operational events, and (9) external events. The team determined that the delamination was caused by the combination and interplay of: (1) certain design features of the containment structure, (2) the type of concrete utilized (material characteristics), and (3) the acts of de-tensioning and cutting the containment structure for creating the opening. It was also noted that the immediate technical root cause of the delamination event was the redistribution of stresses, as a result of the containment opening activities, resulting in additional stress

FIG. 5B.1
DELAMINATION IN CONTAINMENT WALL AT CRYSTAL RIVER 3.
SOURCE: "CRYSTAL RIVER UNIT #3 CONTAINMENT DELAMINATION UPDATE," PROGRESS ENERGY PRESENTATION TO NRC, ROCKVILLE, MARYLAND, NOVEMBER 20, 2009 (HTTP://PBADUPWS.NRC.GOV/DOCS/ML0932/ML093240292.PDF)

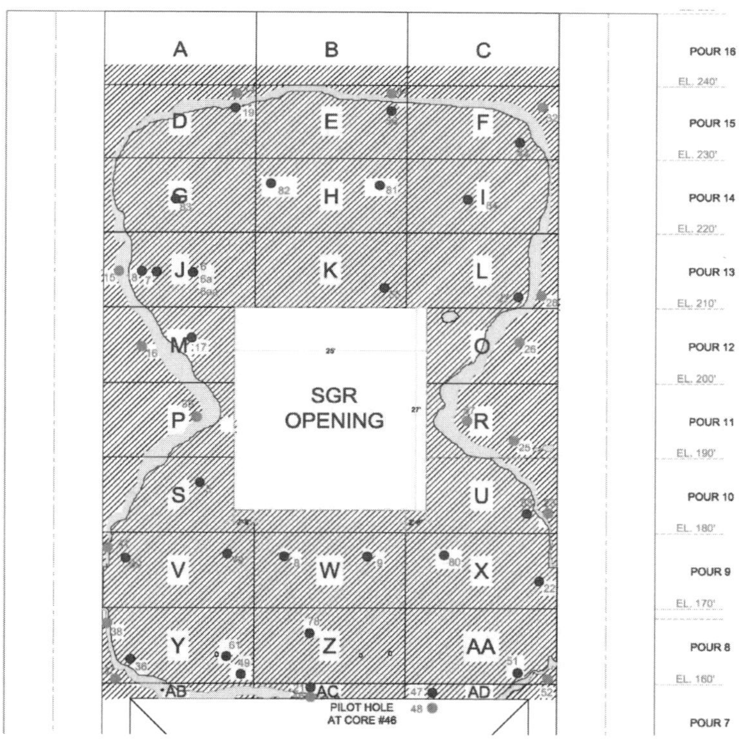

FIG. 5B.2
EXTENT OF WALL DELAMINATION AT CRYSTAL RIVER 3 DETERMINED BY IMPULSE-RESPONSE TESTING AND VALIDATED BY CONCRETE CORING.
SOURCE: "CRYSTAL RIVER UNIT #3 CONTAINMENT INVESTIGATION AND REPAIR," PROGRESS ENERGY PRESENTATION TO NRC, ROCKVILLE, MARYLAND, SEPTEMBER 2, 2010 (HTTP://PBADUPWS.NRC.GOV/DOCS/ML1025/ML102581096.PDF)

beyond original containment design. The condition exceeded the fracture capacity/tensile strength of the concrete resulting in cracking along the high stress plane connecting the horizontal tendons. As the cracking propagated and joined, the delamination occurred over a wide area.

Shield building concrete cracking — Davis-Besse Nuclear Power Station

At the Davis-Besse Nuclear Power Station a temporary access opening was placed in the shield building (reinforced concrete) and containment vessel (steel) to permit removal of the old reactor vessel head and reinstallation of a replacement vessel head. On October 10, 2011, during construction of the shield building access opening using hydrodemolition to remove the concrete, subsurface cracking was discovered near the outer rebar mat [4]. Figure 5B.3 presents an example of the shield building laminar subsurface cracking. Impulse-response testing and concrete core sampling were utilized to investigate the extent of the cracking. Laminar subsurface concrete cracks were identified along the outer rebar mat in the shield building flute shoulders (Fig. 5B.4), at the top of the shield building near the junction with the roof, and at the shield building penetration openings. The root cause of the cracking was attributed to: environmental factors associated with a blizzard in January 1978 (i.e., storm began as rain with winds that drove moisture into the shield building concrete where it froze a day or two later), the lack of an exterior moisture barrier, and the structural design elements of the shield building (e.g., absence of radial reinforcement steel ties or stirrups at intermediate spacing).

FIG. 5B.3
LAMINAR SUBSURFACE CRACKING IN DAVIS-BESSE SHIELD BUILDING CONCRETE. (NOTE THAT THE ACTUAL CRACK WIDTH IS SMALLER THAN SHOWN BECAUSE THE HYDRODEMOLITION PROCESS HAS INCREASED THE CRACK WIDTH). SOURCE: HTTP://PBADUPWS.NRC.GOV/DOCS/ML1217/ML12173A023.PDF

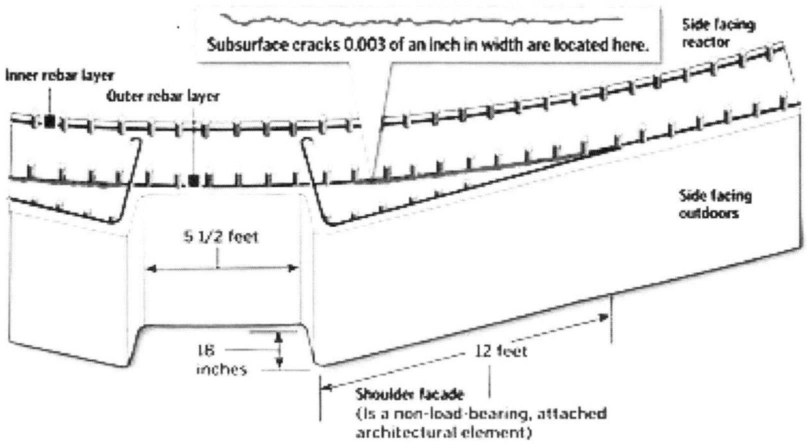

FIG. 5B.4
DAVIS-BESSE SHIELD BUILDING FLUTE AND SHOULDER DETAILS SHOWING SUBSURFACE CRACK LOCATION IN SHOULDER REGION. SOURCE: HTTP://WWW.CLEVELAND.COM/BUSINESS/INDEX.SSF/2012/06/DAVIS-BESSE_STORY.HTML

Both finite-element modeling and experimental studies were used to support the root cause conclusion. Upon repair of the opening, a sealant system will be applied to the exterior of the shield building, and the inspection procedure will be updated to include examination of the shield building exterior sealant system.

Alkali-silica reactions — Seabrook Station

As part of the license renewal activities and its assessment of plant structures at Seabrook Station, the utility performed inspections of safety-related structures and identified the presence of groundwater and visual indications (i.e., pattern cracking) that could be from the occurrence of alkali-silica reactions in the concrete [5].

FIG. 5B.5
CRACKING OBSERVED IN EXTERIOR WALLS OF SEABROOK STATION CONTROL BUILDING/B-ELECTRICAL TUNNEL. SOURCE: A. CHIARAMIDA, "CONCRETE PROBLEM REVEALED," DAILY NEWS, NEWBURYPORT, MASSACHUSETTS, MARCH 16, 2012 (HTTP://WWW.NEWBURYPORTNEWS.COM/LOCAL/X1690519673/CONCRETE-PROBLEM-REVEALED)

The most extensive ASR conditions were in the Control Building/B-Electrical Tunnel (Fig. 5B.5). After the concrete degradation was identified, over 30 concrete core samples were taken and sent to test laboratories where the presence of ASR was confirmed by petrographic examination. Property test results from the concrete cores indicated that the unconfined concrete modulus of elasticity and compressive strength had been reduced by up to 47% and 22%, respectively, relative to reference values that had been obtained in 1979 when the tunnel had been constructed. The occurrence of ASR has been confirmed by petrography in five other structures and visual features indicate its potential occurrence in 14 additional structures [6]. The utility has planned several activities to quantify the significance of the occurrence of ASR in the concrete structures at Seabrook Station: accelerated expansion testing to determine remaining activity of aggregate, conduct extent of conditions inspections at a number of locations, full-scale structural features testing (e.g., anchorage and shear tests) to indicate structural impact of ASR, determination of crack limits and acceptance criteria, implementation of a plant-specific ASR monitoring program, and determination of the impact of alkali-silica reactions on seismic response.

5B.2.3 Steel Liner

A recent review of available information sources such as NRC inspection reports, licensee in-service inspection reports, operational experience, and NRC informational notices indicates that although corrosion of the concrete containment steel liners has occurred most frequently at the concrete/steel interface where the liner becomes embedded in the concrete floor, corrosion has also occurred at other locations in the liner, including adjacent to the concrete side of containment, primarily due to the presence of an embedded foreign material [7].

In October 2009 at Salem Nuclear Generating Station Unit 2 the licensee inspected the containment moisture barrier (i.e., silicone seal between the concrete floor and containment liner) and found heavy corrosion on the containment liner within 15.2 cm (6 in.) of the concrete floor [8]. Figure 5B.6 presents similar containment liner (knuckle plate) corrosion near the intersection of the concrete floor and liner that was identified in Unit 1 during an inspection performed in the spring of 2010. This area of the containment was considered inaccessible because it was normally covered by an insulation package that consisted of a layer of sheet

FIG. 5B.6
SALEM GENERATING STATION UNIT 1 CONTAINMENT LINER KNUCKLE PLATE DEGRADATION. SOURCE: AULUCK, R., "CONTAINMENT DEGRADATION ISSUES WITH U.S. NUCLEAR POWER PLANTS — LESSONS LEARNED," TRANSACTIONS SMIRT 21, PAPER ID# 614, NEW DELHI, INDIA, NOVEMBER 2011.

metal, a layer of plastic sheeting, and a layer of insulation. As a result of this discovery, the licensee decided to enhance inspections of the containment liner in this region [i.e., 15.2 cm (6 in.) above the concrete floor and random inspections at several other areas covered by the insulation package]. Ultrasonic testing was also performed at 440 locations over the bottom 15.2 cm (6 in.) of the cylindrical portion of the containment liner. Since the minimum plate thickness measurement was 16.7 mm (0.667 in.) and the design-required minimum wall thickness was 10.9 mm (0.43 in.), the licensee determined that the liner remained operational. The source of moisture that caused corrosion at the joint between the liner and concrete floor was service water leakage from the containment fan coil units and associated piping.

On April 23, 2009, during a refueling outage at Beaver Valley Power Station Unit 1 the licensee performed a visual examination of the interior reactor containment building steel liner [7]. At a containment elevation of 227.4 m (746 feet), the licensee identified an area approximately 76.2 mm (3 in.) in diameter that exhibited blistered paint. Collapse of the blister during further inspection revealed a protruding rust product underneath. The licensee then cleaned this area revealing a rectangular area of approximately 25.4 mm (horizontal) by 9.5 mm (vertical) (1 in. by 3/8 in.) that penetrated through the entire liner plate thickness (Fig. 5B.7). The licensee removed the corroded section of the liner and discovered a partially decomposed piece of wood approximately 50.8 mm by 101.6 mm by 152.4 mm (2 in. by 4 in. by 6 in.) embedded in the concrete behind the liner (Fig. 5B.8). The wood was left behind as a result of inadequate housekeeping and quality assurance practices during the original construction of the containment wall in the early 1970s. The licensee determined that the cause of the through-wall liner corrosion was a pitting-type corrosion (rust) originating from the concrete side caused by foreign material (wood) that was in contact with the containment carbon steel liner. Corrective actions included removing the embedded wood, grouting the concrete area that was displaced by the wooden debris, and welding a new section of steel plate to the liner to replace the previously removed portion of the liner.

FIG. 5B.7
PERFORATION OF CONTAINMENT LINER AT BEAVER VALLEY POWER STATION UNIT 1 OBSERVED AFTER THE PAINT BLISTER AND CORROSION PRODUCTS WERE REMOVED. SOURCE: "BV1 CONTAINMENT LINER PLATE HOLE," CR 09-57762, FIRSTENERGY CORPORATION, AKRON, OHIO, JUNE 4, 2009

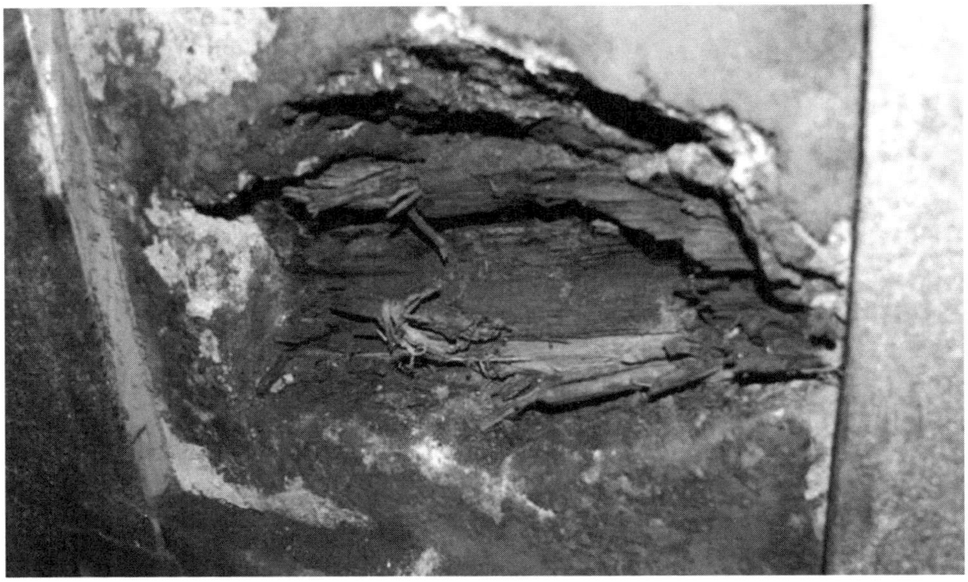

FIG. 5B.8
WOOD EMBEDDED IN THE CONCRETE THAT WAS DISCOVERED AT BEAVER VALLEY POWER STATION UNIT 1 AFTER A SECTION OF THE CONTAINMENT LINER WITH THROUGH-THICKNESS CORROSION WAS REMOVED. SOURCE: "BV1 CONTAINMENT LINER PLATE HOLE," CR 09-57762, FIRSTENERGY CORPORATION, AKRON, OHIO, JUNE 4, 2009.

5B.3 STEEL CONTAINMENTS

Steel containment degradation has occurred in the drywell of BWR free-standing steel containments, steel cylinder of PWR ice-condenser primary containments, and the torus of BWR Mark I containments. Examples representing typical degradation occurrences identified in each of these structures is provided below.

5B.3.1 BWR Free-Standing Steel Containment

In the early 1980's water was discovered leaking out the top of the sand bed through an annulus around the torus vent line at Oyster Creek Nuclear Generating Station [9, 10]. Inspection of this area during a refueling outage revealed that water from the reactor cavity was leaking down and around the outside of the drywell, through the insulation in the air gap, and into the sand bed. It was discovered that several of the drains in the sand bed were clogged allowing water to saturate the sand and corrode the outside surface of the exposed carbon steel drywell shell (Fig. 5B.9). Ultrasonic thickness measurements revealed that the shell thickness in some local areas had been reduced from 29.3 mm to 20.3 mm (1.15 to 0.80 in.), with a minimum acceptable shell thickness of 18.8 mm (0.74 in.). Attempts to stop corrosion through cathodic protection were unsuccessful. Holes were drilled through the biological shield concrete, the sand in the sand bed region removed, the area cleaned, and an epoxy coating applied to the containment. The containment peak pressure was reduced from 427 kPa to 303 kPa and detailed analytical assessments indicated that the minimum required drywell thickness was now 13.7 mm. Periodic visual examinations, ultrasonic thickness measurements of the drywell shell above the sand bed, and inspections for leakage from the reactor cavity were implemented to manage aging.

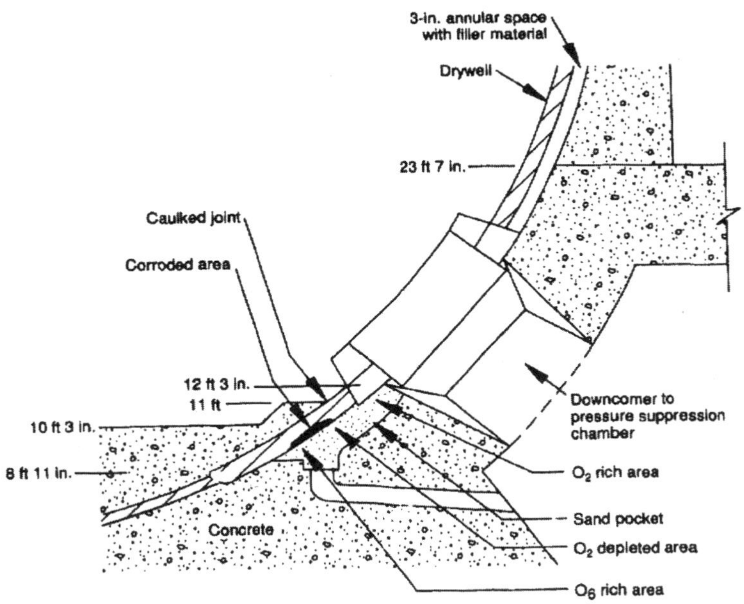

FIG. 5B.9
OYSTER CREEK DRYWELL BASE SHOWING CORRODED AREA ON OUTSIDE SURFACE NEAR SAND POCKET REGION.
SOURCE: V. N. SHAH, S. K. SMITH, AND U. P. SINHA, "INSIGHTS FOR AGING MANAGEMENT OF LIGHT WATER REACTOR COMPONENTS – METAL CONTAINMENTS," NUREG/CR-5314, U.S. NUCLEAR REGULATORY COMMISSION, WASHINGTON, D.C., MARCH 1994.

5B.3.2 Steel Cylinder of PWR Ice-Condenser Primary Containments

In August of 1989, significant base metal corrosion of the outer surface of the steel shell of the McGuire Unit 2 containment (Fig. 5B.10) was discovered during a pre-integrated leak rate test inspection [11]. A significant amount of water in the annular space of the concrete floor was found as a result of the drains not functioning. The corrosion occurred where boric acid deposits had been deposited from leaking instrument line compression fittings. This area did not have a sealant at the joint between the containment and the concrete floor so the corrosive environment could contact the steel shell below the concrete floor. Damaged areas were weld repaired and recoated, accessible areas between the containment and concrete floor were sealed, and the annulus was inspected weekly to identify and remove any standing water or boric acid deposits. The Unit 1 containment at McGuire was found to have similar degradation as well as corrosion of the inside surface of the coated containment shell under the ice condenser at upper and lower floors where a cork material was located. The two ice condenser containments of similar design at Catawba also experienced coating damage and base metal corrosion at the outer surfaces of the steel shells at the intersection of the steel shell and the concrete floor.

5B.3.3 Torus of BWR MK I Plants

Since 1974, 3,800 coating repairs comprising 13.5 m^2 (145 ft^2), or 1.1% of the torus surface area below the waterline, have been made at Cooper Nuclear Station [5, 12]. A total of approximately 2200 of the coating repairs have been made to the torus shell since 2001. Out of the 2200 locations, eighteen have been reported to have pits where the nominal steel shell thickness has been reduced by greater than 10% of the nominal shell thickness [13]. Figure 5B.11 provides an example of torus corrosion at Cooper Nuclear Station. Visual

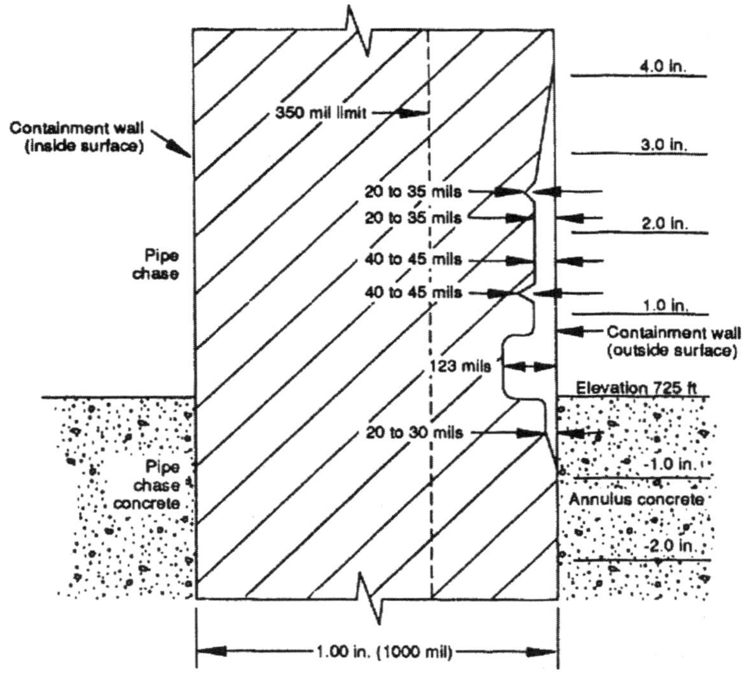

FIG. 5B.10
STEEL CONTAINMENT CORROSION AT MCGUIRE NUCLEAR STATION UNIT 2.
SOURCE: V. N. SHAH, S. K. SMITH, AND U. P. SINHA, "INSIGHTS FOR AGING MANAGEMENT OF LIGHT WATER REACTOR COMPONENTS – METAL CONTAINMENTS," NUREG/CR-5314, U.S. NUCLEAR REGULATORY COMMISSION, WASHINGTON, D.C., MARCH 1994

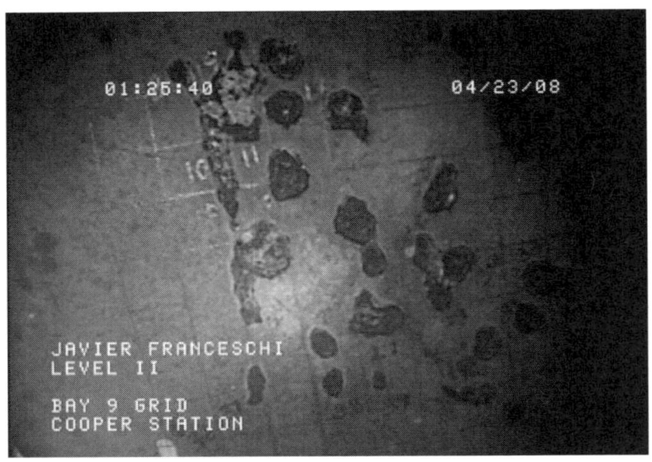

FIG. 5B.11
EXAMPLE OF CORROSION IN COOPER NUCLEAR STATION TORUS BAY 9.
SOURCE: "OFFICIAL TRANSCRIPT OF PROCEEDINGS, ADVISORY COMMITTEE ON REACTOR SAFEGUARDS PLANT LICENSE RENEWAL SUBCOMMITTEE." WORK ORDER NO. NRC-237. NEAL R. GROSS AND CO., INC., COURT REPORTERS AND TRANSCRIBERS: WASHINGTON, D.C., MAY 5, 2010 (HTTP://PBADUPWS.NRC.GOV/DOCS/ML1013/ML101390285.PDF)

inspections of the coating surfaces for adverse coating conditions such as flaking, peeling, blistering, discoloration, and other signs of distress are performed during each refueling outage. Divers visually inspect coatings on the torus below the waterline. Repairs are typically made by applying an epoxy that is intended to arrest the pitting only in areas that experience localized zinc coating failures where the pit depths exceed a threshold [(generally 0.76 to 1.27 mm) (30 to 50 mils)] [12]. However, coating at pits that do not exceed the threshold is not required, but the pitting is monitored at the next inspection (3 years later) for growth. This process, however, has not been totally successful in that it can result in localized galvanic corrosion that can yield higher and unpredictable corrosion rates (pitting) than that of general corrosion. Also, it has contributed to the amount of sludge and corrosion product collecting in the suppression pool which can further increase the corrosion rate. As part of the license renewal application, the applicant within three years of entering the period of extended operation will recoat wetted portions of the torus, sludge will be removed from the wetted portion of the torus and the surface inspected every refueling outage until the torus is recoated, and an analysis will be performed following each torus inspection to demonstrate that the projected pitting of the torus up to the time the torus is recoated will not result in reduction of the wall thickness below minimum accepted values.

At Duane Arnold Energy Center numerous areas of zinc depletion of the torus coating and minor pitting and other indications were identified while conducting inspections (Fig. 5B.12). The torus shell interior surface was initially coated in 1973 and recoated in 1985. Since 1995, over 15,000 repairs have been made to the torus coating [14]. All of these repairs have been made below the water line, primarily at the very bottom of the shell area. This represents about 5% of the underwater torus area. Only one pit had degradation that exceeded the maximum allowable pit depth of 1.35 mm (53 mils) [10% of 13.56 mm (0.534 in.), the nominal shell thickness]. That pit measured 1.42 mm (56 mils) in depth and 6.35 mm (0.25 in.) in diameter. The pit was dispositioned in the corrective action program through an engineering evaluation that determined the pit was acceptable without repair. The current scope of the coatings program includes inspection of the interior and exterior surfaces of the suppression chamber (torus), vent lines, and downcomers, and the interior and accessible exterior surfaces of the drywell [15]. Visual inspections are performed during each refueling cycle. Prior to startup from the first refueling outage during the period of extended operation, the torus will be recoated.

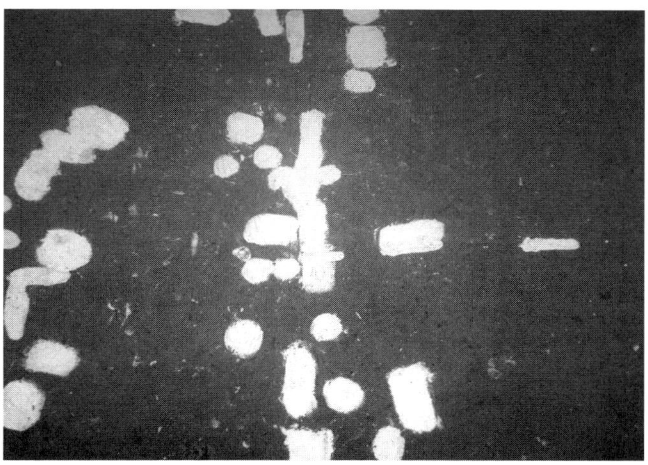

FIG. 5B.12
DUANE ARNOLD ENERGY CENTER TORUS CORROSION AND COATING REPAIRS. SOURCE: THE APPLICANT (DUAC) PRESENTATION TO ACRS SUBCOMMITTEE DATED JAN. 20, 2010

REFERENCES

[1] Letter from D. Skay, U.S. Nuclear Regulatory Commission, to C. H. Cruse, Calvert Cliffs Nuclear Power Plant, "Calvert Cliffs Nuclear Power Plant, Unit Nos. 1 and 2 – Containment Tendon Long-Term Corrective Action Plan (TAC Nos. MA7782 and MA7783," September 21, 2011 (http://pbadupws.nrc.gov/docs/ML0125/ML012530249.pdf).

[2] "Degradation of Prestressing Tendon Systems in Prestressed Concrete Containments," Information Notice No. 99-10, U.S. Nuclear Regulatory Commission, Washington, D.C., April 13, 1999.

[3] "Root Cause Assessment Crystal River 3 Containment Concrete Delamination," Non-Proprietary Version, Performance Improvement International, Oceanside, California, August 10, 2010.

[4] Letter from S. A. Reynolds, U.S. Nuclear Regulatory Commission, to B. Allen, Davis-Besse Nuclear Power Station, "Davis Besse Nuclear Power Station – Inspection to Evaluate the Root Cause Evaluation and Corrective Actions for Cracking in the Reinforced Concrete Shield Building of the Containment System (05000346/2012009 (DRS)," June 21, 2012 (http://pbadupws.nrc.gov/docs/ML1217/ML12173A023.pdf).

[5] Auluck, R., "Containment Degradation Issues With U.S. Nuclear Power Plants – Lessons Learned," Transactions SMiRT 21, Paper ID# 614, New Delhi, India, November 2011.

[6] "Impact of Alkali Silica Reaction on Seabrook Concrete Structures," Nextera Energy, Seabrook, New Hampshire presentation to Advisory Committee on Reactor Safeguards, Rockville, Maryland, April 23, 2012 (http://pbadupws.nrc.gov/docs/ML1211/ML121160422.pdf).

[7] Dunn, D. S., Pulvirenti, A. L., and Hiser, M., "Containment Liner Corrosion Operating Experience Summary Technical Letter Report – Revision 1," U.S. Nuclear Regulatory Commission, Washington, D.C., August 2, 2011 (http://pbadupws.nrc.gov/docs/ML1120/ML112070867.pdf).

[8] "Containment Liner Corrosion," Information Notice No. 2010-12, U.S. Nuclear Regulatory Commission, Washington, D.C., June 18, 2010.

[9] Lipford, B. L. and Flynn, J. C., "Drywell Corrosion Stopped at Oyster Creek," *Power Engineering*, Vol. 97, No. 11, November 1993, pp. 47–50.

[10] "Degradation of Steel Containments," Information Notice 86-99, U.S. Nuclear Regulatory Commission, Washington, D.C., December 8, 1986.

[11] Shah, V. N., Smith, S. K., and Sinha, U. P., "Insights for Aging Management of Light Water Reactor Components," NUREG/CR-5314, Vol. 5, U.S. Nuclear Regulatory Commission, Washington, D.C., March 1994.

[12] U.S. Nuclear Regulatory Commission. — "Safety Evaluation Report Related to the License Renewal of Cooper Nuclear Station." NUREG-1944. U.S. Nuclear Regulatory Commission: Washington, D.C., October 2010.

[13] O'Grady, B. J., Nebraska Public Power District, letter to U.S. Nuclear Regulatory Commission, December 21, 2009.

[14] Costanzo, C. R., NextEra Energy Duane Arnold, LLC, letter to U.S. Nuclear Regulatory Commission, December 14, 2009.

[15] U.S. Nuclear Regulatory Commission. — "Safety Evaluation Report Related to the License Renewal of Duane Arnold Energy Center." NUREG-1955. U.S. Nuclear Regulatory Commission, Washington, D.C., November 2010.

CHAPTER

6

CONTAINMENT STRUCTURE TESTING, MODELING, AND DEGRADATION

Jason Petti

Numerous studies have been conducted at Sandia National Laboratories and around the world to examine the severe accident behavior of containments through containment structural testing and modeling. Large-scale testing has been conducted on both prestressed concrete containments as well as on various steel shell designs. Since the containment includes the entire pressure boundary and not just the steel or concrete shell, some additional testing has been conducted on components (e.g., bellows, hatches). One of the goals of these test programs was to provide data for the calibration of computational models. Therefore, the accompanying computational analyses efforts for a number of the testing programs are also discussed. In addition, a summary of work examining the effects of containment degradation is provided. Various forms of degradation (e.g., liner corrosion) have been observed throughout the U.S. fleet of nuclear power plants. The effects of degradation on the severe accident response are examined in several studies using both deterministic and probabilistic analyses. These efforts describe the effect containment degradation has on the severe accident failure pressures, risks, and consequences.

Keywords: containment degradation, liner corrosion, containment testing, concrete containments, steel containments, severe accident response, containment fragility, fragility curves

6.1 INTRODUCTION

This chapter summarizes past efforts in containment structural testing, modeling, and examining degradation effects under severe accident conditions. Previous chapters provide details related to the design of containments under design bases conditions using the appropriate requirements established in the U.S. Code of Federal Regulations, ASME Boiler & Pressure Vessel Code [1], and U.S. NRC Regulatory Guides. Since severe accidents are perceived to have a low probability of occurrence, they are not considered in the design basis. However, the probability of a severe accident is not zero as shown through the events at the Three Mile Island Nuclear Power Plant as well as the recent events in Japan. A subsequent chapter provides the history of severe accident considerations and also includes details on severe accident progression and modeling. The structural behavior of the containment under such severe accident conditions is addressed in this chapter through both large-scale testing and computer modeling. Tests and analyses conducted on both reinforced concrete and prestressed concrete containments will be summarized as well as tests and analyses performed with various steel shell designs. Since the containment includes the entire pressure boundary and not just the primary steel or concrete shell, some additional discussion is also provided on component (e.g., bellows, hatches) testing and modeling. This work was the basis of the recent guidelines provided by the U.S. NRC for estimating the structural integrity of containments under above design basis internal pressures [2]. This chapter also includes a summary of work examining the effects of containment degradation. As described in a Chapter 5, various forms of degradation (e.g., liner corrosion) have been observed throughout the U.S. fleet of nuclear power plants. The procedures for assessing and repairing degradation are also described in Chapter 5. As part of the assessment process, the effect that the degradation has on the design basis of the containment can be determined through analysis. This is discussed briefly in this chapter; however, the main focus centers on the effects of degradation on the severe accident response. Several studies are described that have examined the effect containment degradation has on the severe accident failure pressures, risks, and consequences.

6.2 EARLY ESTIMATES OF CONTAINMENT STRUCTURAL RESPONSE TO SEVERE ACCIDENTS

As discussed in an earlier chapter, containments are designed to resist the conditions (e.g., pressurization up to the design pressure and design accident temperatures) present during design accidents. However, containments are not designed to resist the conditions (e.g., pressurization and temperatures above the design pressure and temperatures) for beyond design basis accidents. Therefore, the conditions present within the containment during a severe accident leads to a structural response that is increasingly uncertain. This is due to the highly nonlinear behavior of the containment that is induced by the severe loadings.

A comprehensive examination of severe accident risks was performed under the NUREG-1150 [3] study. This study employed probabilistic risk assessments (PRAs) in examining the severe accident response at five US plants: Zion (PWR with prestressed containment), Surry (PWR with a reinforced concrete containment), Sequoyah (PWR with a steel shell containment), Peach Bottom (Mark I BWR with a steel shell containment), and Grand Gulf (Mark III BWR with a reinforced concrete containment). For each plant, a detailed assessment of the potential severe accidents was conducted that included modeling of numerous plausible accident progressions. In some cases, accident progressions led to core damage and a pressurization and heat-up of the containment. Therefore, a significant component of the NUREG-1150 study was to assess the containment's structural behavior during pressurization and heat-up. Since limited work had been conducted in assessing the behavior of the containments to severe loadings, expert panels were assembled to provide their opinions as to the performance of the containments under the severe accident loadings. Due to the nature of the PRA, the experts were required to provide the containment responses as the probability of containment failure versus pressure (also designated as the fragility curve). Containment failure in NUREG-1150 was defined as a leak failure or greater. A leak was defined as a breach in the containment's pressure boundary (e.g., a tearing of the steel shell in steel containments, a tearing of the steel liner in concrete containments, or the initiation of other pathways for the leakage of material to the outside environment) that was on the order of 0.028 m^2 (0.3 ft^2) and would not depressurize the containment in less than 2 hours. Conditional failure probabilities were also provided by the experts to include the estimates for

TABLE 6.1
NUREG-1150 CONTAINMENT STATIC FAILURE PRESSURE RANGES, 5TH TO 95TH PERCENTILE [3]

Plant	Design Pressure, MPa (psi)	Failure Pressure Range, MPa (psi)
Zion	0.32 (47)	0.74–1.24 (108–180)
Surry	0.31 (45)	0.66–1.03 (95–150)
Sequoyah	0.076 (11)	0.28–0.66 (40–95)
Peach Bottom	0.39 (56)	0.83–1.20 (120–174)
Grand Gulf	0.10 (15)	0.26–0.50 (38–72)

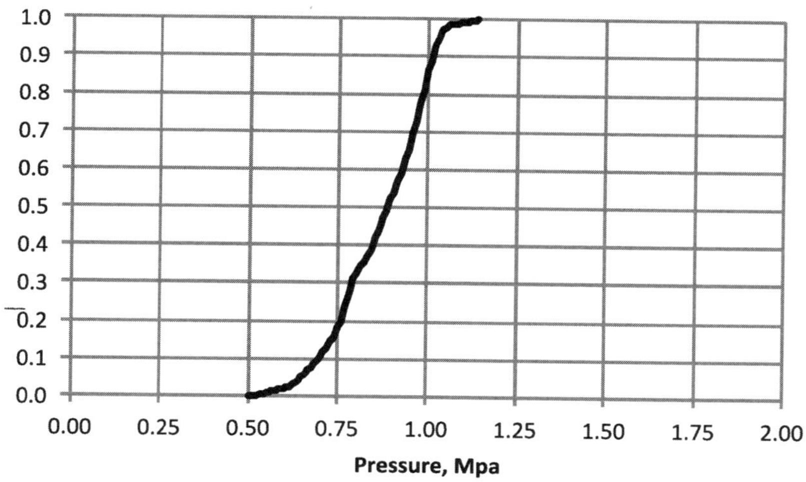

FIG. 6.1
SURRY IPE FRAGILITY CURVE [5]

larger ruptures (breaches larger than 0.028 m^2 that would depressurize the containment in less the 2 hours) and catastrophic rupture failures (significant sections of the containment removed). The methods used by the experts varied considerably from simple hand calculations to more detailed computer programs and analyses. Due to the limited knowledge of containment response to severe accident loads and the different approaches and assumptions made by the expert panel members, there was a wide variation in the estimates. Therefore, the expert panel estimates were combined in order to define a single "best estimate" probability of failure versus pressure curve for use in the PRA. Table 6.1 summarizes the failure pressure ranges as used in the NUREG-1150 study.

In addition to the NUREG-1150 expert estimates, the plants also performed their own analyses of the containment capacity under the Individual Plant Evaluations (IPEs) [4]. As with the NUREG-1150 expert estimates, the IPE estimates of containment failure versus pressure varied considerably from plant to plant. The estimates differed in the level of sophistication, the techniques used, and the assumption made for the failure definitions. Figure 6.1 illustrates the fragility curve developed for the Surry plant subjected to static pressurization. In addition to Surry, NUREG/CR-6338 [5] and NUREG/CR-6906 [6] include all of the available IPE fragilities for containments in the US fleet.

6.3 LARGE-SCALE CONTAINMENT AND COMPONENT TESTING

Since the containment is the last barrier between the release of radioactive material and the environment during a severe accident, a thorough understanding of its behavior during such events is critical. Therefore, the US NRC conducted a study [7] with Sandia National Laboratories in the early 1980s to develop a plan for building the tools needed to predict the ultimate capacity of steel and concrete containments under severe accident conditions. Among other findings, the study concluded that little experimental data was available examining the behavior of containments under severe conditions. This then led to a series of tests on both steel and concrete containments at Sandia. This section summarizes those tests along with a number of tests performed at other facilities. While some of the designs tested at Sandia and at other facilities are of containments not built in the United States, the insights gained through these tests are primarily used for understating the failure mechanisms and for improving modeling techniques. In addition to the individual test reports referenced in this chapter, Hessheimer and Dameron [6] provide a more detailed summary of these tests in NUREG/CR-6906.

6.3.1 Containment Testing Purpose

As mentioned previously, testing of containments was necessary for the development methods for predicting the containments behavior and the ultimate capacity during severe accidents. Due to the high costs of testing and other limitations, tests were required to be performed at a smaller scale than actual containments are built. Since there are many different containment designs throughout the US fleet and the world, scaling down versions of any specific design was typically not considered. However, the scaled tests were designed in order to insure that general characteristic features were maintained and that unrealistic failure modes were not introduced [7]. The primary goal of the experimental program was to provide data for the benchmarking of computational analysis models and techniques. The intent being that after the computational tools were sufficiently developed using the experimental data, analyses of full scale containments under similar conditions could be conducted. The use of the experimental data in the benchmarking of the codes and computational tools greatly increases the confidence in the modeling predictions. The analyses of the tests are discussed in more detail later in this chapter.

6.3.2 Containment Tests

Under severe accident conditions, the primary loading applied to the containment is internal pressurization. This pressurization is caused by the generation of steam or other gases within the containment due to damage to the core. These events also include a significant increase in the temperature within the containment. The pressures and temperatures that would be present during a severe accident can far exceed the design pressure and temperatures. If these conditions are sufficiently high, a breach of the containment boundary will occur, leading to a leak or rupture. In designing the tests, decisions were made on whether or not to include both pressure and temperature loadings. In order to simplify the tests, most of the tests were conducted using pressure loading only. The temperature loadings were judged to be sufficiently accounted for in the computational models [6].

In addition to obtaining data for the benchmarking of computational tools, the tests also allowed for the study of the failure modes that could occur in the containment. As discussed earlier, the failures could range from leaks caused by a tearing of the liner or a penetration seal failure to a larger rupture failure where a significant breach is induced. In order for the tests to be capable of reflecting these failures under similar mechanisms as in the full-scale containments, the designs were scaled in order to not remove or introduce failure modes. The tests were typically scaled geometrically (e.g., 1:10 scale or 1:4 scale). They were also built using the same materials as would be used in the full-scale containments. A thorough discussion of the scaling and failure issues considered during these tests is provided in NUREG/CR-6906. It should also be noted that some representative penetrations or hatch openings were included in many of the tests, but not to the extent or level seen in actual full-scale containments. A number of separate effects penetration and component tests are also briefly described.

6.3.2.1 Reinforced Concrete Containment Tests. The 1:6 scale reinforced concrete containment tested at Sandia National Laboratories in 1987 is the only large-scale reinforced concrete containment test where detailed information has been made publicly available [8]. The containment model was designed and constructed by United Engineers using the ASME code [1]. The test was of a PWR design with a design pressure of 0.32 MPa (46 psi). Figure 6.2 illustrates some of the details for the 1:6 scale containment including several penetrations. Figure 6.3 provides images of the containment construction including the rebar and penetration details. In order to obtain data on the containment's performance, over 1200 channels of instrumentation were used to monitor mainly strains and displacements during pressurization. After the low-pressure Structural Integrated Test and Integrated Leak Rate Tests, the model was pressurized to 1.0 MPa (145 psi) where the leak rate exceeded 5000% mass/day. This leak rate was sufficiently high that the pressurization system could no longer increase the pressure. The post-test inspection revealed numerous tears in

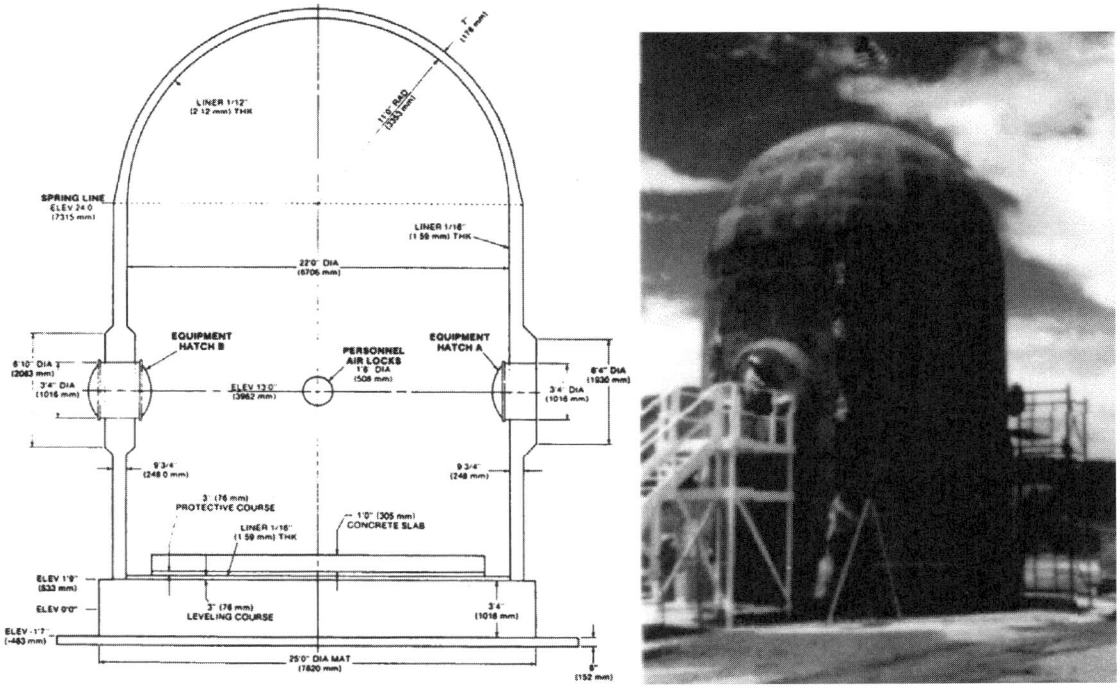

FIG. 6.2
DETAILS OF THE 1:6 SCALE REINFORCED CONCRETE CONTAINMENT MODEL [9]

FIG. 6.3
CONSTRUCTION IMAGES FOR THE 1:6 SCALE REINFORCED CONCRETE CONTAINMENT MODEL [6]

FIG. 6.4
LINER TEAR NEAR A PENETRATION FOR THE 1:6 REINFORCED CONCRETE CONTAINMENT [9]

the liner near the liner anchor studs surrounding penetrations. Figure 6.4 shows the most significant tear that extended approximately 56 cm (22 in.) along a thickened plate near a penetration [9]. Other results from the test include maximum mid-height rebar and liner strain between 1.5% and 2% and peak liner strains of 8%.

6.3.2.2 Prestressed Concrete Containment Tests. Significantly more prestressed concrete containments have been tested than typical reinforced concrete containments. NUREG/CR-6906 [6] provides a listing and at least a short discussion of approximately 10 prestressed concrete containments tested around the world. These tests include the 1:10 scale Sizewell-B model tested by the Central Electricity Generating Board in the United Kingdom [10, 11] and the 1:14 scale CANDU model tested in Canada [12–15] among others [16, 17]. This section will focus on the resent 1:4 scale prestressed concrete containment vessel (PCCV) tested at Sandia National Laboratories as part of a cooperative program between the U.S. NRC and the Nuclear Power Engineering Corporation (NUPEC) of Japan [18].

The PCCV model was an actual scaled version of the Ohi-3 Pressurized Water Reactor containment in Japan. Therefore, the test model was constructed in order to approximate the response of the actual containment in terms of global behavior, details near penetrations, and the steel liner. Figure 6.5 illustrates the details of the PCCV test model. As part of the construction process, nearly 1500 transducers were placed within the containment in order to monitor the behavior of the containment during testing. Figure 6.6 shows the PCCV under construction as well as the completed model. After approximately 3 1/2 years of construction, the PCCV model was completion in June of 2000.

The testing program for the PCCV included several different levels of static pressurization tests. These included the System Functionality Test to 0.5 times the design pressure (design pressure = 0.39 MPa) and

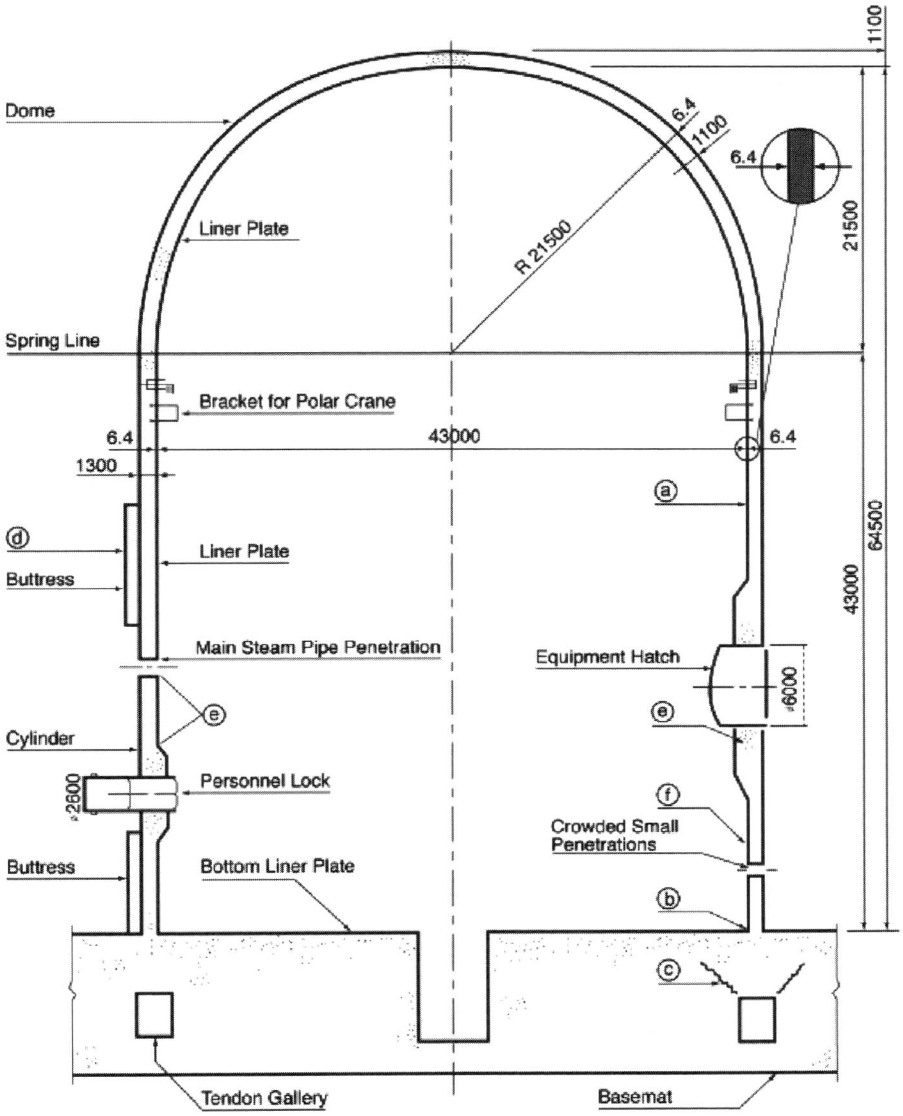

FIG. 6.5
NUPEC/NRC/SANDIA PCCV MODEL DETAILS [18]

the Structural Integrity and Integrated Leak Rate Test to 1.125 times the design pressure for 1 hour and then 0.9 times the design pressure for 24 hours. The leak rate during the Integrated Leak Rate Test was measured at less than 0.1% mass/day which was considered "leak tight." This was followed by the Limit State Test to assess the behavior of the containment to beyond design basis pressures that could result from a severe accident. During the Limit State Test, a leak rate of 0.48% mass/day was observed at 1.5 times the design pressure and dropped to 0.003% mass/day at 2 times the design pressure. The leak rate then increased to 1.63% mass/day at 2.5 times the design pressure with the average hoop strain at 0.18%. At approximately 2.4 times, the design pressure, the acoustic sensors indicated that an event occurred near the equipment hatch likely leading to the high leak rate. The containment was further pressurized to approximately 3.3 times the design pressure leading to a leak rates far exceeding 100% mass/day. The test was terminated when the pressurization

FIG. 6.6
NUPEC/NRC/SANDIA PCCV UNDER CONSTRUCTION AND COMPETED [18]

system could no longer maintain the pressure due to the high leak rate. The maximum local liner strains were approximately 6.5% with an average hoop strain of 0.4%. Figure 6.7 shows two of the liner tears that occurred near the equipment hatch. In addition to these tears, several others were observed but there was no indication of any tendon or rebar damage. A final Structural Failure Mode Test was conducted in order to achieve a fully inelastic failure of the containment. This required that an elastomeric membrane be placed in the interior of the containment against the liner so the containment could be filled with water. Only a small volume (~3%) of the containment was left unfilled in order to be pressurized. Figure 6.8 shows the failure of the containment after reaching a peak pressure of 3.63 times the design pressure with a maximum hoop strain of 1.65%.

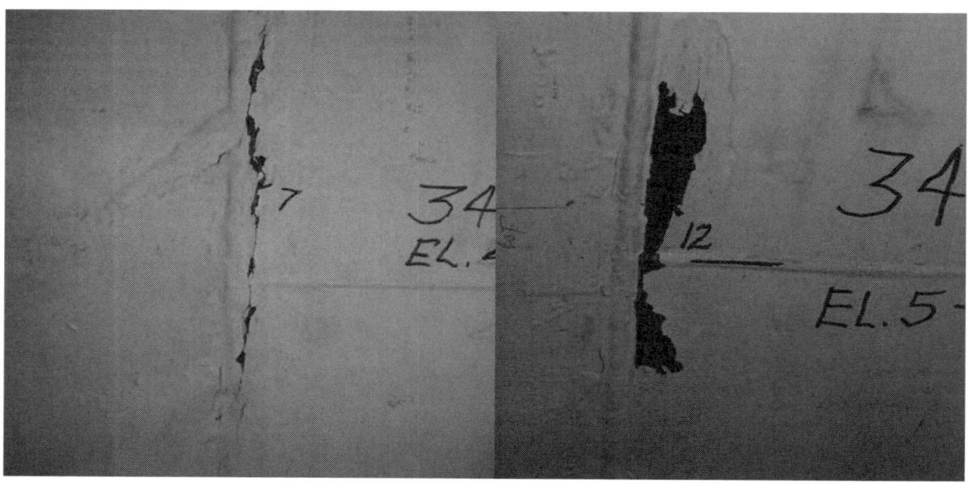

FIG. 6.7
NUPEC/NRC/SANDIA PCCV LIMIT STATE TEST [18]

FIG. 6.8
NUPEC/NRC/SANDIA PCCV AT STRUCTURAL FAILURE [18]

6.3.2.3 Steel Containment Tests. Similar to the concrete containment testing programs, a series of steel containments were tested at Sandia during the 1980s and 1990s. The initial tests were conducted on a 1:32 scale (~1.65 m or 65 in. height and ~1.17 m or 46 in. diameter with a 1.22 mm or 0.048 in. wall thickness) containment consisting of a cylindrical shell with a hemispherical dome [19, 20]. Four of the 1:32 scale tests were conducted with examples shown in Fig. 6.9. Two of the tests (designated SC-0 and SC-1) included

FIG. 6.9
NRC/SANDIA 1:32 SCALE STEEL CONTAINMENTS TESTED AT SANDIA.
SC-1 (TOP TWO IMAGES), SC-3 BOTTOM IMAGE [6]

no stiffening elements or penetrations. One (SC-2) included stiffeners attached to the outside surface of the cylinder and one (SC-3) was constructed with penetration details. The two plain steel containments (SC-0 and SC-1) failed at 0.93 and 0.76 MPa with the first catastrophically rupturing and the later failing due to leakage from a tear in a weld. The stiffened model (SC-2) failed catastrophically at 0.97 MPa after being repaired when leakage occurred during the initial test. The model that included penetrations (SC-3) failed catastrophically at 0.83 MPa with the initiation near one of the penetrations.

The 1:32 scale tests were followed by a larger 1:8 scale steel containment with dimensions and the constructed model shown in Fig. 6.10. This containment had a design pressure of 0.27 MPa (40 psi) and included stiffening elements and penetration features common in PWR ice condensers and Mark III BWRs [21]. The containment failed catastrophically at a pressure of 1.34 MPa (195 psi) due to a tear initiating in a stiffener detail near one of the penetrations [22]. Since the containment failed at nearly five times the design pressure, fragments of the containment traveled over 1500 ft after rupturing.

Similar to the 1:4 scale PCCV model, a cooperative sponsorship between the U.S. NRC and NUPEC of Japan was formed to conduct a test of a 1:10 scale steel containment at Sandia. The containment was constructed at 1:10 scale geometrically, but the thickness was fabricated at 1:4 scale [23]. The higher thickness was required in order to facilitate construction. Figure 6.11 illustrates the details of the model and Fig. 6.12 shows the final model after final construction. The model was designed to resemble a Mark II BWR in Japan. The design pressure for the test model was 0.78 MPa with failure occurring at 4.66 MPa. Failure was initiated due to a large tear that developed at a detail near the equipment hatch as shown in Fig. 6.13. In addition, Fig. 6.14 illustrates another significant tear that was found to have initiated near the middle stiffening ring.

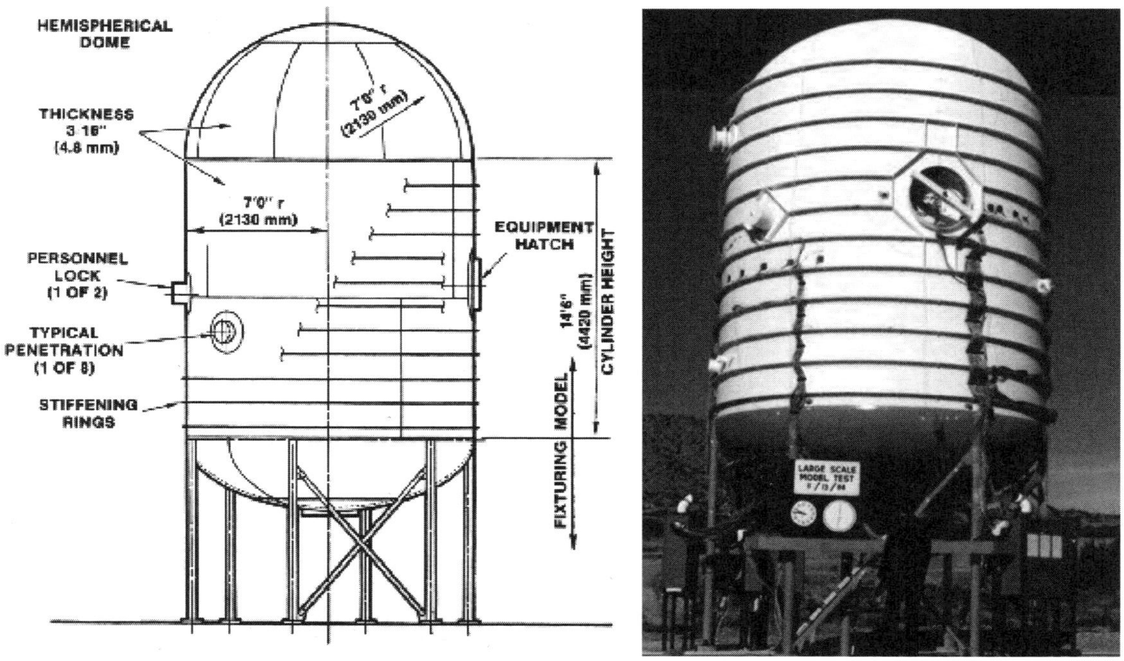

FIG. 6.10
NRC/SANDIA 1:8 SCALE STEEL CONTAINMENT DETAILS AND TEST MODEL [6]

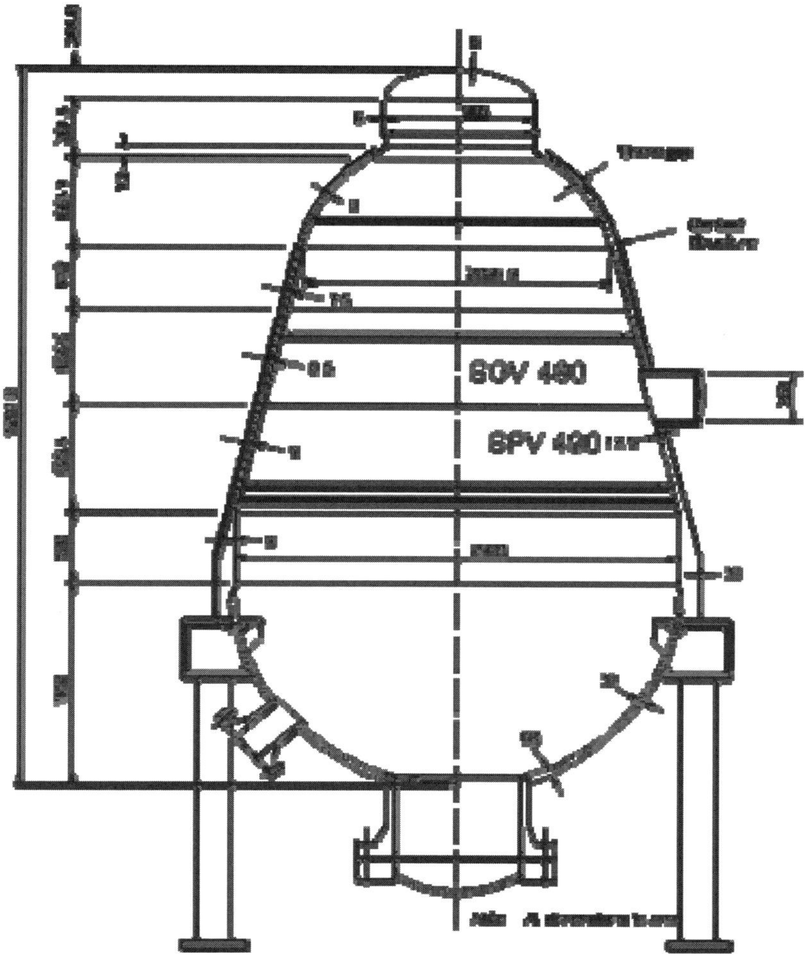

FIG. 6.11
NUPEC/NRC/SANDIA 1:10 SCALE STEEL CONTAINMENT DETAILS [23]

6.3.2.4 Containment Component Testing. In addition to the main structure of the containment, tests have been conducted on other components that comprise the pressure boundary. These components include the penetrations that enter and exit from the main containment structure. Penetrations are typically surrounded by expansion bellows. The bellows, as shown in Fig. 6.15, are placed around pipe penetrations in both steel and concrete containments. The bellows are designed to be capable of significant deformation without tearing in order to allow relative movement between the containment shell and piping without damaging the pipe or the containment structure. The extent of the deformation that typical bellows are able to withstand were studied in a series of tests at Sandia [24, 25]. Since the bellow is part of the pressure boundary, the tests subjected the bellows to axial displacement, lateral displacement, as well as internal pressure. Since severe accidents also cause increases in temperature, the temperature was also varied during the bellows tests. An example of a tested bellow is shown in Fig. 6.16. The testing program also included a number of artificially

FIG. 6.12
NUPEC/NRC/SANDIA 1:10 SCALE STEEL CONTAINMENT MODEL [23]

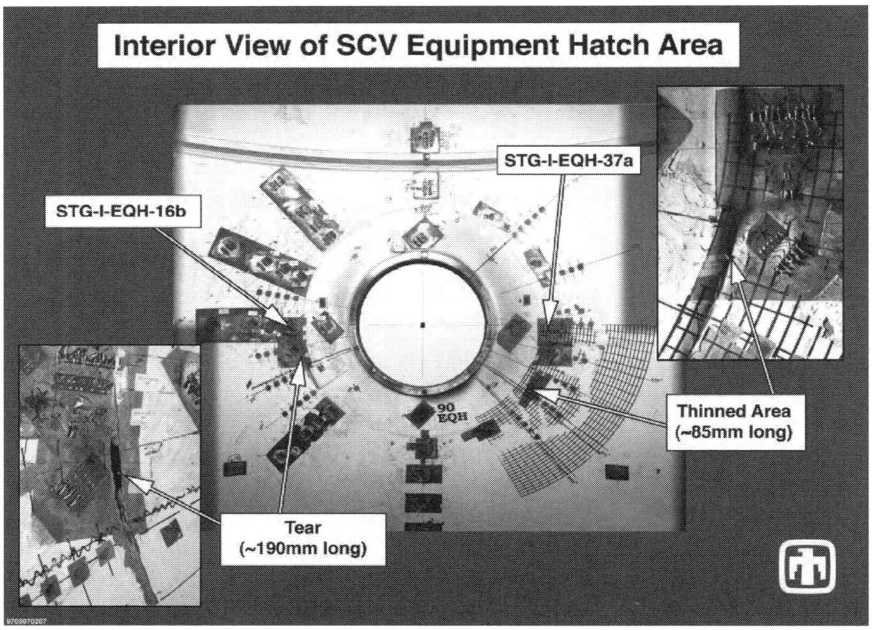

FIG. 6.13
NUPEC/NRC/SANDIA 1:10 SCALE STEEL CONTAINMENT TEARS
NEAR EQUIPMENT HATCH [6]

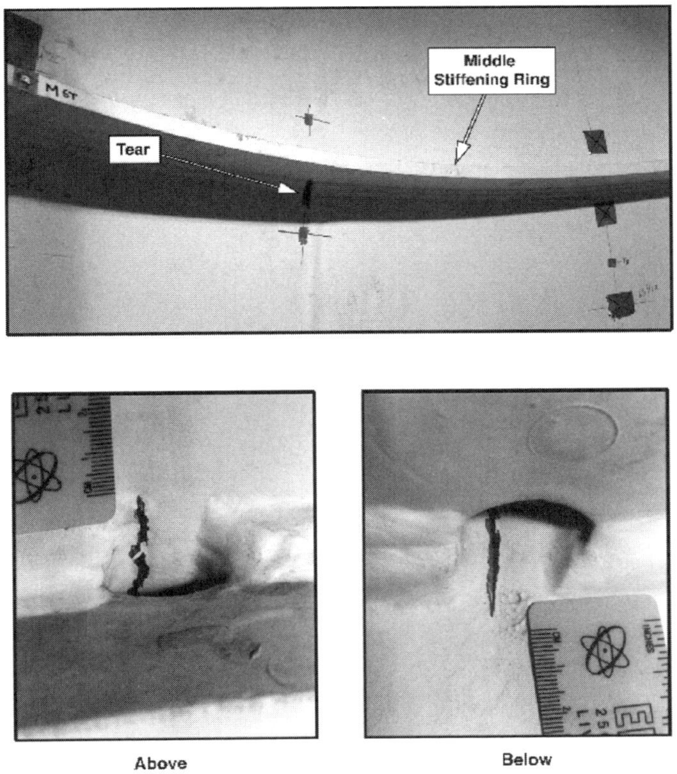

FIG. 6.14
NUPEC/NRC/SANDIA 1:10 SCALE STEEL CONTAINMENT RAT HOLE TEARS [23]

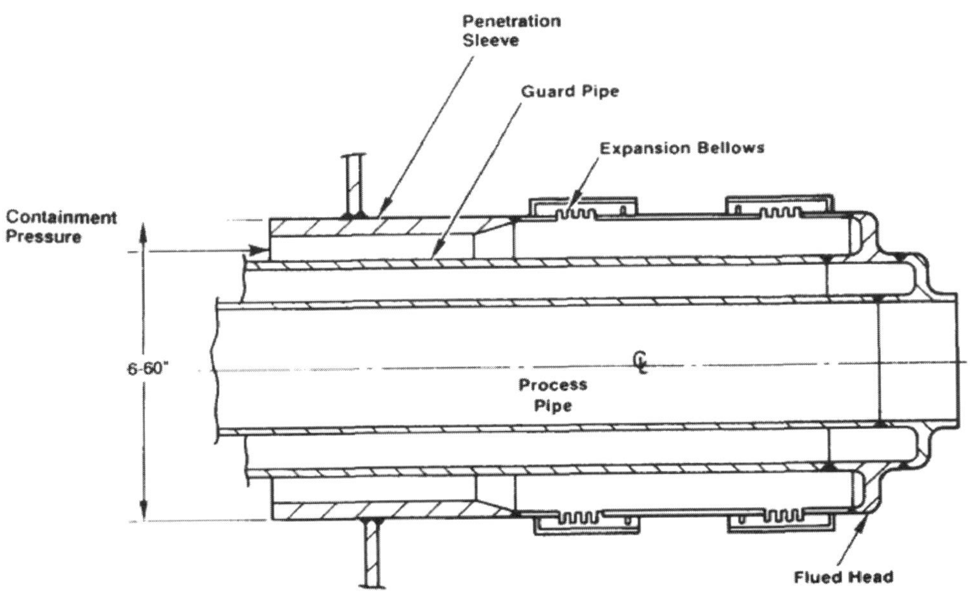

FIG. 6.15
BELLOWS LAYOUT EXAMPLE [24]

FIG. 6.16
PRE AND POST TEST BELLOWS EXAMPLE [24]

corroded bellow test articles to assess the degraded performance. The results of the testing showed that the as-new bellow performed as expected with no leakage prior to full compression while the corroded bellows did show that, in some cases, premature leaking was possible.

In addition to the bellows tests, a number of other components have been tested over the years. These include the seals in personnel airlock doors [26, 27], compression seals and gaskets [28, 29], electrical penetration assemblies [30], inflatable seals [31], equipment hatch [32], and the drywell head [33]. The references have been provided for each of these tests to enable further research into the specific areas. In addition, NUREG/CR-6906 provides a summary of some of these component tests.

6.4 CONTAINMENT SEVERE ACCIDENT MODELING AND INSIGHTS

As part of the testing programs discussed in this chapter, many included a parallel effort that studied methods to analyze these test with computational tools. This process often included pretest analyses followed by posttest analyses that took into account the lessons and observations obtained during the test. In addition, several of the testing programs included international Round Robin exercises that allowed groups from around the world to conduct pre and posttest analyses. These Round Robin programs also had the advantage of each group sharing and learning from each other during the testing and analysis process. As discussed earlier in this chapter, the testing and subsequent analyses have enabled the development of computational techniques and methods for predicting the severe accident behavior of containment vessels. This section summarizes the efforts and results for some example analysis programs conducted along with the testing.

6.4.1 Concrete Containment Analyses

Both pretest [34] and posttest [35, 36] analyses and Round Robins were conducted for the 1:6 scale reinforce concrete containment test discussed previously [8, 9]. The pretest analyses were relatively good in predicting the global response of the containment; however, the location of the liner tearing was more challenging for the analysts to predict. This is due to the computational models lacking local details that are often the initiation sites for liner tearing. The posttest analyses in Fig. 6.17 were able to confirm the high strains near the large tear shown in Fig. 6.4.

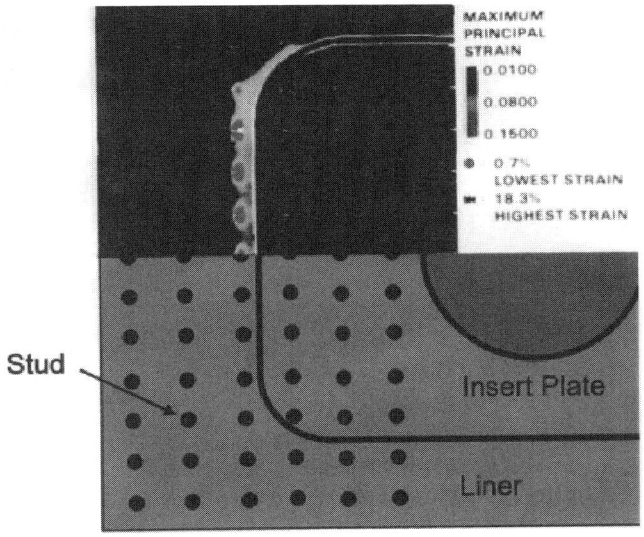

FIG. 6.17
SANDIA REINFORCE CONCRETE POSTTEST LINER ANALYSIS AT 1.0 MPA (145 PSIG) [6]

Similarly, pretest [37, 38] and posttest [39, 40] analyses and Round Robins were conducted for the 1:4 scale PCCV test [18]. Seventeen groups from around the world participated in the pretest Round Robin program [38]. Predictions up to yielding (approximately 2.5 times the design pressure, or 1.0 MPa) were similar but varied considerably within the inelastic range. The average failure pressure predicted was approximately 3.6 times the design pressure, or 1.46 MPa, with half predicting a functional failure (e.g., leakage through a liner tear) and half predicting structural failure (e.g., rebar or tendon rupture). The posttest analysis by Sandia [39] thoroughly examined the response of the containment including exploring the causes of the failure shown in Fig. 6.8. Figure 6.18 illustrates the 3D finite element model developed for the PCCV and the prediction of the failure location (e.g., dark region to the left of the equipment hatch opening). It was discovered that the additional hoop reinforcement surrounding the equipment hatch opening terminated in the region of the failure. This caused that

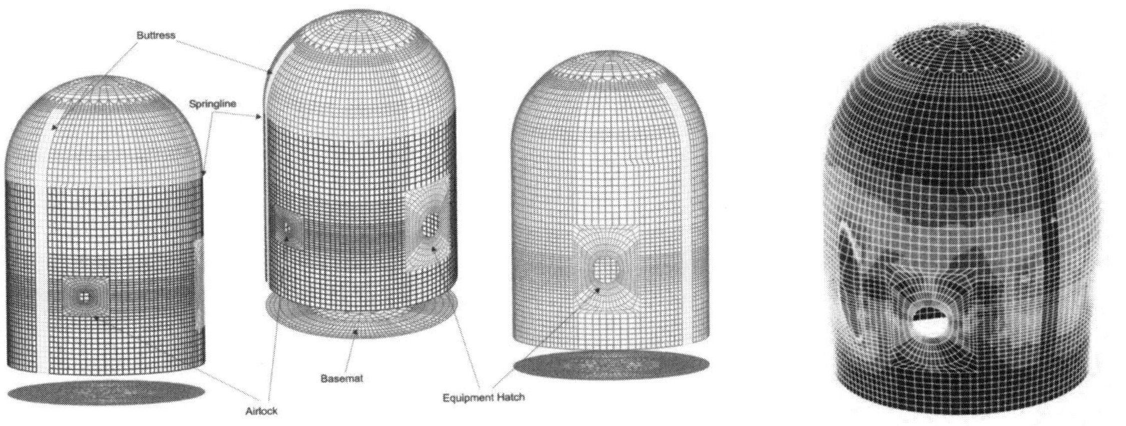

FIG. 6.18
SANDIA PCCV POST-TEST FINITE ELEMENT MODEL AND ANALYSES RESULTS [39]

region to experience a rise in strain that led to the failure initiating at that location. No formal posttest Round Robin was conducted, though a number of groups participated in the OECD/NEA/CSNI International Standard Problem 48 on Containment Capacity. Posttest analyses of the PCCV were included as part of the ISP [40].

6.4.2 Steel Containment Analyses

The 1:32 scale steel containments tests shown in Fig. 6.9 were analyzed by Horchel [20] with pretest predictions proving to be fairly consistent with the test results [19]. The analyses predicted failure of SC-0 and SC-1 at 0.91 MPa (tests at 0.93 and 0.76 MPa, respectively), SC-2 at 0.95 MPa (test at 0.93 MPa and 0.97 MPa for retest), and SC-3 at 0.84 MPa (test at 0.83 MPa).

For the 1:8 scale steel containment [21] tested in Fig. 6.10, pretest [41] analyses predicted failure of the containment by ovalization of the equipment hatch and leakage through the seals. Posttest analyses [42] were able to predict the high strains in the stiffener geometry near the equipment hatch that lead to catastrophic rupture prior to the ovalization predicted in the pretest analysis.

The 1:10 scale (1:4 scale thickness) steel containment in Figs. 6.11 through Fig. 6.14 likewise was subject to pretest [43, 44] and posttest [45–47] analyses efforts, but also included a Round Robin program. The predictions for the pretest Round Robin [44] were all in the range or higher than the actual failure pressure, or much higher in some cases (exceeding twice the failure pressure). None of the estimates correctly predicted the failure location with most predicting material failure at a thinned liner or buckling of the drywell head. In general, the posttest analysis Round Robin [46] showed improved correlation to the test results; however, some of the local regions of high strain in the test (e.g., weld seams) were still not predicted since those local details could not be included in the models. The finite element model used in the Sandia posttest analysis is illustrated in Fig. 6.19 alongside an image of the test model.

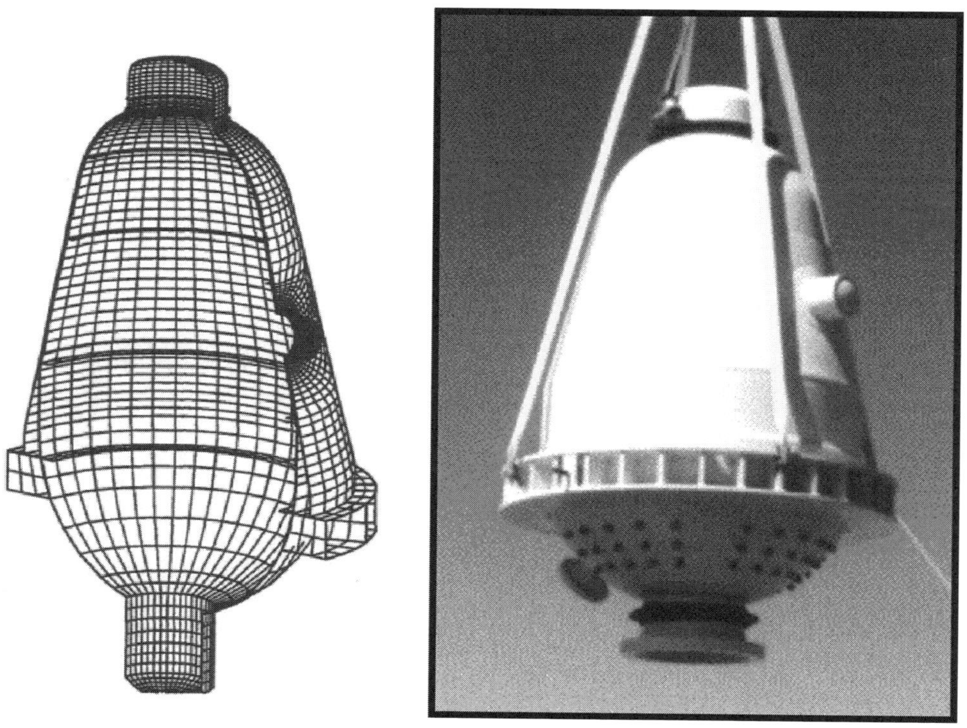

FIG. 6.19
SANDIA STEEL CONTAINMENT FINITE ELEMENT MODEL [23, 46]

6.4.3 Probabilistic Modeling of Containment Severe Accident Response

The analysis efforts described previously were deterministic in that they computed the response of the containment given one set of assumptions. The assumptions used for each analysis included the level of detail in the model, the material properties, the failure criterion used for steel and concrete, and the boundary conditions among others. Often, there is a large amount of uncertainty in many of these assumptions, and "best estimate" values are used to assess the containment's behavior. However, studies have attempted to account for these uncertainties through probabilistic modeling. As discussed earlier in this chapter, fragility curves were developed for use in probabilistic risk assessments [3] and in the IPE [4] studies. Following those efforts, more sophisticated research programs explored methods to develop fragility curves using computational finite element analyses. One study by Ellingwood and Cherry [48] examined the development of fragility curves using these techniques. Ellingwood and Cherry accounted for two types of uncertainty in their study. The inherent randomness for the most part focuses on material property uncertainty, while the knowledge-based uncertainty centers on finite element modeling assumptions, the failure criterion, and the extrapolation of material test data for use in the models. The uncertainties were defined for these parameters using test data and engineering judgment as available. A number of techniques were available to develop the fragility curves including classical Monte Carlo; however, Monte Carlo approaches would require a large number of samples to be developed. For each sample, a complete finite element analysis must be performed. Since this would not be practical, Ellingwood and Cherry utilized a Latin Hypercube sampling (LHS) [49] technique to develop the set of input parameters for each finite element analysis. The use of the LHS greatly reduced the number of samples required in order to develop the fragility curve. Ellingwood and Cherry performed 14 finite element analyses of a PWR Ice Condenser with a steel shell containment. In each of these analyses, the material properties and other uncertainty parameters were varied as defined by the LHS process. After conducting the finite element analyses and computing the failure (e.g., leak or rupture) pressure for each analysis, the failure pressures were rank ordered and plotted on the log-normal plot shown in Fig. 6.20. The

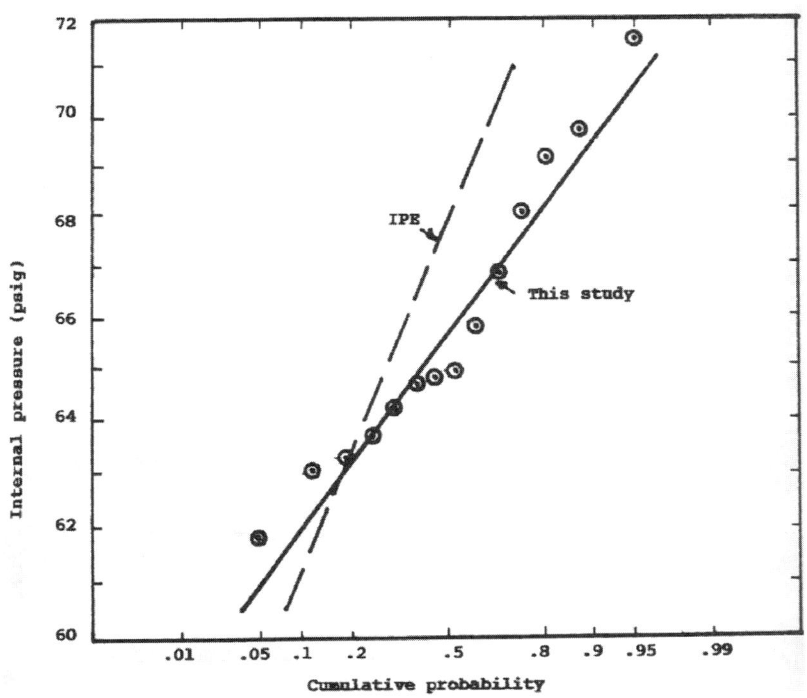

FIG. 6.20
FRAGILITY CURVES FOR A PWR ICE CONDENSER [48]

plot also includes the IPE fragility developed for the plant (e.g., Sequoyah) used as the basis for the PWR Ice Condenser finite element model. For most of the probability range, the Ellingwood and Cherry estimates show higher failure probabilities at the same pressure relative to the IPE values.

This method of developing fragility curves can be, and has been, used for input into PRA models in order to include the uncertainty in the containment performance in the severe accident assessment. The following section includes a description of a study that uses this technique to develop containment fragility curves to assess the effects of degradation (e.g., corrosion) on the severe accident response.

6.5 EFFECTS OF CONTAINMENT DEGRADATION ON ITS SEVERE ACCIDENT RESPONSE

Given the advancements in containment computational analysis made possible by the testing described in this chapter, it was then reasonable to the study more complex containment conditions. One of these conditions includes the wide range of potential degradation mechanisms that have been observed in aging containments. Many of the containments in the US fleet have been operating for 30–40 years. Over this period of time, various types of degradation have been observed (e.g., liner corrosion, concrete cracking, stress loss in tendons, etc). Chapter 5 summarizes many on the discovered degradation instances in the US fleet and the programs that have been ongoing to manage the response to degradation. Finite element modeling is often used to assess the effects of degradation on the design basis response of the containment. One example is shown in Fig. 6.21 where the BWR Mark I containment at the Oyster Creek plant was analyzed under design accidents in its actual degraded state [50]. The Oyster Creek containment experienced significant corrosion in the steel shell near the sandbed region. The model results shown are for a design accident where the internal pressure was increased to the design pressure and stresses were estimated in the corroded regions to ensure that the ASME code allowable stresses were not exceeded. However, the internal pressure during severe accidents can greatly exceed the design pressure. This section summarizes several studies that have explored the effects of degradation on the containment capacity during a severe accident. Two studies have also studied the effects of the change in containment capacity due to degradation on the risks and consequences during severe accidents.

6.5.1 Examples of Deterministic Modeling

An extensive study was performed by Cherry and Smith [51] where they examined the effects of degradation on the ultimate capacity of four typical containment types. These four containments were also studied in NUREG-1150 and include a PWR Ice Condenser steel containment, a BWR Mark I steel containment, a PWR reinforced concrete containment, and a PWR prestressed concrete containment.

Since the dominant failure mechanism considered in the study focused on the tearing of the steel liner or steel shell, testing was conducted to determine the effect of corrosion on the failure strain. This testing consisted of measuring the failure strain in a series of pristine and corroded coupons. It was determined that the local stress concentrations caused by the roughness of the corroded surface reduced the strain to failure to approximate 50% of the coupons without corrosion. This effect was important in developing the failure criterion used in the finite element modeling.

The degradation in this study was hypothetical in nature but was placed in areas of know corrosion in actual plants, placed in areas of high stress, or placed in regions where water is present. In the models, the liner or the shell was thinned uniformly to represent the reduction in thickness due to the corrosion. However, this thinning did not include the surface roughness effect. Therefore, the failure criterion in the corroded regions was reduced by half based on the test results. A number of other uncertainties or adjustments were made to the failure criterion based on the large scale containment tests and modeling considerations. Recall that the large scale tests showed numerous cases of liner or shell tearing at local features or welds. These local details

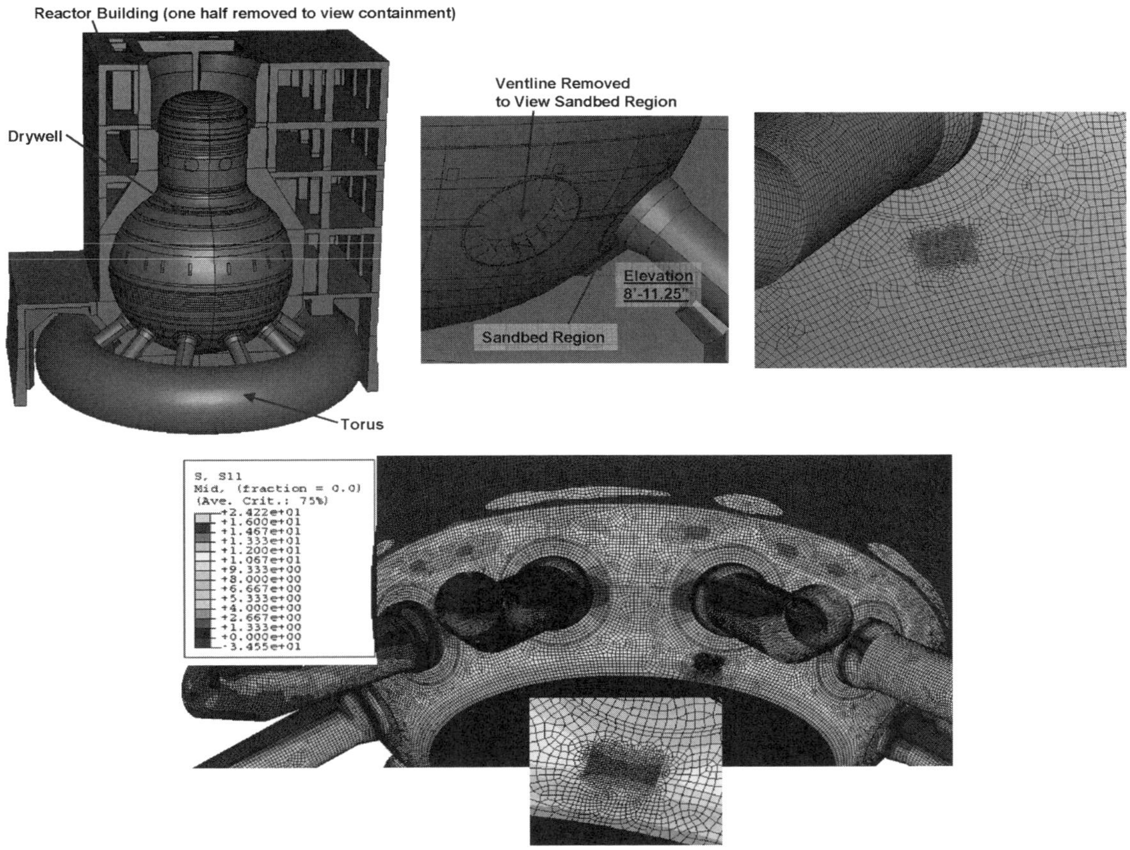

FIG. 6.21
CORRODED STEEL SHELL ANALYSIS OF THE OYSTER CREEK DRYWELL WITH HOOP STRESS SHOWN (IN KSI) FOR A DESIGN ACCIDENT (HIGH ELEMENT DENSITY RECTANGULAR REGION REPRESENTS AN AREA OF HIGH LOCALIZED CORROSION) [50]

are well beyond the resolution of features in the finite element models of the containments. Therefore, the failure strain was also reduced by a factor (0.5) to account for the lack of these features. The baseline failure strain uses results from a uniaxial tension test. So, an addition reduction factor (between 0.61 and 0.69) was used to account for the biaxial stress state in the actual containment liner or shell. Finally, a factor (1.0) was used to account for the uncertainty and variability in the material properties. These were the "best estimate" reduction factors used by Cherry and Smith. They also performed analyses to determine the "lower bound" and "upper bound" containment capacity. The lower and upper bound reduction factors were based on test data when available and on engineering judgment when insufficient data existed. The summaries of the degraded containment analyses in the following sections discuss only the best estimate failure pressures. The full lower and upper bound predictions are included in Cherry and Smith [51].

6.5.1.1 PWR Ice Condenser Steel Containment. The first containment type examined was a PWR Ice Condenser with a cylindrical steel shell and a hemispherical dome (53 m or 173 ft high with a 35 m or 115 ft diameter). The steel shell is made of steel plates of varying thickness (1.27 cm or 0.5 in. to 3.81 cm or 1.5 in.) welded together. The containment also includes numerous ring stiffeners welded perpendicular

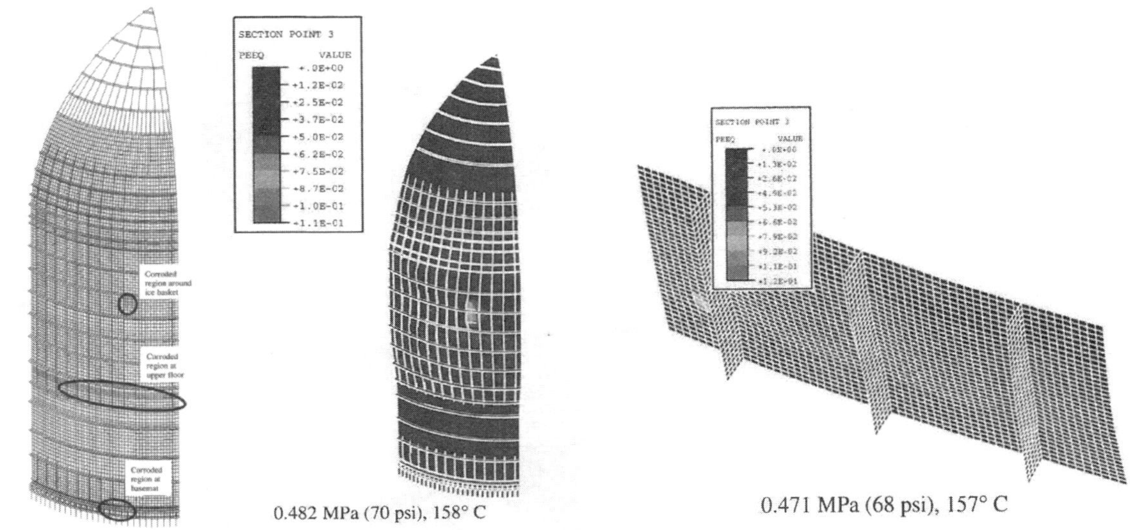

FIG. 6.22
PWR ICE CONDENSER CONTAINMENT FINITE ELEMENT MODEL WITH EQUIVALENT PLASTIC STRAIN RESULTS FOR THE ENTIRE CONTAINMENT WITH NO CORROSION AND THE EQUIVALENT PLASTIC STRAIN IN A CORRODED REGION NEAR THE UPPER FLOOR [51]

to the main shell. The finite element was constructed as a 53° slice of the steel shell as shown in the far left side of Fig. 6.22. For this containment, three different corrosion locations were chosen including the ice basket region, the upper floor region, and the shell-basemat junction. These locations are the site of corrosion experienced at actual ice condensers or are regions that have water present. In addition, each location was modeled with two different levels of nominal corrosion (e.g., thickness reduced by hypothetical corrosion at levels of 10%, 25%, 50%, or 65% based on the case). A baseline case with no corrosion was also examined. The loads applied to the containment included internal pressure as well as raising the temperature in accordance with the saturated steam temperature-pressure relationship. The increase in temperature leads to both thermal expansion in the steel as well as changes to the stress–strain relationship. Figure 6.22 illustrates the equivalent plastic strains in the entire containment with no corrosion at a given pressure-temperature (Full color images of the strains are provided in [51]). In this case, the discontinuity created at the thickness change of several steel plates led to a strain concentration near the ice basket region. The figure also shows a submodel used to model the corrosion near the upper floor location. Again, a strain concentration is observed at the transition from one plate thickness to another thickness, or to a reduced thickness corroded region in this case. The previously discussed failure criterion was applied to the results of the analysis to compute the pressure where tearing of the steel shell would initiate. For the case with no corrosion, failure was predicted at 0.472 MPa (68.4 psi) (best estimate prediction). The case with corrosion at the ice basket region was shown to be very sensitive to slight reductions in thickness since corrosion with only a 10% reduction in thickness caused the failure pressure to drop to 0.410 MPa (59.5 psi). In order to achieve similar drops in failure pressure with corrosion at the upper floor and basemat locations, thickness reductions in the range of 50% to 65% were required.

6.5.1.2 BWR Mark I Steel Containment. The second containment examined by Cherry and Smith was a typical steel BWR Mark I. The Mark I consists of a drywell connected to the wetwell (or suppression pool) by ventlines. The drywell examined here has a 20 m (65 ft) diameter spherical section below a 12-m

(39 ft) diameter cylindrical section which is covered by the drywell head. The torus shaped wetwell was 9 m (30 ft) in diameter with the torus diameter of approximately 34 m (112 ft). The steel plates for the drywell ranged from 1.91 to 7.3 cm (0.75 to 2.87 in.) and from 1.53 to 2.86 cm (0.60 to 1.13 in.) in the wetwell. The finite element model of the drywell was constructed as a 45° segment and the wetwell was represented with a 22.5° segment. As with the PWR ice condenser, three hypothetical locations of corrosion were assumed and include the sandbed region (similar to Oyster Creek), the knuckle region (connection between spherical and cylindrical sections of the drywell), and the waterline in the wetwell. Again, these are either regions where corrosion has been observed or are locations thought to be susceptible to corrosion due to the potential presence of water. Reductions of 50% and 65% of the thickness were examined at each location. Figure 6.23 illustrates the finite element model and the detailed knuckle region for modeling corrosion. The containment was both pressurized and subjected to increased temperatures based on the saturated steam versus pressure relationship. It should be noted that the most likely leak location for the Mark I containment is the drywell cylinder to drywell head bolted interface. Extension of the bolts due to internal pressurization would open a gap leading to release of material from inside the containment. For this study, the estimate of drywell head leakage was between 0.69 MPa (100 psi) (if intentionally vented) and 0.72 MPa (104 psi) (if caused by actual extension of the head bolts). Therefore, the failure pressures reported exceeding the drywell head leakage pressure neglect this failure mode and would be the failure pressure caused by a tear/rupture in the steel shell if the drywell head leak was prevented. For the case with no corrosion, failure was predicted to exceed 1.39 MPa (200 psi) (best estimate prediction) (failure not reached at maximum pressure analyzed). The sandbed corrosion was least sensitive to corrosion with the failure pressures predicted at 1.17 MPa (170 psi) and 0.855 MPa (124 psi) for 50% and 65% corrosion, respectively. The wetwell waterline corrosion was more sensitive to corrosion with the failure pressures predicted at 0.786 MPa (114 psi) and 0.545 MPa (79 psi) for 50% and 65% corrosion, respectively. Finally, the knuckle corrosion was most sensitive to corrosion with the failure pressures predicted at 0.731 MPa (106 psi) and 0.531 MPa (77 psi) for 50% and 65% corrosion, respectively. Figure 6.23 shows the equivalent plastic strains in the knuckle region with 50% corrosion at a given pressure-temperature (Again, full color images are provided in [51]).

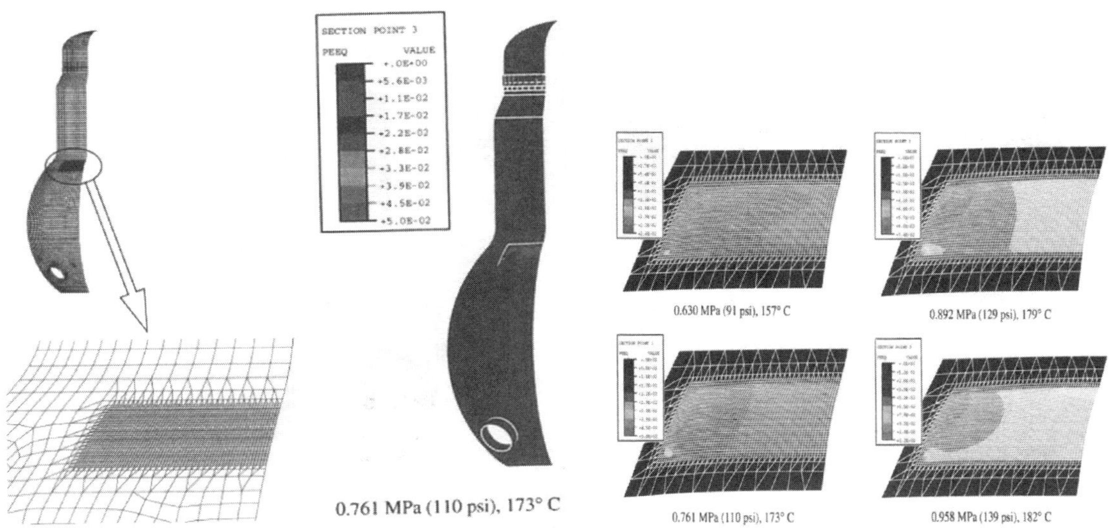

FIG. 6.23
BWR MARK I CONTAINMENT FINITE ELEMENT MODEL WITH EQUIVALENT PLASTIC STRAIN RESULTS FOR THE ENTIRE CONTAINMENT AND IN A DETAILED AREA WITH CORROSION IN THE KNUCKLE REGION [51]

6.5.1.3 PWR Reinforced Concrete Containment. The third containment examined by Cherry and Smith was a typical PWR with a reinforced concrete containment including a steel liner. The inside diameter of the containment was 38.4 m (126 ft) with the inside height of 56.4 m (185 ft). The concrete wall thickness was 1.37 m (4.5 ft). The steel liner varies from 6.35 mm (0.25 in.) (floor) to 12.7 mm (0.5 in.) (dome) with a 9.5 mm (0.375 in.) thickness on the walls. The liner was attached to the concrete with studs welded to the liner and embedded in the concrete. A simple axisymmetric model was used to represent the containment in the finite element model as shown in Fig. 6.24. Two corrosion locations were assumed with one at the mid-height of the cylinder (high stress location) and the second at the basemat-wall junction (corrosion observed at this location) (a third location near the equipment hatch was also examined but failure was not predicted to occur at that location). For these locations, submodels of the liner were constructed. Tearing of the liner was assumed to be sufficient for failure since it was assumed that at the tearing pressures the concrete would be cracked and would provide little resistance to material released from inside the containment. Reductions of 25% and 50% of the liner thickness were examined at each location (10% was also examined but failure was not achieved for the best estimate cases). Again, the containment was both pressurized and subjected to increased temperatures based on the saturated steam versus pressure relationship. However, due to the thick concrete wall, a temperature gradient was modeled through the thickness where the liner was heated more than the outer surface of the containment. For the case with no corrosion, failure was predicted to exceed 1.03 MPa (150 psi) (best estimate prediction). The failure pressure was not very sensitive to basemat corrosion with the pressure falling to about 0.99 MPa (143 psi) at 25% corrosion (only 0.02 MPa additional drop at 50% corrosion). Mid-height corrosion at 25% had little effect on the failure pressure, but 50% corrosion did drop the failure pressure to 0.88 MPa (127 psi). Figure 6.24 illustrates the containment deformed shape and the equivalent plastic strains in the mid-height 50% corrosion submodel at a given pressure-temperature.

6.5.1.4 PWR Prestressed Concrete Containment. The final containment examined by Cherry and Smith was a typical PWR with a prestressed concrete containment including a steel liner. The inside diameter of the containment was 42.7 m (140 ft) with the inside height 64.6 m (212 ft). The wall thickness was 1.07 m (3.5 ft) in the cylinder and 0.81 m (2.67 ft) in the dome with the steel liner at 6.35 mm (0.25 in.). The liner

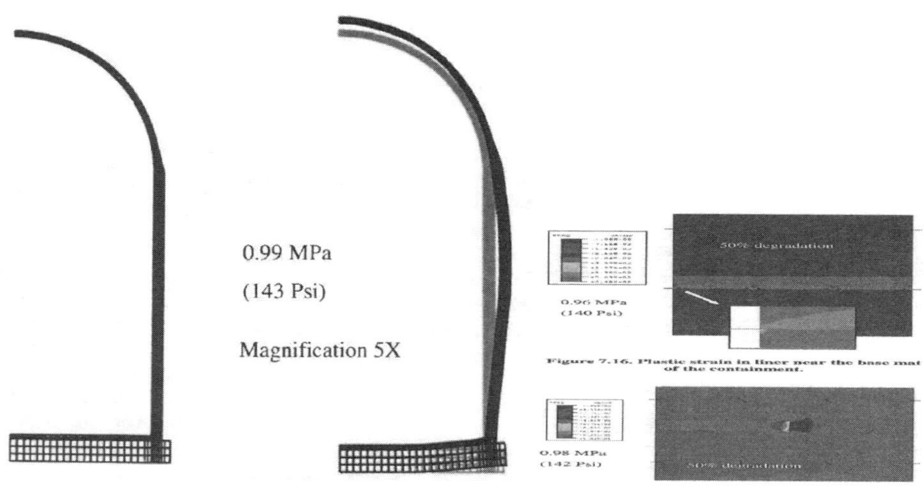

FIG. 6.24
PWR WITH REINFORCED CONCRETE CONTAINMENT FINITE ELEMENT MODEL AND DEFORMED SHAPE WITH EQUIVALENT PLASTIC STRAIN RESULTS FOR MID-HEIGHT 50% CORROSION LINER SUBMODEL [51]

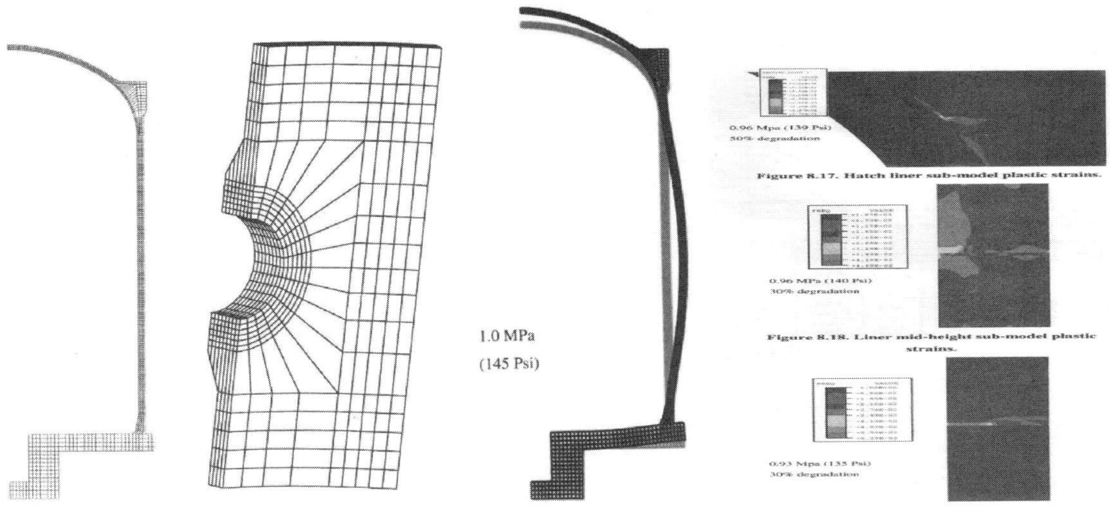

FIG. 6.25
PWR WITH PRESTRESSED CONCRETE CONTAINMENT FINITE ELEMENT MODEL, HATCH SUBMODEL, DEFORMED SHAPE, AND EQUIVALENT PLASTIC STRAIN RESULTS FOR MID-HEIGHT 30% CORROSION LINER SUBMODEL [51]

was attached to the concrete with angles and channels welded to the liner and embedded in the concrete. Again, a simple axisymmetric model was used to represent the containment in the finite element model as shown in Fig. 6.25. The figure also shows the equipment hatch submodel that was also constructed. Two corrosion locations and submodels were assumed with one at the mid-height of the cylinder and the second at the basemat-wall junction (a third location near the equipment hatch was also examined but failure was not predicted to occur at that location prior to the case with no corrosion). Reductions of 30% and 50% of the liner thickness were examined at each location. Again, the containment was both pressurized and subjected to increased temperatures based on the saturated steam versus pressure relationship with a through-wall gradient. For the case with no corrosion, failure was predicted at 0.99 MPa (143 psi) (best estimate prediction). The failure pressure was not very sensitive to 30% mid-height corrosion with no change in failure pressure. At 50% mid-height corrosion, the failure pressure dropped to 0.93 MPa (135 psi). Basemat corrosion at 30% dropped the failure pressure to 0.94 MPa (137 psi) and 50% corrosion dropped the failure pressure to 0.92 MPa (134 psi). Figure 6.25 illustrates the containment deformed shape and the equivalent plastic strains in the mid-height 30% corrosion submodel at given pressure-temperatures.

As a follow-on to NUREG/CR-6707 by Cherry and Smith [51], Smith [52] explored the effects of tendon degradation on the PWR prestressed concrete containment. In order to include the tendon degradation, a 3D finite element model was constructed that included explicitly modeled tendons as shown in Fig. 6.26. The model included a 30° segment of the containment. In all, seventeen different tendon degradation cases were examined. They included hypothetical scenarios that assumed uniform loss of prestressing, grouped tendon loss of prestressing, distributed tendon loss of prestressing, and grouped and distributed tendon cross-sectional area loss. The study concluded that loss of prestressing or area in the vertical tendons had little effect on the failure pressure (e.g., leak pressure). Loss of prestressing in the hoop tendons also had little effect on the failure pressure. The cases that did lower the failure pressure significantly were those that reduced the area (by 50%) of a given percentage (~10%) of the hoop tendons or where a certain percentage (10 to 20%) of the hoop tendons were removed completely. For the cases examined, failure pressure reductions on the order of 30% were achieved for a number of the hoop degradation cases.

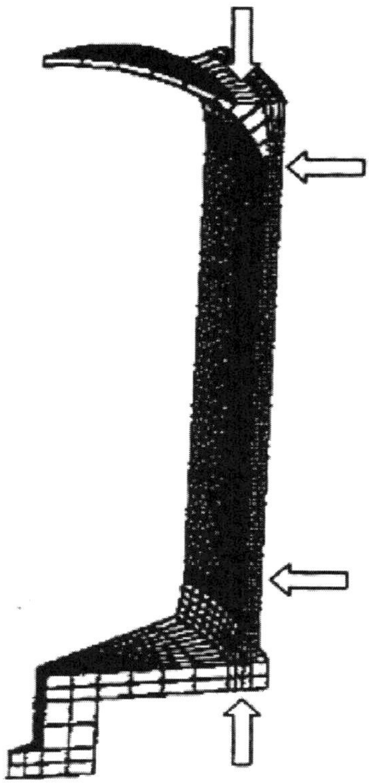

FIG. 6.26
PWR WITH PRESTRESSED CONCRETE CONTAINMENT FINITE ELEMENT MODEL [52]

6.5.2 Probabilistic Analysis of Degradation Effects

Expanding on the deterministic analyses, several studies have examined the effects of containment degradation on the failure pressure using probabilistic methods in order to account for uncertainties. As discussed earlier, the work by Ellingwood and Cherry [48] included a method for developing fragility curves (e.g., probability of failure versus pressure). That study also included analyses of degradation in the PWR ice condenser. Fragility curves for the containment with and without hypothetical corrosion in various locations were developed as shown in Fig. 6.27. The figure illustrates the fragility curves for three different corrosion cases compared to the case with no corrosion. These comparisons provide a more comprehensive comparison of the failure probability that can potentially illustrate responses of the containment that could be missed when performing deterministic "best estimate" comparisons only.

Building on both the work by Cherry and Smith [51] on modeling containment degradation and Ellingwood and Cherry [48] on probabilistic modeling, Spencer et al. [53] developed degraded fragility curves for various cases using the same four containments studied in Cherry and Smith. Due to increases in computation resources, Spencer et al. performed 30 analyses (using the LHS method) for each scenario as compared to the 14 performed by Ellingwood and Cherry. As will be discussed in a subsequent section, the fragility curves developed by Spencer et al. were used as input into a Probabilistic Risk Assessment (PRA) analysis. For that analysis, additional input was required to assess the potential release of material from within the containment. That input included a further breakdown of the fragility curve into distinct failure modes: leak, rupture, and catastrophic rupture. The distinctions between these failure modes were discussed earlier in this chapter. Depending on the containment, the corrosion case, and the location within the containment being assessed, leaks could either initiate well before rupture or leaks could immediately transition into a rupture of

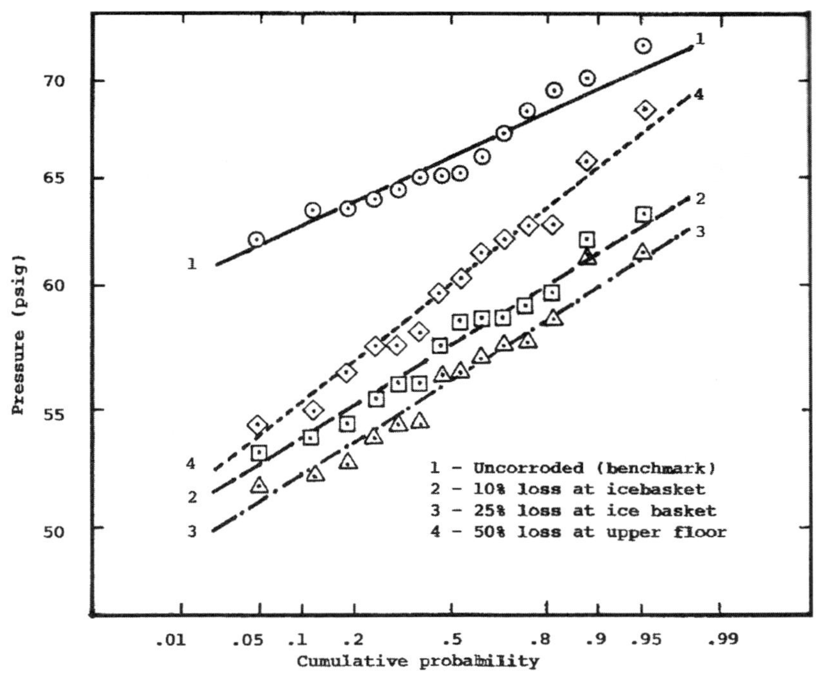

FIG. 6.27
DEGRADED FRAGILITY CURVES FOR A PWR ICE CONDENSER [48]

the containment (typical in steel containments at high pressures). Spencer et al. computed individual statistically independent fragility curves for each potential failure location and for each of the failure modes (e.g., leak, rupture, catastrophic rupture). Probability theory was then used to combine the individual fragilities into a single cumulative probability of failure fragility. The cumulative probability of failure fragility curves reflect the different modes in several cases where leak failures dominate at low pressures and rupture and catastrophic ruptures dominate at higher pressures as would be expected. Brief summaries of the analyses and results are presented below for the PWR ice condenser steel containment, the BWR Mark I steel containment, the PWR reinforced concrete containment, and the PWR prestressed concrete containment.

6.5.2.1 PWR Ice Condenser Steel Containment. The PWR ice condenser model is similar to the model created by Cherry and Smith [51]; however, the level of mesh refinement was significantly higher, especially in the areas of the corroded regions. The corrosion locations examined are shown in Fig. 6.28 and were slightly modified with cases including 25% and 50% ice basket corrosion, 25% and 50% basemat corrosion, and 50% and 65% upper floor corrosion. In addition, a seventh degradation case was examined to assess bellows corrosion. Numerous bellows are attached to the containment shell surrounding the various penetrations entering and exiting the containment. For this case, the bellows were not explicitly modeled. They were assessed using a combination of the displacements of the containment shell near the bellow location and the corroded bellows testing previously discussed in this chapter. The model developed would predict leakage if the steel shell displacement near the bellow exceeded a critical value based on the test results. Figure 6.29 shows the fragility curves developed from the 30 analyses for each scenario (lines fit to the 30 analysis values). The 50% ice basket and 65% upper floor cases show the largest shift in fragility. However, at the lower probabilities, the basemat corrosion cases have the potential to leak at very low pressures due to the bending that occurs at that location. At higher pressures/probabilities, rupture dominates and failure in the ice basket and upper floor locations increases in probability.

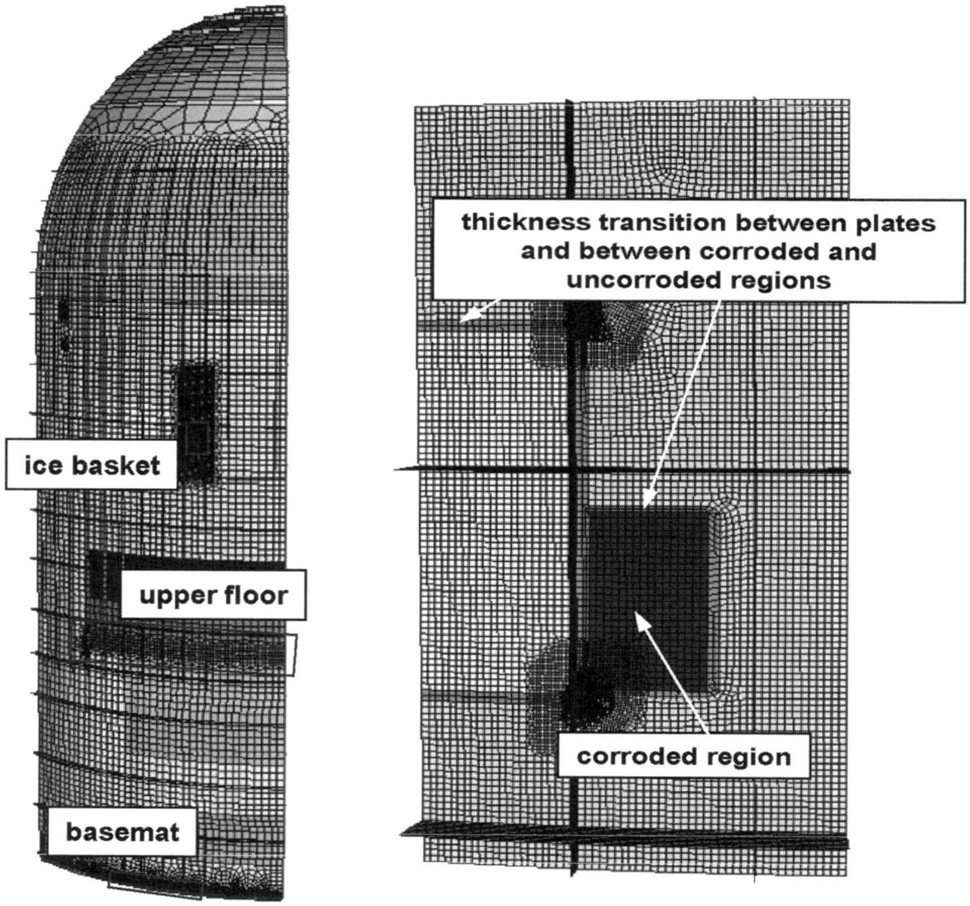

FIG. 6.28
PWR ICE CONDENSER FINITE ELEMENT MODEL WITH CORRODED LOCATIONS IDENTIFIED AND THE DETAILED REGION OF CORROSION NEAR THE ICE BASKET [53]

6.5.2.2 BWR Mark I Steel Containment. As with the ice condenser, the BWR Mark I model created by Spencer et al. is similar to the one created by Cherry and Smith [51], but has higher mesh refinement. In addition, numerous detailed penetration models were created to examine the potential for failure in these locations. Again, the corroded cases examined were similar to the case studied by Cherry and Smith and included 25% and 50% corrosion at the knuckle, sand bed region, and the wetwell waterline. As with the PWR ice condenser, an additional case of bellows corrosion was added. However, the corrosion was explicitly modeled in the bellow surrounding the ventline that connects the drywell to the wetwell for the Mark I containment. Figure 6.30 shows the 22.5° section finite element model of the entire containment as well as the detailed models used for the knuckle and bellow corrosion cases. The potential for leakage through the drywell head gasket was also included using an external program that computed the extension in the bolts holding the drywell head to the drywell. In general, leak failure through the head gasket dominates the no corrosion case fragility shown in Fig. 6.31. The head gasket leak failure also dominates most of the corrosion case fragilities as observed by the lack of a difference between the curves. Rupture failures initiated in the corroded regions do begin to influence the fragility curves as most obviously demonstrated by the 50% knuckle and 50% wetwell corrosion cases.

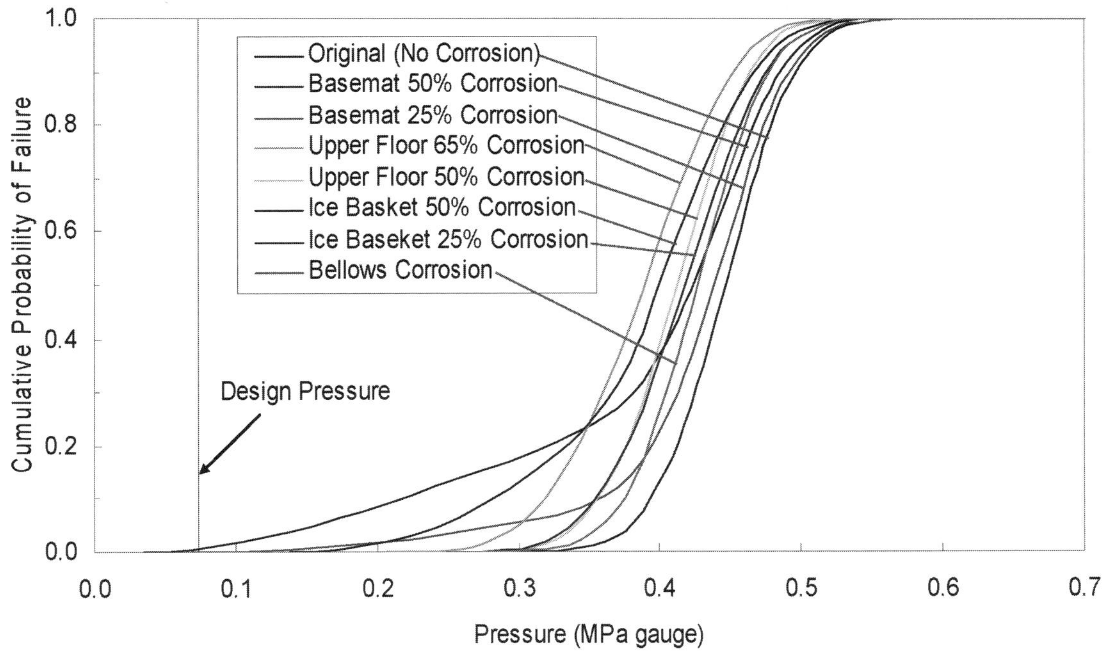

FIG. 6.29
PWR ICE CONDENSER FRAGILITY CURVES FOR VARIOUS DEGRADATION CASES [53]

6.5.2.3 PWR Reinforced Concrete Containment. The axisymmetric finite element model used by Cherry and Smith essentially remained unchanged in the Spencer et al. study as shown in Fig. 6.32 along with the deformed shape. One addition was made to the analysis methodology that was not included in the Cherry and Smith study. The stress concentrations at features not modeled in detail in the axisymmetric model were included using magnification factors developed by Tang et al. [54]. Tang et al. provided the magnification factors to be multiplied by the strain in the axisymmetric liner to estimate the increase that could be expected if the actual stress concentration feature were present. The stress concentration magnification

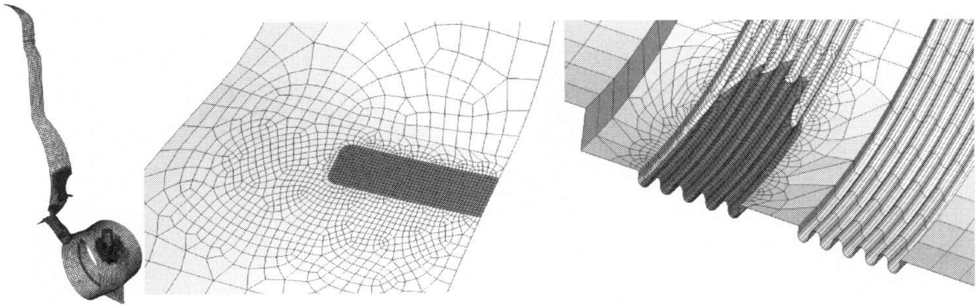

FIG. 6.30
BWR MARK I STEEL CONTAINMENT FINITE ELEMENT MODEL AND THE DETAILED REGION OF CORROSIONS NEAR THE KNUCKLE AND BELLOWS (CORRODED REGIONS IN DARK GREY) [53]

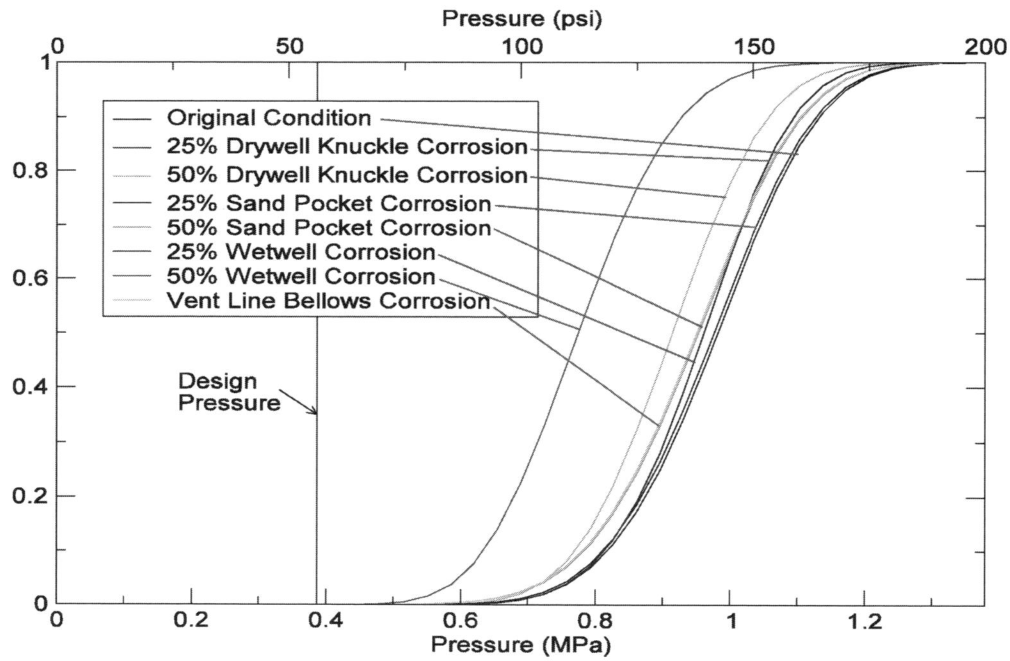

FIG. 6.31
BWR MARK I STEEL CONTAINMENT FRAGILITY CURVES FOR VARIOUS DEGRADATION CASES [53]

factors included the steam penetration, the wall-basemat junction, the equipment hatch, and the springline. The corrosion cases examined were similar in location to Cherry and Smith but with new submodels created for the mid-height (shown in Fig. 6.32) and basemat cases. For each location, 50% and 65% corrosion cases were analyzed. Figure 6.33 illustrates the fragility curves where the mid-height corrosion cases showed more of an effect than the basemat cases. Since the tearing of the liner due to corrosion only leads to a leak in the

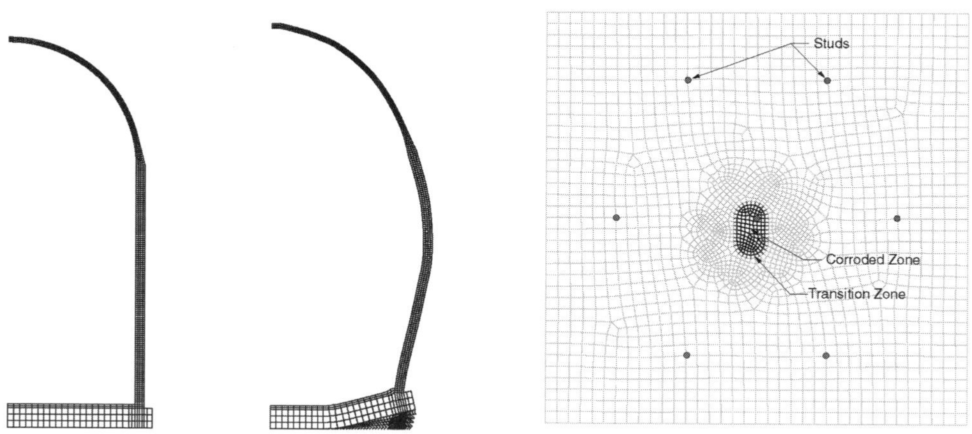

FIG. 6.32
PWR WITH A REINFORCED CONCRETE FINITE ELEMENT MODEL AND DEFORMED SHAPE (MAGNIFIED 10X) AT 1.05 MPA (152 PSI) AND THE LINER SUBMODEL FOR THE MID-HEIGHT CORROSION [53]

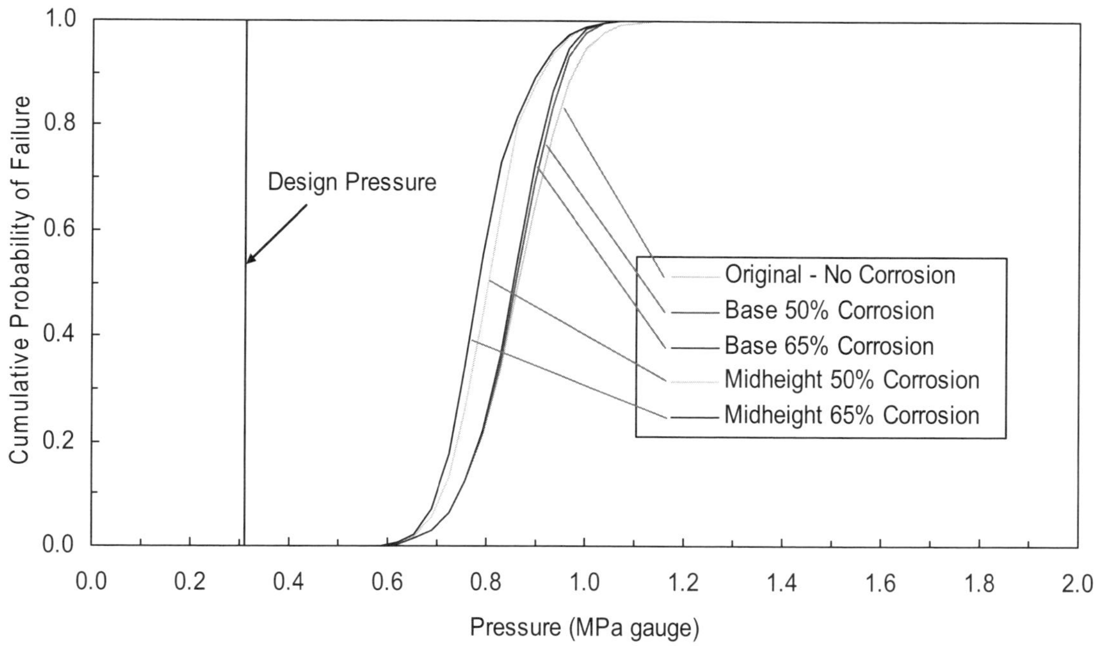

FIG. 6.33
PWR WITH REINFORCED CONCRETE CONTAINMENT FRAGILITY CURVES FOR VARIOUS DEGRADATION CASES [53]

containment (cracked concrete is still held together by the rebar), leak is the dominate failure mode for the reinforced concrete containment. However, the rupture criteria used in PRA models defines rupture as a tear reaching a certain area of opening as discussed earlier in this chapter. Therefore, the rupture failure does provide a small influence especially at the higher pressures as the liner tears grow.

6.5.2.4 PWR Prestressed Concrete Containment. The PWR with a prestressed concrete containment examined by Spencer et al. uses essentially the same finite element model as used by Smith [52] in the follow-on study to Cherry and Smith [51] as shown in Fig. 6.34 along with the deformed shape. As with the reinforced concrete containment analyses, magnification factors developed by Tang et al. [54] for prestressed concrete containments were used to include the potential failure at stress concentrations not modeled in detail. The corrosion cases examined were also similar in location to the Cherry and Smith study but with new submodels created for the mid-height (shown in Fig. 6.34) and basemat cases. For each location, 50% and 65% corrosion cases were analyzed. In addition, two tendon degradation cases were examined based on the Smith study. Those two cases were reducing the prestressing force by 50% in 20% of the hoop tendons grouped at the mid-height and reducing the area by 50% in 20% of the hoop tendons grouped at the mid-height. Figure 6.35 illustrates the fragility curves where the mid-height and basemat corrosion cases showed little difference from the no corrosion case. This is due to the wall-basemat magnification factor-based location tearing prior to failure in the corroded locations in most cases. As with the reinforced concrete containment, tearing of the liner only leads to a leak in the containment (cracked concrete is still held together by the rebar), and leak is the dominate failure mode for the corrosion cases. However, the fragility curve is affected by the tendon degradation cases, especially for the 50% tendon area reduction case since the tendon degradation cases increase the likelihood of a rupture type failure at lower pressures. This will be discussed further in the next section on the PRA modeling.

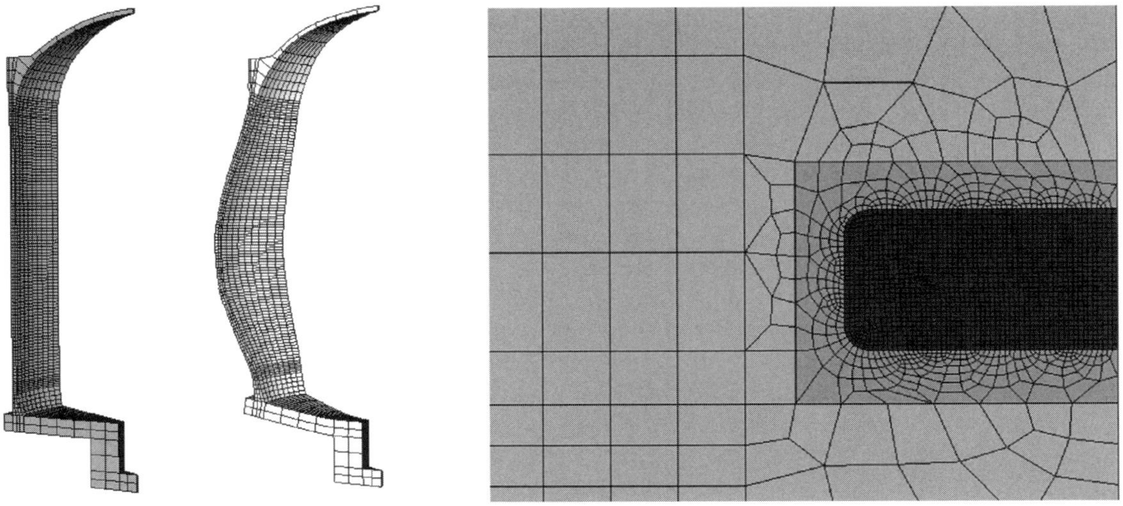

FIG. 6.34
PWR WITH A PRESTRESSED CONCRETE CONTAINMENT FINITE ELEMENT MODEL AND THE DEFORMED SHAPE (MAGNIFIED 5X) AT 1.08 MPA (157 PSI) AND THE DETAILED CORROSION REGION FOR THE LINER SUBMODEL [53]

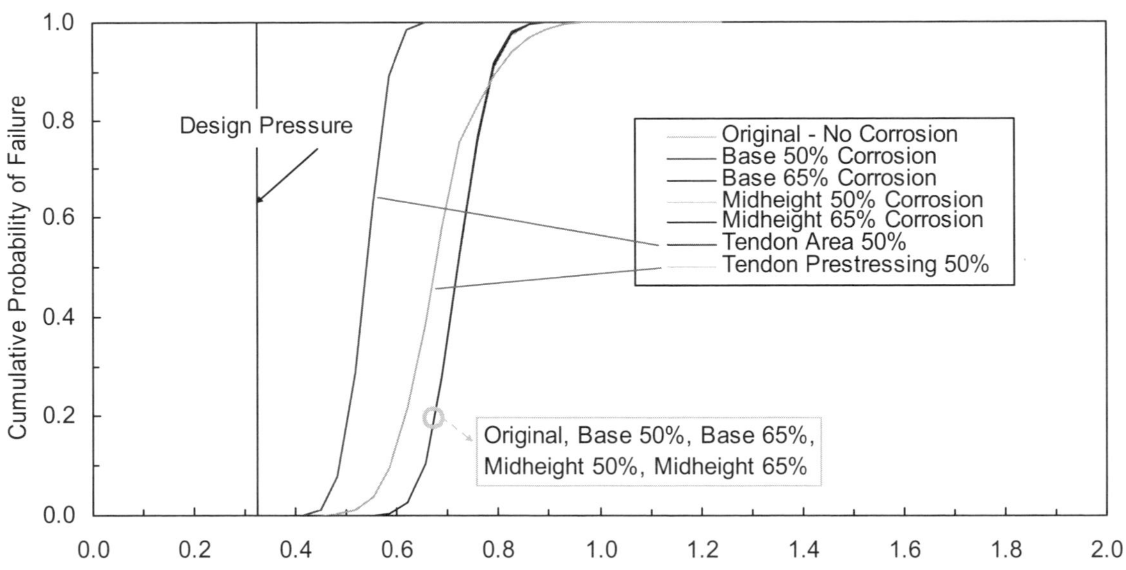

FIG. 6.35
PWR WITH PRESTRESSED CONCRETE CONTAINMENT FRAGILITY CURVES FOR VARIOUS DEGRADATION CASES [53]

6.5.3 Risk-Informed Assessment of Degraded Containments

A major component of the work by Spencer et al. [53] was to assess a potential metric for assessing containment degradation. Specifically, the metric examined was the large early release frequency (LERF). LERF is computed with a PRA analysis in order to determine the frequency of a large early release of material from inside the containment to the outside environment during a severe accident. PRA analyses will be discussed in more detail in the next chapter. For this study, the PRA models developed for the NUREG-1150 study were used to compute LERF for each of the containments discussed previously when in their original no corrosion states and each of the degraded states. The difference in LERF was then computed to see how it was affected by the hypothetical degradation cases. This was potentially a convenient metric for assessing containment degradation since the U.S. NRC's Regulatory Guide 1.174 [55] provides acceptance guidelines for changes in LERF due to modifications to the plant's licensing basis. Regulatory Guide 1.174 was developed for determining the acceptance of other changes to the plant and not for corrosion. However, the Spencer et al. study was tasked with assessing the potential for using the established ΔLERF limits for degradation.

As discussed earlier, the conditional probability of failure curves were computed in addition to the cumulative probability of failure fragility curves for each containment degradation case. The conditional probability of failure curves include individual curves for each the probability of a leak, a rupture, and a catastrophic rupture as a function of pressure. Each of these conditional curves represents the probability of that mode of failure occurring given that a failure occurs. Figure 6.36 shows the conditional probability of failure curves

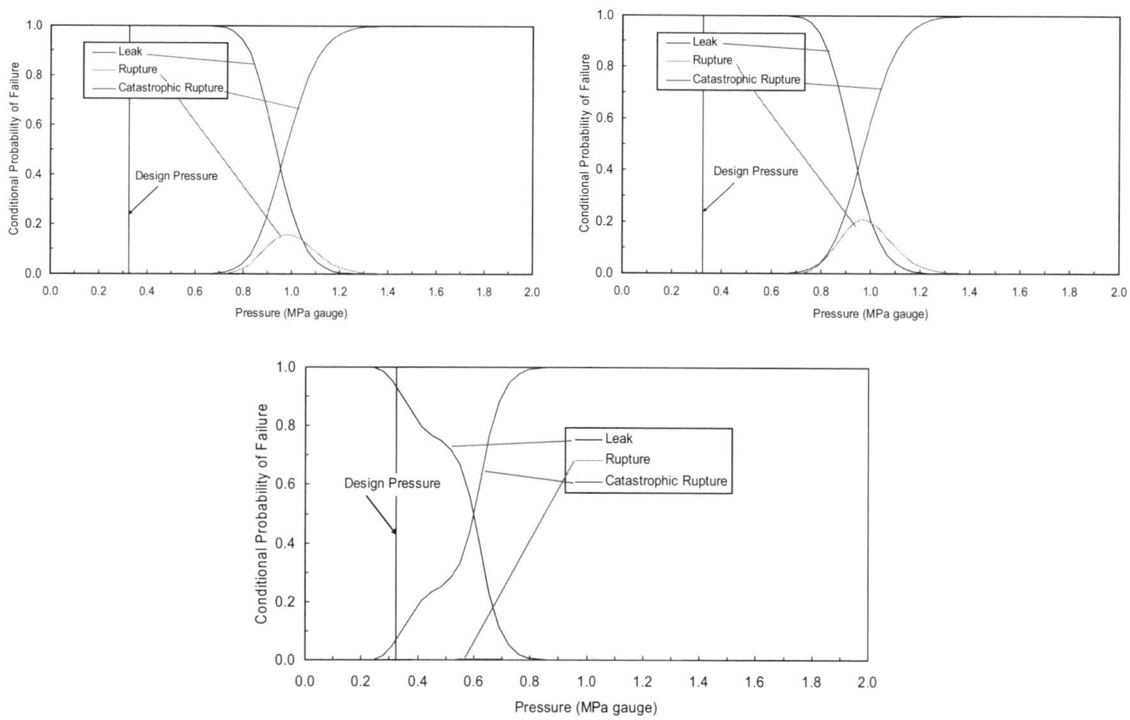

FIG. 6.36
PWR WITH PRESTRESSED CONCRETE CONTAINMENT CONDITIONAL FAILURE PROBABILITY CURVES FOR THE NO CORROSION (UPPER LEFT), 65% MID-HEIGHT CORROSION (UPPER RIGHT), AND 50% TENDON AREA REDUCTION (BOTTOM) CASES [53]

TABLE 6.2 PWR WITH PRESTRESSED CONCRETE CONTAINMENT ΔLERF (RELATIVE TO THE NO CORROSION CASE) [53]

Case	ΔLERF
50% Corrosion near basemat	–8.79*E* – 9
50% Corrosion at mid-height	–8.79*E* – 9
65% Corrosion near basemat	–8.79*E* – 9
65% Corrosion at mid-height	–8.79*E* – 9
50% Tendon Area Reduction	5.47*E* – 7
50% Tendon Prestressing Loss	1.17*E* – 9

for the prestressed concrete containment analysis by Spencer et al. for the no corrosion, 65% mid-height corrosion, and 50% tendon area reduction cases. The differences in the no corrosion case and the 65% mid-height corrosion case are minimal. However, the 50% tendon area reduction case shows the catastrophic rupture probability increasing significantly at very low pressures. This reflects the fact that the reduction in tendon area has an effect on the global capacity of the containment and not just the leak pressure as caused by liner tearing. Table 6.2 provides the ΔLERF values computed for the prestressed concrete containment for each of the degradation cases. The liner corrosion cases actually show decreases in LERF (LERF for the corrosion case is lower than the no corrosion case), a small increase for the 50% tendon prestressing force reduction, and a large increase for the 50% tendon area reduction. The decrease in LERF was possible due to the binning process use in the PRA analysis. Leak failures in the PRA are placed into the small early release frequency (SERF) while ruptures and catastrophic rupture failures are binned into LERF. Therefore, degradation cases that only affected the leak probability showed decreases in LERF and increases in SERF. Cases that affected the rupture and catastrophic rupture probability led to increases in LERF.

The Spencer et al. study demonstrated that LERF is not the metric to assess containment degradation, at least not for liner corrosion degradation cases. The study did show the potential for using SERF or the total early release frequency (LERF plus SERF); however, no current acceptance guidelines exist for these metrics.

6.5.4 Containment Degradation Effects on Severe Accident Consequences

Since LERF is only a surrogate for the actual release consequences during a severe accident, additional analyses were performed to study the effects of containment degradation beyond LERF. The study by Petti et al. [56] coupled the PWR reinforced concrete structural finite element results with a series of deterministic consequence analyses. These consequence analyses employed two codes under ongoing development at Sandia National Laboratories for the U.S. Nuclear Regulatory Commission: MELCOR and MACCS. MELCOR [57] is a consequence code that models the progression of a severe accident in light water reactors. The accident progression includes the thermal-hydraulic response in the reactor coolant system, reactor cavity, containment, and confinement building. It also models the core heat-up, core-concrete reactions, hydrogen production and combustion, and fission product release. MACCS [58] uses the output from MELCOR to predict the offsite consequences due to the fission product release by taking into account the surrounding geographic, demographics, and meteorological data at the plant.

The Petti et al. study used the Spencer et al. PWR reinforced concrete containment finite element results for the mid-height corrosion cases at 50% and 65% since these led to earlier leaks than the basemat corrosion cases. In addition, two new cases were created by assuming that 10 locations of mid-height corrosion existed at each level of corrosion. This was simply approximated by multiplying the computed crack area opening

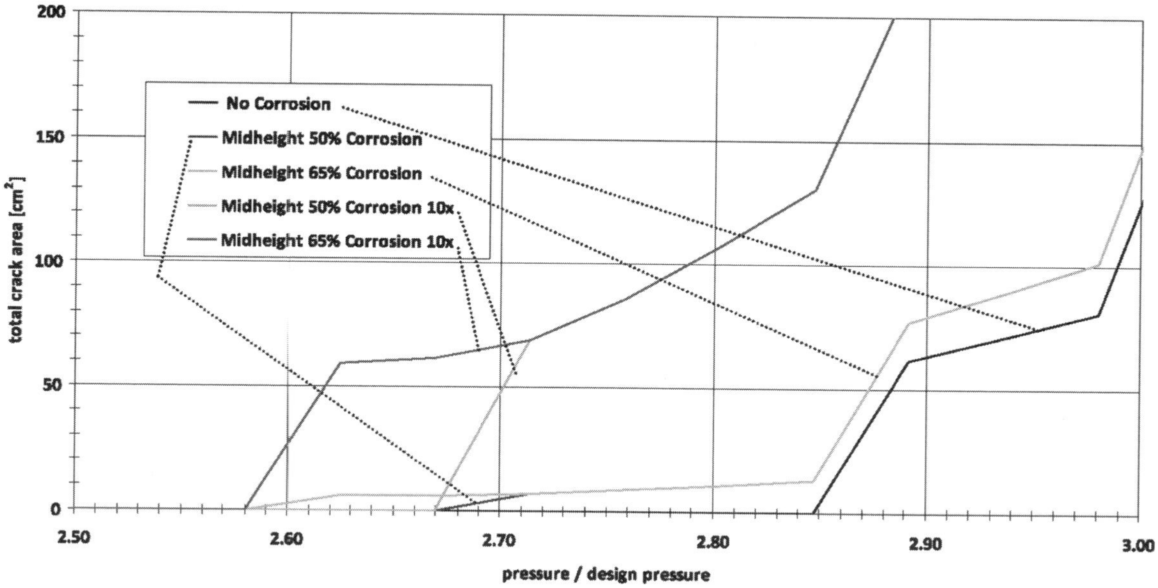

FIG. 6.37
PWR REINFORCED CONCRETE CONTAINMENT CRACK OPENING AREA WITH PRESSURE CURVES [56]

for the original single corrosion location cases by 10. This resulted in tears initiating in the liner at the same pressure, but opening to much larger areas at the same pressures. The crack open area versus pressure curves are illustrated in Fig. 6.37. These areas were input into the containment model within the MELCOR code.

The MELCOR code employed the crack open area curves to assess two different accident scenarios: a short-term station blackout (STSBO) and a long-term station blackout (LTSBO). Both accident progression sequences were modeled through containment pressurization and release through the containment opening based on the crack opening versus pressure relationship for each of the degradation cases. Figure 6.38 illustrates the containment pressure versus time response predicted by MELCOR for the STSBO and LTSBO. The pressure responses show the initial pressurization, leveling off after initiating a tear in the liner, and then depressurization as material is released to the outside environment as the accident progresses.

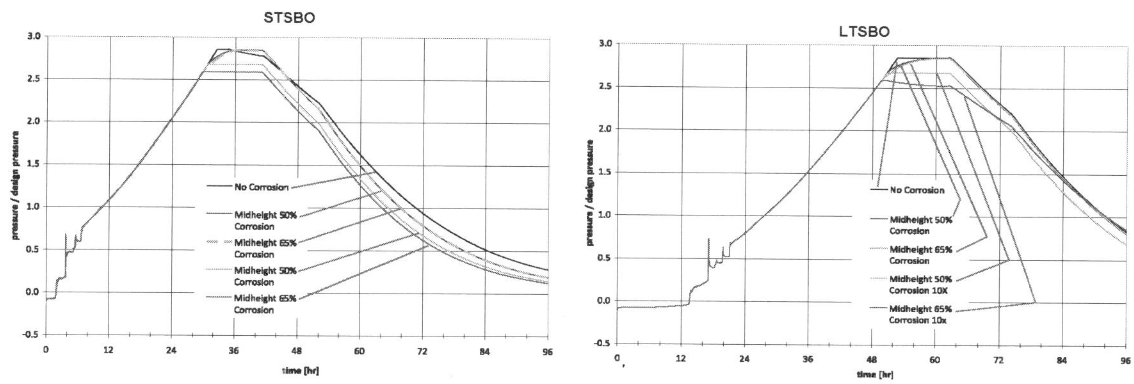

FIG. 6.38
PWR REINFORCE CONCRETE CONTAINMENT STSBO AND LTSBO PRESSURE VS. TIME RESPONSES [56]

TABLE 6.3
STSBO AND LTSBO HEALTH CONSEQUENCES [56]

Corrosion Case	STSBO — Individual Latent Cancer Fatality Risk (Percent of No Corrosion Case)	LTSBO — Individual Latent Cancer Fatality Risk (Percent of No Corrosion Case)
Midheight 50% Corrosion	123.8%	103.3%
Midheight 65% Corrosion	123.8%	103.3%
Midheight 50% Corrosion 10×	170.3%	122.1%
Midheight 65% Corrosion 10×	161.3%	115.3%

The fission products released through the breached containment for each case were then input into the MACCS code to compute the offsite consequences. In this case, the Individual Latent Cancer Fatality Risk was the consequence chosen to compare the different degradation cases. Table 6.3 provides the risk values as a percentage of the no corrosion case for each of the degradation cases. A number of conclusions were provided by Petti et al. [56] including the risk being proportional to the source term release to the environment, that more corrosion area and deeper corrosion in most, but not all, cases resulted in higher risk, that lower peak pressures are associated in most, but not all, cases with higher risk, that higher maximum containment failure area in most, but not all, cases resulted in higher risk, and that lower initial containment failure pressure in most, but not all, cases resulted in higher risk. In addition, a number of observations were counterintuitive. Some of these observations occurred when having a leak at a lower pressure but not having a higher consequence (mid-height 50% corrosion 10× versus mid-height 65% corrosion 10× for example). The containment failure does occur at a lower pressure (and earlier time) for the 65% corrosion case which caused early venting of the containment. This early venting actually led to a lower containment pressures later in the accident. This lower pressure did not increase the crack open area to a value as high the value reached in the 50% corrosion case. The combination of a smaller crack area and lower containment pressure for the 65% corrosion case resulted in less source term being released to the environment and hence a lower consequence value. Another observation was that even small areas of corrosion can lead to early leaks in the containment that can depressurize the containment. This depressurization can occur prior to the inventory of fission products being fully developed within the containment. Even though the amount of time that leakage is occurring through the containment is longer than say the no corrosion case, the pressure within the containment is lower later in the accident when the most risk significant material is produced. This means that material will not be pushed out of the containment as quickly as it would in the no corrosion case and leads to lower consequences. This is in no way a general conclusion for accidents in any type of containment, just a potential scenario under a given set of conditions. The lack of consistent trends from this study ultimately showed that degradation in containments should be examined in a case-by-case basis whether examining the failure pressures, fragility curves, or coupling those results to a PRA or consequence analysis.

REFERENCES

[1] ASME Boiler and Pressure Vessel Code, Section III, Division I-Subsection NE, Nuclear Power Plant Components-Class MC Components and Section III, Division 2, Code for Concrete Reactor Vessels and Containments, The American Society of Mechanical Engineers, New York, NY.

[2] U.S. Nuclear Regulatory Commission, 2010, Regulatory Guide 1.216, "Containment Structural Integrity Evaluation for Internal Pressure Loadings above Design-Basis Pressure," U.S. Nuclear Regulatory Commission, Washington, DC.

[3] U.S. Nuclear Regulatory Commission, 1990, "Severe Accident Risks: An Assessment or Five U.S. Nuclear Power Plants," NUREG-1150, U.S. Nuclear Regulatory Commission, Washington, DC.

[4] U.S. Nuclear Regulatory Commission, 1989, "Individual Plant Examination: Submittal Guidance," NUREG-1335, U.S. Nuclear Regulatory Commission, Washington, DC.

[5] Pilch, M. M., Allen, M. D., and Klamerus, E. W., 1995, "Resolution of the Direct Containment Heating Issues for All Westinghouse Plants with Large Dry Containments or Subatmospheric Containments," NUREG/CR-6338, SAND95-2381, Sandia National Laboratories.

[6] Hessheimer, M. F., and Dameron, R. A., 2006, "Containment Integrity Research at Sandia National Laboratories," NUREG/CR-6906, SAND2006-2274P, Sandia National Laboratories, Albuquerque, NM.

[7] Blejwas, T. E., Dennis, A. W., Woodfin, R. L., and von Riesemann, W. A., 1982, "Background Study and Preliminary Plans for a Program on the Safety Margins of Containments," NUREG/CR-2549, SAND82-0324, Sandia National Laboratories, Albuquerque, NM.

[8] Horschel, D. S., 1988, "Design, Construction, and Instrumentation of a 1:6-Scale Reinforced Concrete Containment Building," NUREG/CR-5083, SAND88-0030, Sandia National Laboratories, Albuquerque, NM.

[9] Horschel, D. S., 1992, "Experimental Results from Pressure Testing a 1:6-Scale Nuclear Power Plant Containment," NUREG/CR-5121, SAND88-0906, Sandia National Laboratories, Albuquerque, NM, January.

[10] Dameron, R., and Rashid, Y. R., 1991, "Post-Test Report on Testing and Analysis of the Sizewell-B 1:10-Scale Containment Model," NUREG/CR-5825, SAND92-7064, ANATECH Research Corp., San Diego, CA.

[11] Dameron, R. A., Rashid, Y. R., and Parks, M. B., 1991, "Comparison of Pre-Test Analyses with the Sizewell-B 1:10 Scale Prestressed Concrete Containment Test," *Proceedings of SMiRT 11*, Paper H10, Tokyo, Japan.

[12] MacGregor, J. G., Simmonds, S. H., and Rizkalla, S., 1980, "Test of a Prestressed Concrete Secondary Containment Structure," University of Alberta, Dept. of Civil Engineering Structural Engineering Report No. 85, (also *Proceedings of the 5th International Conference on Structural Mechanics in Reactor Technology*, Scibac, Berlin, Germany, 1979, Papers J3/2, J3/5 and J 4/2).

[13] Atchison, I. J., Asmis, G. J. K., and Campbell, F. R., 1979, "Behaviour of Concrete Containment under Over-Pressure Conditions," *Proceedings of SMiRT 5*, Paper No. 5312, Berlin, Germany.

[14] Rizkalla, S., Simmonds, S. H., and MacGregor, J. G., 1979, "A Test of a Model of a Thin Walled Prestressed Concrete Secondary Containment Structure," *Proceedings of SMiRT 5*, Paper No. 5412, Berlin, Germany.

[15] Murray, D. W., Chitnuyanondh, L., and Wong, C., 1979, "Modelling and Predicting Behavior of Prestressed Concrete Secondary Containment Structures Using BOSOR5," *Proceedings of SMiRT 5*, Paper 5315, Berlin, Germany.

[16] Appa Rao, T. V. S. R., 1975, "Behaviour of Concrete Nuclear Containment Structures Up to Ultimate Failure With Special Reference to MAPP-1 Containment," Inelastic Behaviour, Report 4-SM-THEME75, Madros, India: Structural Engineering Research Centre.

[17] Donten, K., Knauff, M., Sadowski, A., and Scibak, W., 1980, "Tests on a Model of Prestressed Reactor Containment," Archiwum Inzynierii Ladowej, vol. XXVI, no. 1, pp. 231–245 (also *Proceedings of the 5th International Conference on Structural Mechanics in Reactor Technology*, Scibac, Berlin, Germany, 1979, Paper J 4/8).

[18] Hessheimer, M. F., Klamerus, E. W., Rightley, G. S., Lambert, L. D., and Dameron, R. A., 2003, "Overpressurization Test of a 1:4-Scale Prestressed Concrete Containment Vessel Model", NUREG/CR-6810, SAND2003-0840P, Sandia National Laboratories, Albuquerque, NM.

[19] Horschel, D. S., and Clauss, D. B., 1984, "The Response of Steel Containment Models to Internal Pressurization," Structural Engineering in Nuclear Facilities, J. J. Ucciferro: Editor, Raleigh, North Carolina, September 1984, 1, pp. 534–553.

[20] Horschel, D. S., and Blejwas, T. E., 1983, "An Analytical Investigation of the Response of Steel Containment Models to Internal Pressurization," *Proceedings of the 7th International Conference on Structural Mechanics in Reactor Technology*, Chicago, Illinois, Paper J 6/4, pp. 297–304.

[21] Reese, R. T., and Horschel, D. S., 1985, "Design and Fabrication of a 1:8-Scale Steel Containment Model," NUREG/CR-3647, SAND84-0048, Sandia National Laboratories, Albuquerque, NM.

[22] Koenig, L. N., 1986, "Experimental Results for a 1:8-Scale Steel Model Nuclear Power Plant Containment Pressurized to Failure," NUREG/CR-4216, SAND85-0790, Sandia National Laboratories, Albuquerque, NM.

[23] Luk, V. K., Hessheimer, M. F., Rightley, G. S., Lambert, L. D., and Klamerus, E. W., 1999, "Design, Instrumentation, and Testing of a Steel Containment Vessel Model," NUREG/CR-5679, SAND98-2701, Sandia National Laboratories, Albuquerque, NM.

[24] Lambert, L. D., and Parks, M. B., 1994, "Experimental Results from Containment Piping Bellows Subjected to Severe Accident Conditions, Volume 1: Results from Bellows Tested in 'Like-New' Conditions," NUREG/CR-6154, SAND94-1711, Sandia National Laboratories, Albuquerque, NM.

[25] Lambert, L. D., and Parks, M. B., 1995, "Experimental Results from Containment Piping Bellows Subjected to Severe Accident Conditions, Volume 2: Results from Bellows Tested in Corroded Conditions," NUREG/CR-6154, SAND94-1711, Sandia National Laboratories, Albuquerque, NM.

[26] Clauss, D. B., 1987, "An Evaluation of the Leakage Potential of a Personnel Air Lock Subject to Severe Accident Loads," *Proceedings of the 9th International Conference on Structural Mechanics in Reactor Technology*, J, Lausanne, Switzerland, pp. 147–152.

[27] Julien, J. T., and Peters, S. W., 1989, "Leak and Structural Test of a Personnel Air Lock for LWR Containments Subjected to Pressure and Temperature Beyond Design Limits," NUREG/CR-5118, SAND88-7155, Sandia National Laboratories, Albuquerque, NM.

[28] Brinson, D. A., and Graves, G. A., 1988, "Evaluation of seals for Mechanical Penetrations of Containment Buildings," NUREG/CR-5096, SAND88-7016, Sandia National Laboratories, Albuquerque, NM.

[29] Bridges, T. L., 1987, "Containment Penetration Elastomer Seal Leak Rate Tests," NUREG/CR-4944, SAND87-7118, Sandia National Laboratories, Albuquerque, NM.

[30] Clauss, D. B., 1989, "Severe Accident Testing of Electrical Penetration Assemblies," NUREG/CR-5334, SAND89-0327, Sandia National Laboratories, Albuquerque, NM.

[31] Parks, M. B., 1991, "Leakage behavior of inflatable seals subject to severe accident conditions," Nuclear Engineering and Design, No. 137, pp. 175–186.

[32] Parks, M. B., Walther, H. P., and Lambert, L. D., 1991, "Experiments to Determine the Leakage Behavior of Pressure-Unseating Equipment Hatches," *Proceedings of the 11th International Conference on Structural Mechanics in Reactor Technology*, Tokyo, Japan, Paper No. FO 3/4, pp. 49–54.

[33] Kulak, R. F. et al., 1985, "Structural Response of Large Penetrations and Closures for Containment Vessels Subjected to Loadings Beyond Design Basis," NUREG/CR-4064, SAND84-7177, ANL-84-41, Sandia National Laboratories, Albuquerque, NM.

[34] Clauss, D. B., 1987, "Round Robin Pretest Analyses of a 1:6-Scale Reinforced Concrete Containment Model Subject to Static Internal Overpressurization," NUREG/CR-4913, SAND87-0891, Sandia National Laboratories, Albuquerque, NM.

[35] Weatherby, J. R., 1990, "Posttest Analysis of a 1:6-Scale Reinforced concrete Reactor Containment Building," NUREG/CR-5476, SAND89-2603, Sandia National Laboratories, Albuquerque, NM.

[36] Clauss, D. B., 1989, "Round Robin Analysis of the Behavior of a 1:6-Scale Reinforced Concrete Containment Model Pressurized to Failure: Posttest Evaluations," NUREG/CR-5341, SAND89-0349, Sandia National Laboratories, Albuquerque, NM.

[37] Dameron, R. A., Zhang, L., Rashid, Y. R., and Vargas, M. S., 2000, "Pretest Analysis of a 1:4-Scale Prestressed Concrete Containment Vessel Model," NUREG/CR-6685, SAND2000-2093, ANATECH Corporation, San Diego, CA and Sandia National Laboratories, Albuquerque, NM.

[38] Luk, V. K., 2000, "Pretest Round Robin Analysis of a Prestressed Concrete Containment Vessel Model," NUREG/CR-6678, SAND00-1535, Sandia National Laboratories, Albuquerque, NM.

[39] Dameron, R. A., Hanson, B. E., Parker, D. R., and Rashid, Y. R., 2003, "Posttest Analysis of a 1:4 Scale Prestressed Concrete Containment Vessel Model," NUREG/CR-6809, SAND2003-0839P, ANA-01-0330, ANATECH Corporation, San Diego, CA and Sandia National Laboratories, Albuquerque, NM.

[40] Hessheimer, M. F., and Mathet, E., 2005, "International Standard Problem 48, Containment Capacity, Synthesis Report," Organization for Economic Co-operation and Development, Nuclear Energy Agency, Committee on the Safety of Nuclear Installations, NEA/CSNI/R(2005)5, Paris, France.

[41] Clauss, D. B., 1985, "Pretest Predictions for the Response of a 1:8-Scale Steel LWR Containment Building Model to Static Overpressurization," NUREG/CR-4137, SAND85-0175, Sandia National Laboratories, Albuquerque, NM.

[42] Clauss, D. B., 1985, "Comparison of Analytical Predictions and Experimental Results for a 1:8-Scale Steel Containment Model Pressurized to Failure", NUREG/CR-4209, SAND85-0679, Sandia National Laboratories, Albuquerque, NM.

[43] Porter, V. L., Carter, P. A., and Key, S. W., 1999, "Pretest Analysis of the Steel Containment Vessel Model," NUREG/CR-6516, SAND96-2877, Sandia National Laboratories, Albuquerque, NM.

[44] Luk, V. K., and Klamerus, E. W., 1998, "Round Robin Pretest Analysis of a Steel Containment Vessel Model and Contact Structure Assembly Subject to Static Internal Pressurization," NUREG/CR-6517, SAND96-2899, Sandia National Laboratories, Albuquerque, NM.

[45] Van Den Avyle, J. A., and Eckelmeyer, K. H., 1999, "Posttest Metallurgical Evaluation Results for the Steel Containment Vessel (SCV) High Pressure Test," SAND98-2702, Sandia National Laboratories, Albuquerque, NM.

[46] Luk, V. K., and Klamerus, E. W., 2000, "Round Robin Posttest Analysis of a Steel Containment Vessel Model," NUREG/CR-5678, SAND98-700, Sandia National Laboratories, Albuquerque, NM.

[47] Ludwigsen, J. S., Luk, V. K., Hessheimer, M. F., and Costello, J. F., 2000, "Posttest Analyses of the Steel Containment Vessel Model," NUREG/CR-6649, SAND99-2954, Sandia National Laboratories, Albuquerque, NM.

[48] Ellingwood, B. R., and Cherry, J. L., 1999, "Fragility Modeling of Aging Containment Metallic Pressure Boundaries," NUREG/CR-6631, ORNL/SUB/99-SP638V, Oak Ridge National Laboratory, Oak Ridge, TN.

[49] Iman, R. L., and Conover, W. J., 1980, "Small Sample Sensitivity Analysis Techniques for Computer Models with an Application to Risk Assessment," *Communications in Statistics Part A — Theory and Methods*, A9(17), pp. 1749–1842.

[50] Petti, J. P., 2007, "Structural Integrity Analysis of the Degraded Drywell Containment at the Oyster Creek Nuclear Generating Station," SAND2007-0055, Sandia National Laboratories, Albuquerque, NM.

[51] Cherry, J. L., and Smith, J. A., 2001, "Capacity of Steel and Concrete Containment Vessels with Corrosion Damage," NUREG/CR-6706, SAND2000-1735, Sandia National Laboratories, Albuquerque, NM.

[52] Smith, J. A., 2001, "Capacity of Prestressed Concrete Containment Vessels with Prestressing Loss," SAND2001-1762, Sandia National Laboratories, Albuquerque, NM.

[53] Spencer, B. W., Petti, J. P., and Kunsman, D. M., 2006, "Risk-Informed Assessment of Degraded Containment Vessels," NUREG/CR-6920, SAND2006-3772P, Sandia National Laboratories, Albuquerque, NM.

[54] Tang, H. T., Dameron, R. A., and Rashid, Y. R., 1995, "Probabilistic Evaluation of Concrete Containment Capacity for Beyond Design Basis Internal Pressures," *Nuclear Engineering and Design*, 157, pp. 455–467.

[55] U.S. Nuclear Regulatory Commission, 2002, Regulatory Guide 1.174, "An Approach for Using Probabilistic Risk Assessment in Risk-Informed Decisions on Plant-Specific Changes to the Licensing Basis," U.S. Nuclear Regulatory Commission, Washington, DC.

[56] Petti, J. P., Kalinich, D. A., Jun, J., Graves III, H., and Pires, J., 2009, "Effects of Liner Degradation on the Severe Accident Consequences at a PWR Plant with a Reinforced Concrete Containment," *Proceedings from SMiRT20*, Division IV, Paper 1786, Espoo, Finland.

[57] Gauntt, R. et al., 2005, "MELCOR Computer Code Manuals," NUREG/CR-6119, Vol. 2, Rev. 3, SAND2005-5713, Washington, DC.

[58] Chanin, D. I. et al., 1990, "MELCOR Accident Code System (MACCS), Vol. I: User's Guide," NUREG/CR-4691, SAND86-1562, Sandia National Laboratories, Albuquerque, NM.

CHAPTER

7

CONTAINMENT SYSTEM CHALLENGES UNDER SEVERE ACCIDENTS

Dana Powers, Shawn Burns and Hansraj Ashar

The Nuclear Regulatory Commission (NRC) defines three types of reactor accidents that have to be addressed to ensure public health and safety; (1) design basis accidents, (2) beyond design basis accidents and (3) severe accidents. A design basis accident is a postulated accident that a nuclear facility must be designed and built to withstand without loss to the systems, structures, and components necessary to ensure public health and safety. Loss of coolant accidents (LOCAs) are those postulated accidents that result from the loss of reactor coolant at a rate that is in excess of the capability of the reactor coolant makeup system. LOCAs are from breaks in the reactor coolant pressure boundary, up to and including a break equivalent in size to the double-ended rupture of the largest pipe of the reactor coolant system. The containment structures are required to withstand the LOCA loads, as well as the loads imposed by the design basis natural phenomena, without exceeding conservatively calculated acceptance criteria.

Beyond design basis accidents is the term used as a way to discuss accident sequences that are possible but were not fully considered in the design process because they were judged to be too unlikely. In that sense, they are considered beyond the scope of design basis accidents. As the regulatory process strives to be as thorough as possible, "beyond design basis" accident sequences are analyzed to fully understand the capability of a design.

In a generic way, the NRC defines severe accident in nuclear power plants as a type of accident that may challenge safety systems at a much higher level than estimated. Beyond design basis, natural phenomena are addressed in Chapter 8 of this book.

Chapter 1 of this book describes two types of light water reactors used in the U. S. nuclear power plants; Pressurized Water Reactors (PWRs), and Boiling Water Reactors (BWRs). Severe accident is loosely defined as an event when uncovery of the reactor core is imminently expected. For the PWRs, it is assumed that the onset of the core damage occurs at uncovery of the top of active fuel without possibility of recovery. For the BWRs, it is assumed that onset of core damage would occur when the water level is less than 0.6 m (2 ft) above the bottom of the active fuel.

This chapter principally describes (1) severe accident phenomena that load the containment, (2) source terms developed during and after the core melt, and (3) the regulatory requirements and guidelines provided to address severe accident threats.

Keywords: Reactor Safety Studies, Severe Accident, Three Mile Island Accident, Chernobyl Accident, Fukushima Accident Loads on Containment, Core Debris Interaction, Containment Heating, Source Terms, Iodine Dilemma

7.1 INTRODUCTION

Severe reactor accidents that go well beyond the design basis of a nuclear reactor involve extensive reactor fuel degradation and even extensive melting. Core debris may be expelled from the reactor coolant system. Though reactor containments are not designed to withstand the rigors of all severe accidents, they are the last physical barrier preventing the release of radioactive material from the plant into the public environment. The worth of modern reactor containments for protection of the public was demonstrated well during the accident at Three Mile Island. Equally so, the dire consequences of prompt failure of this final barrier was demonstrated by the reactor accident at Chernobyl. It is too early to know the exact roles containment played in the reactor accidents at Fukushima Daiichi. Current analyses certainly suggest that the containments and engineered safety systems within these containments were important, while they were functional in limiting the extent of radionuclide release to the environment [1].

Severe accidents are expected to be very rare events [2], but there is now ample evidence that they can occur. In this chapter, severe accident processes that threaten the containment integrity are examined. Also, in this chapter, processes within the containment that affect the inventory of radioactive material suspended in the containment atmosphere and available for release should there be a loss of structural integrity are examined.

Accident phenomena that threaten containment integrity early in accident progression before emergency response measure can be fully implemented are, of course, of primary concern. Three such accident phenomena are examined here:

- Combustible gas explosions
- Explosive interactions of core debris with water
- Direct containment heating

In addition, the interaction of core debris with structural concrete represents a longer term threat to structural integrity and is also considered in this chapter.

Radionuclides released from containment will be present as gases, primarily noble gases, and aerosolized, solid particles. Isotopes of iodine can transform between these two physical forms and is particular consequential should it escape the containment. In this chapter, physical and chemical processes that affect the radioactive gases and aerosol concentration in containment will be examined with particular attention given to the behavior of radioactive iodine.

7.2 HYDROGEN COMBUSTION

Pressurization as a result of hydrogen combustion has long been recognized as a threat to containment integrity under the conditions of reactor accidents [3]. The threat was highlighted by the hydrogen combustion event during the accident at Three Mile Island (see Fig. 7.1). Of course, the far more damaging hydrogen combustion events during the reactor accidents at Fukushima Daiichi have raised awareness of hydrogen combustion even more.

7.2.1 Hydrogen Sources

Under reactor accident conditions, the predominant source of hydrogen comes from the steam oxidation of zirconium alloy cladding on the reactor fuel:

$$Zr + 2H_2O \rightarrow ZrO_2 + 2\ H_2 + \text{heat}$$

The rate of this reaction is limited by the formation of the oxide product and follows parabolic kinetics [4]. The rate does increase exponentially with temperature up to the point it is limited by the mass transport of steam to the cladding. In typical predictions of reactor accidents, the mass transport of steam becomes

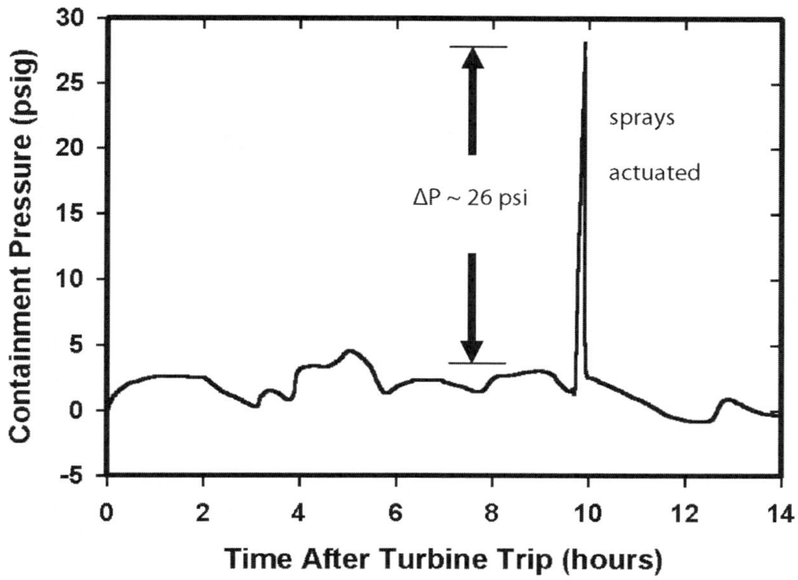

FIG. 7.1
CONTAINMENT PRESSURE DURING THE ACCIDENT AT THREE MILE ISLAND

limiting when peak fuel cladding temperatures are about 2100 K. That is, vigorous hydrogen production can take place before there is substantial fuel degradation and melting within the reactor core.

Oxidation of other metals such as stainless steel within the reactor core and especially structures above the reactor core can augment hydrogen production [5]. Water corrosion of aluminum and zinc alloys in containment and water radiolysis are not significant sources of hydrogen in comparison to those coming from steam oxidation of zirconium alloy and stainless steel during core degradation [6].

Steam oxidation of fuel cladding and other metals within the reactor containment is often predicted in severe accident analyses to be incomplete by the time core debris has relocated to the lower plenum of a reactor vessel. Penetration of the reactor vessel by the core debris can lead to additional phases of combustible gas generation during core debris interactions with water and structural concrete in the containment. In the case of core debris interactions with concrete, combustible carbon monoxide, CO, is produced along with hydrogen. These additional sources of combustible gas released to the reactor containment will be discussed further in later sections of this chapter. Here the focus is on the modes of hydrogen combustion and the loads combustion can impose on reactor containments.

7.2.2 Modes of Combustion

Reactor accident analyses treat the reaction of hydrogen with air in the reactor containment:

$$2\,H_2 + O_2 \rightarrow 2\,H_2O + \text{heat}$$

These analyses usually consider three modes of hydrogen combustion:

- Standing flames: Combustible gas is ignited at the point that it emerges from the reactor coolant system or other source location and burns as a jet.
- Deflagrations: Combustible gas burns throughout a volume along a reaction front that propagates at a velocity less than the speed of sound in the unburned gas.
- Detonations: Combustible gas burns throughout a volume along a reaction front behind a shock wave that propagates at a velocity greater than the speed of sound in the unburned gas.

Standing flames can develop at vent points in the reactor coolant system or in reactor buildings of boiling water reactors at vent points in the reactor containment [7, 8]. Typically, standing flames are not found to produce significant containment pressurization because the rate of heat release is slow in comparison to the natural heat removal capability of the containment or reactor building. Standing flames can impose high heat loads on adjacent structures and equipment which can have ramifications on the continued progression of a reactor accident. The effects of standing flames are very specific to the reactor design and are not pursued further here.

7.2.2.1 Deflagrations. Typically, accident analyses predict that hydrogen accumulates in reactor containments throughout the period of core degradation within the reactor coolant system. Further accumulation of hydrogen and other combustible gases in the containment can occur when core debris is expelled from the reactor coolant system and interacts with water or attacks concrete. Once minimal concentrations of combustible gas accumulate in containment, it is possible to initiate combustion throughout the gas volume at a rate that is faster than the rate of heat removal. This leads to pressurization of the containment as observed in the Three Mile Island accident.

The minimal concentrations of hydrogen necessary for deflagrations are called "ignition limits." Because deflagrations propagate by convective heat transfer, ignition limits depend on the direction of flame propagation. At room temperature, upward propagation of the flame requires there to be more than 4 volume percent hydrogen. Downward propagation in a direction opposed to the buoyancy of the heated gas requires about 9 volume percent hydrogen. It is widely thought that the combustion event during the Three Mile Island accident corresponded to the reaction of about 8 volume percent hydrogen. These minimal concentrations of hydrogen necessary to sustain a volumetric combustion do depend on temperature (see Fig. 7.2). At about 800 K, hydrogen will react spontaneously with oxygen in air at any concentration.

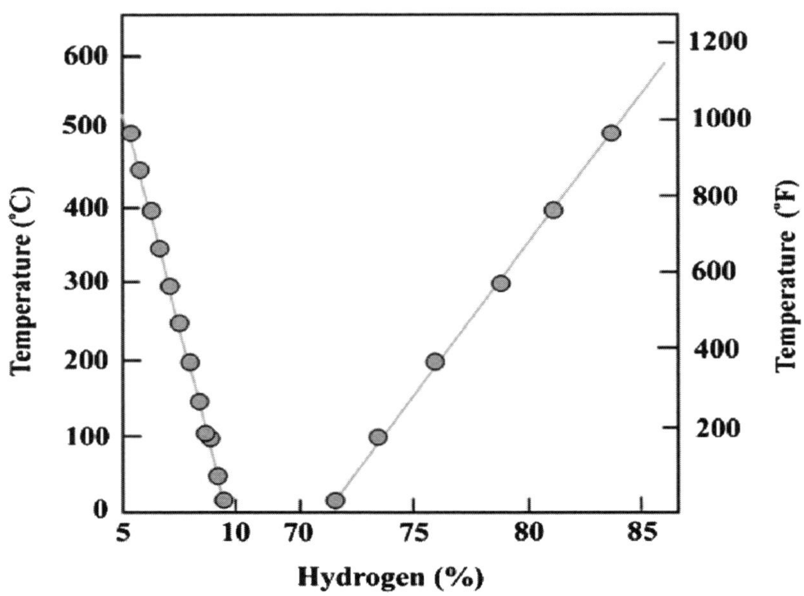

FIG. 7.2
TEMPERATURE DEPENDENCE OF THE IGNITION LIMITS FOR HYDROGEN–AIR MIXTURES FOR DOWNWARD PROPAGATING FLAMES AT ATMOSPHERIC PRESSURE

As discussed in Section 7.5, core debris interactions with concrete can generate large amounts of carbon monoxide which is a combustible gas. The lower flammability limit for carbon monoxide in air at room temperature is about 12.5 volume percent [9]. This lower flammability limit decreases to about 10.5 volume percent at 600 K. Hustad and Sønju [10] suggest that the lower flammability limit for mixtures of hydrogen and carbon monoxide can be estimated based on the principle of LeChatlier:

$$\mathrm{LFL(mix)} = \frac{\mathrm{P(H_2) + P(CO)}}{\mathrm{P(H_2)/LFL(H_2) + P(CO)/LFL(CO)}}$$

where LFL(j) is the lower flammability limit of gas j alone in air and P(j) is the partial pressure of gas j in the mixture. Wierzba and Kilchyk [11] obtained experimental results that validate this estimation procedure.

Ignition limits also depend on the noncombustible content of a gas mixture. The Mark I and Mark II containments of General Electric boiling water reactor designs are deliberately purged with nitrogen to limit the concentration of oxygen below that necessary to sustain a deflagration (< 5% oxygen). Steam released to containment can also change ignition limits as shown in Fig. 7.3. Steam at sufficiently high concentrations will render an otherwise combustible gas mixture of hydrogen and air immune to ignition. This immunity arises largely because the heat capacity of the atmosphere is high enough that any reaction of hydrogen with air will not produce local temperatures sufficient for self-sustained chemical reaction.

The inerting of containment atmospheres by steam can be reversed quite easily by condensation of the steam. In particular, actuation of the containment sprays can rapidly decrease the mole fraction of steam in

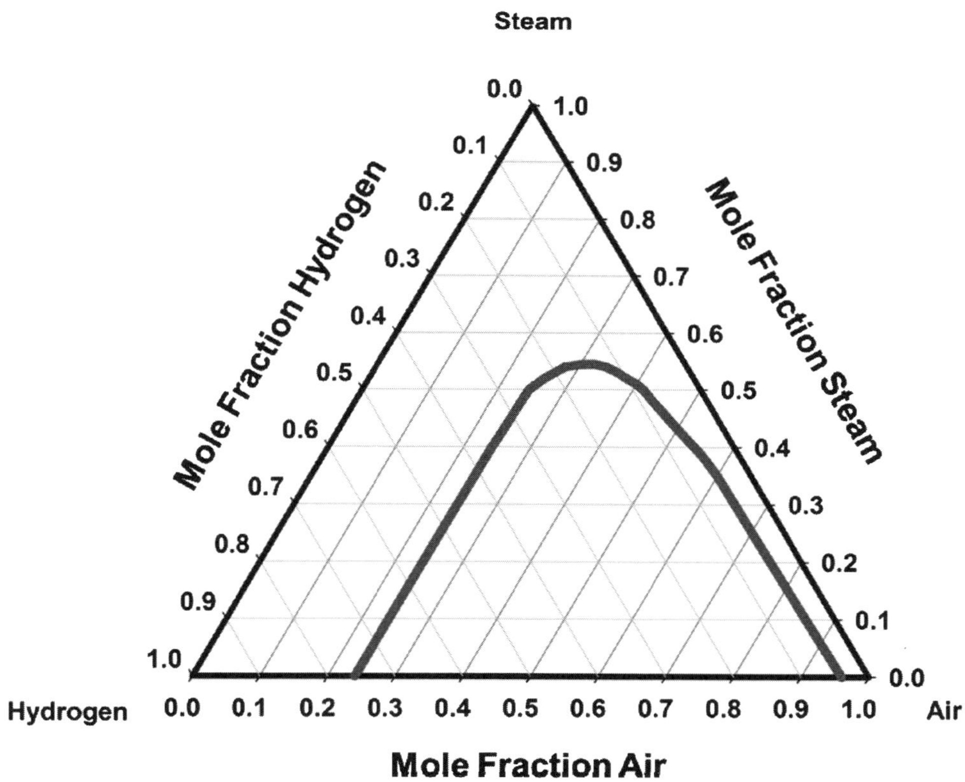

FIG. 7.3
IGNITION LIMITS FOR UPWARD PROPAGATION OF FLAME IN A MIXTURE OF HYDROGEN, AIR, AND STEAM AT ABOUT 370 K

the containment atmosphere. Sufficient steam condensation can change the gas composition so it falls into a region that is flammable and can sustain a volumetric combustion of the hydrogen.

At temperatures below about 800 K, gaseous mixtures of hydrogen and air do not ignite spontaneously. Some external source of local energy is required to start a deflagration. The energy required for ignition depends on the hydrogen concentration but in all cases is small — typically less than milliJoules. That is, static electric sparks are sufficient to initiated hydrogen deflagration (see Fig. 7.4). Indeed, the deflagration at Three Mile Island is often attributed to sparking associated with electronic actuation of a valve.

Pressure and temperature loads imposed on reactor containments by deflagrations are quasi-static and do not involve shock loading. The temperature rise and pressurization associated with a deflagration depend on the hydrogen concentration and the completeness of the combustion. Theoretical loads based on complete combustion reach a maximum when the hydrogen concentration in the atmosphere is just sufficient to consume all the oxygen present. For air, this amounts to about 30 volume percent hydrogen. Seldom are such high hydrogen concentrations predicted to be reached in containments during reactor accidents. Complete combustion of 4 volume percent hydrogen can raise containment temperatures from about the boiling point of water to about 800 K. The pressure of the containment atmosphere would be increased to a little more than twice its initial value. A deflagration involving 9 volume percent hydrogen would raise the temperature to about 1200 K and increase the containment pressure by about a factor of 4 [13].

Complete combustion is affected by the levels of atmospheric turbulence which, in turn, can be affected by the combustion event itself. Experimental studies show [14] that combustion will be substantially incomplete for hydrogen concentrations less than about 6 volume percent even when the atmosphere is stirred by fans (see Fig. 7.4). Because deflagrations are incomplete at low hydrogen concentrations, it is not uncommon for reactor accident analyses to predict multiple, sequential deflagration events during periods of rapid hydrogen release to containment.

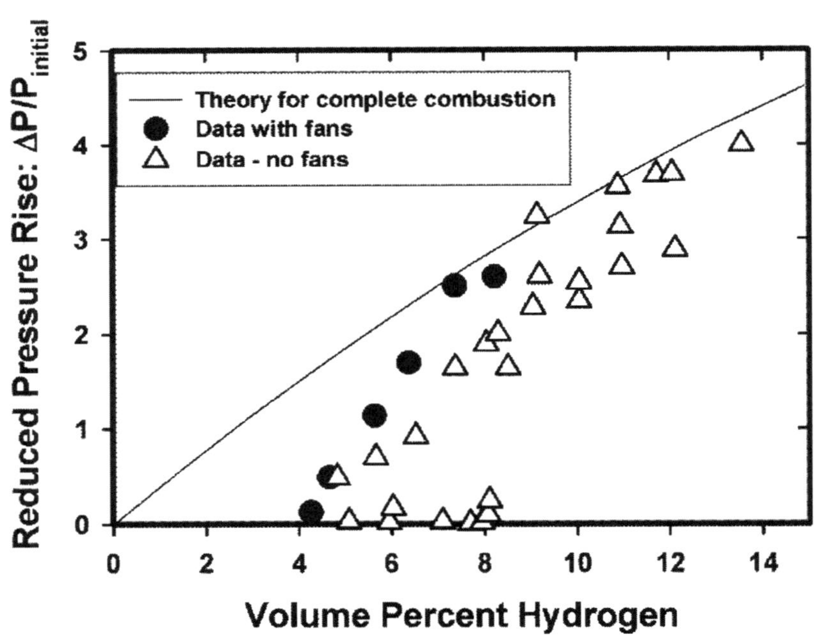

FIG. 7.4

COMPARISON OF PREDICTED PRESSURE RISE FROM AN INITIAL PRESSURE OF $P_{INITIAL}$ DUE TO AN ADIABATIC ISOCHORIC COMBUSTION TO PRESSURE RISES OBSERVED IN EXPERIMENTS WITH STATIC GAS MIXTURES AND MIXTURES STIRRED BY FANS [12,13]

7.2.2.2 Detonations. Detonations of hydrogen–air mixtures occur because of radical chain reactions in the mixtures. Pressurization pulses produce by these rapid, exoergic reactions propagate more rapidly through the high temperature gaseous products than they can through the cool reactants. Consequently, these pulses accumulate to produce a shock wave. The damaging effects of detonation shock waves were dramatically apparent by the events during the accidents at the Fukushima Daiichi reactor installation.

Detonations require higher minimal concentrations of combustible gas to be present than those necessary to sustain deflagrations. In the past, it was assumed that there were fixed composition limits needed for detonations much like the ignition limits for deflagrations. These limits were established by experiments done in small (~2.5 cm) diameter tubes. A commonly quoted minimal concentration for detonations of hydrogen in air is 18% hydrogen. It was found, however, in later experiments with larger facilities that detonation limits depend on geometry as well as composition of the gas. Detonations have been observed in larger pipes at concentrations less than 11%. Detonations are sensitive to other variables including the levels of turbulence in the gas phase. It is common for the purposes of reactor accident analysis to correlate the sensitivity of a gas mixture to the detonation cell size characteristic of the structure of the shock wave produced by a detonation. The detonation cell size has been correlated with gas composition as is shown in Fig. 7.5. A variety of empirical and semi-empirical "rules" are used to compare the detonation cell size to the local geometry to ascertain if a detonation can be sustained in the region. For example, the detonation cell size must be less than the minimum dimension of a duct for a detonation to propagate down the duct [16].

The direct initiation of a detonation requires a very large local energy source as shown in Fig. 7.6. Such large energy sources are not easily identified in reactor containments even under accident conditions. There is, however, another mechanism for inducing hydrogen–air mixtures to detonate. This is the phenomenon of "deflagration to detonation transition" which will be discussed in the next subsection of this chapter.

Detonations can impose shock loading on containments. Thus, the pressure pulse to the containment can be doubled by the reflection of a shock wave from a surface. Pressure pulses in excess of 10 times the initial containment pressure can be imposed on surfaces. Even more damaging pressure pulses can be created when

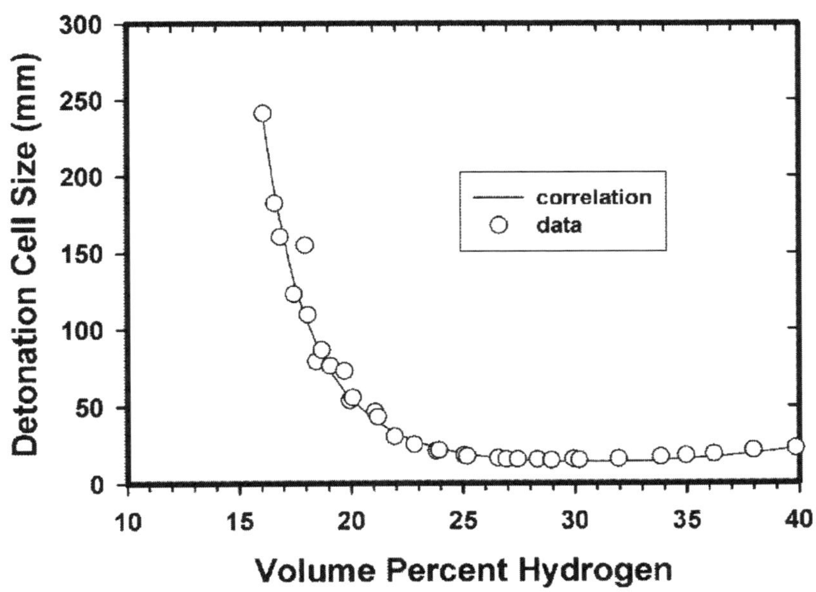

FIG. 7.5
DETONATION CELL SIZE AS A FUNCTION OF THE HYDROGEN CONCENTRATION IN AIR [15]

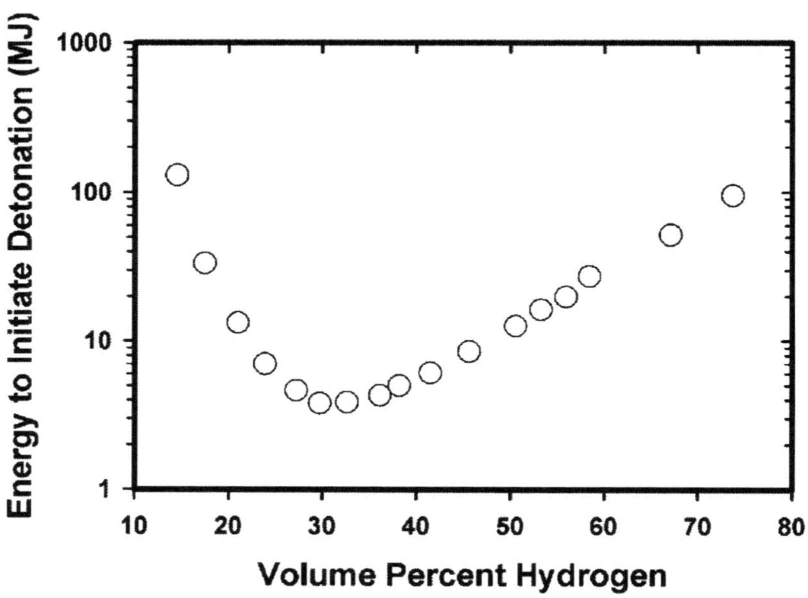

FIG. 7.6
ENERGY REQUIRED TO INITIATE DIRECTLY A DETONATION IN A MIXTURE OF HYDROGEN IN AIR AT ATMOSPHERIC PRESSURE AND ROOM TEMPERATURE [15]

multiple reflected shock fronts combine constructively at a surface. Shock heating can create local temperatures in excess of 2500 K. Calculation of the potential effects of detonations on containment structures is not usually done with systems level accident analysis computer codes. The physics of detonations and the complicating effects of details of containment geometry exceed the level of resolution of these computer codes. Specialized analyses are required to investigate effects of detonations on containments and internal structures [17, 18].

7.2.2.3 Deflagration to Detonation Transitions. Direct initiation of detonations in hydrogen–air mixtures is thought to be unlikely in nuclear reactor containments. On the other hand, deflagrations can evolve into detonations [19]. The evolution comes about because of turbulence in the burning gas. This turbulence, among other things [20], can entrain unburned gases into combustion region of a flame front. This leads to flame acceleration and additional turbulence at the leading edge of the flame front. Acceleration of the flame generates a shock wave sustained by the expansion of the hot, gaseous products of combustion. As the flame accelerates, the shock heating of the reactants becomes more significant. If shock heating is sufficient to initiate rapid chemical reactions of hydrogen with oxygen, the shock front and the chemical reaction zone become closely coupled and there is a transition to detonation.

Turbulence in the gas necessary for flame acceleration can be induced by partial blockages along the flame front pathway [21]. Turbulence can also be induced by the injection of hot gases from a vent into the combustible gas mixture [22]. Prediction of flame acceleration and the transition to detonation is not possible with current reactor accident analysis models. More empirical approaches have been adopted to identify locations and situations where flame acceleration and detonation are possible [23]. Dorofeev et al. [21] suggest that the hydrogen concentration must be such that the characteristic detonation cell size (Fig. 7.5) is at most 1/7 of the characteristic dimension of the combustion pathway. Ciccarelli and Dorofeev [24] suggest much more detailed and complicated methods to identify deflagration-to-detonation hazards.

7.2.3 Hydrogen Combustion Mitigation

Two approaches have been adopted to mitigate threats to containment structural integrity posed by hydrogen combustion [25]. The Mark I and Mark II containments for boiling water reactors are purged with nitrogen to prevent ignition of hydrogen. As demonstrated by the accidents at the Fukushima Daiichi plant, inerting the reactor containment does not prevent hydrogen combustion events in the reactor building should hydrogen leak from either the reactor coolant system or from the containment.

The second approach to mitigation of the hydrogen combustion threats is to burn hydrogen as it is released to reactor containments and prevent accumulations of hydrogen to levels where combustion could threaten the structural integrity of the reactor containment. In the USA, Mark III boiling water reactor containments and ice condenser containments for pressurized water reactors are equipped with glow plug igniters. These glow plugs initiate chemical reaction of hydrogen with oxygen from the air so there is not an accumulation of hydrogen. The glow plugs do require electrical power and special measures have been necessary to assure the availability of electrical power for the igniters in the event of a loss of all power for the plant (Station Blackout).

Canadian CANDU reactors and many types of reactors in Europe are equipped with catalytic hydrogen recombiners. These catalytic systems consist of finely dispersed noble metals (commonly palladium but in some cases platinum or platinum and palladium alloys) on large surfaces within structures to facilitate natural convection over the surfaces [26]. Molecular hydrogen, H_2, absorbs and probably dissociates on the surfaces to produce surface-bound atomic hydrogen. Atomic hydrogen will react readily with gaseous oxygen to form eventually water vapor. Heat evolved by the reaction raises the temperature of the surface to facilitate natural convection of gases over the surfaces.

The catalytic recombiners do not require external power sources. They will begin to catalyze the reaction of hydrogen with air at hydrogen concentrations of about 1%. The combiners are susceptible to "poisoning" [27]. That is, the reactive noble metal particles dispersed on the surfaces can react with other atmospheric gases to either prevent hydrogen absorption or to occlude the surface from interaction with hydrogen. Poisoning of the catalyst surfaces can occur during normal operations and especially during shutdown operations when maintenance activities can introduce vapors and fumes into the containment atmosphere. Sulfur containing vapor species are especially reactive toward catalysts. The catalysts need to be tested regularly for poisoning or degradation during plant shutdowns for maintenance. Some testing has been done to ascertain the susceptibility of catalysts to poisoning during the progression of reactor accidents [26].

7.3 CORE DEBRIS INTERACTIONS WITH COOLANT

Continued progression of reactor accidents beyond the stage of rapid steam oxidation of the zirconium alloy cladding leads to fuel melting. Inevitably this molten core debris will interact with liquid water. Attempts to arrest the accident progression will involve submerging the debris in water. Relocation of the molten debris from the core region will lead to interaction with residual water in the lower plenum of the reactor vessel. If accident management measures cannot arrest the accident, molten core debris will be expelled eventually into the containment where large quantities of water have accumulated from the pressure vessel blowdown and actions of the emergency core cooling system.

7.3.1 Core Debris Quenching

The interaction of hot, but still solid, core debris with water is expected to be benign in the sense that the interaction will not pose a prompt threat to the structural integrity of either the reactor pressure vessel or the reactor containment. Certainly, this was the case during the accident at Three Mile Island [28, 29]. QUENCH tests conducted in Germany also support this supposition [30]. Fragmentation of the core materials can lead to the formation of rubble beds with cooling characteristic quite different than those of intact fuel. The cooling of rubble beds depends on particle size and debris depth [31]. The dependence on debris size is shown in

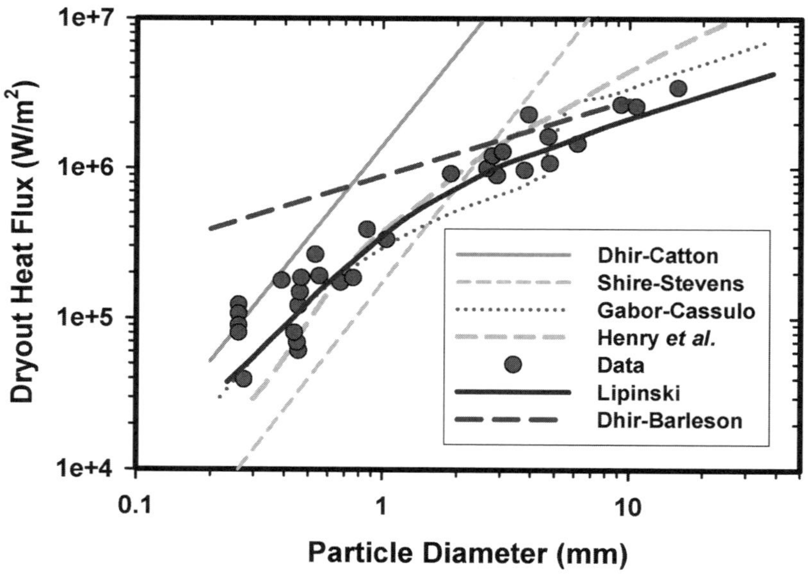

FIG. 7.7
COOLABILITY OF CORE DEBRIS WITH INTERNAL HEAT GENERATION AS A FUNCTION OF DEBRIS PARTICLE SIZE [31, AND REFERENCES THEREIN]

Fig. 7.7. Fortunately, it is expected that quenching hot, solid debris yields rather coarse debris that is easily cooled provided there is a continuing source of coolant.

Molten core debris poses more challenges. Certainly, the accident at Three Mile Island demonstrated that molten core debris is more difficult to quench. If the molten material can be quenched, it need not fragment into fine debris. Sustained cooling requires a continued source or water and means to extract heat from the reactor containment. More problematical is that interactions of high temperature molten materials with liquid water can lead to explosive events that do pose threats to both the reactor coolant system and to the reactor containment. These steam explosions are discussed in the next section.

7.3.2 Explosive Interactions of Core Debris with Water

Steam explosions, also known as "fuel-coolant interactions (FCI)," arise when high temperature molten materials interact with water to produce a shock wave. The phenomena entered the considerations of containment integrity during the preparation of the Reactor Safety Study [32] as a result of explosive magma interactions with ocean water in Iceland [33]. Steam explosions have plagued the ferrous and nonferrous metals industries for a very long time [34]. Buxton and Nelson [35] provide a review of the history of steam explosions including accidents in industrial settings.

An example of an explosive melt interaction with water is shown in Fig. 7.8. The explosion shown in this photograph was produced by dropping a 5-kg mass of molten iron and aluminum oxide produced by a thermitic reaction into a pool of cold water contained within a Plexiglas receiver.

The concern that arose originally in connection with accident progression under severe accident conditions dealt with molten core debris interactions with residual water in the reactor pressure vessel. The concern was that explosive interactions could either rupture the reactor pressure vessel head or generate missiles that would penetrate the reactor containment. Such a containment penetration early in the course of a reactor accident could result in very large releases of radioactivity into the environment. This immediate failure of containment as a result of core debris interactions with water became known as an "alpha-mode containment failure."

FIG. 7.8
STEAM EXPLOSION PRODUCED BY 5 KG OF MOLTEN IRON AND ALUMINUM OXIDE FALLING INTO A LARGE POOL OF WATER

Core debris that penetrates the reactor vessel will cascade into the reactor cavity which may contain water. There is, then, the possibility of steam explosions outside the reactor coolant system. There is less concern about the structural consequences of ex-vessel steam explosions because the reactor cavities are typically robust and numerous obstacles exist along pathways missiles might take to impact the reactor containment structure. Some concern remained that steam explosions might damage the pedestal supports for the reactor vessels in boiling water reactor containments. Collapse of the vessel could fail the containment and lead to release paths for radionuclides from the containment to the environment [37].

Following publication of the Reactor Safety Study, a period of intense investigations of steam explosions was undertaken by the nuclear community [38]. These investigations paralleled those undertaken especially by the aluminum and copper industry following steam explosions in industrial situations that damaged property and in a few cases resulted in the loss of life. Two configurations for molten core debris and water were considered. Most of the research attentions focused on the explosive interactions that were possible when high temperature melts were poured into liquid water. Much less attention was devoted to the so-called alternate contact mode or stratified contact mode [39] in which a water pool overlays a pool of molten core debris. It was established, however, that this alternate contact mode could lead to explosive interactions of molten material and water.

The research established the general phases of melt interactions with water. Photographs in Fig. 7.9 show some of these important phases. High temperature melts streaming into water break into droplets as a result of hydrodynamic instability. The breakup is into droplets that are on the order of 0.01 m in diameter. The droplets are surrounded by a film of steam since droplet temperatures are well into the film-boiling regime for liquid water. Visual evidence indicates that droplets vibrate and oscillate vigorously during this phase of mixing with the larger body of coolant. Of course, heat removal from the droplets by film boiling is low compared to nucleate boiling. Furthermore, radiant heat loss is limited by high temperature sources of the cloud of surrounding droplets also in film boiling. Consequently, this coarsely mixed configuration of large droplets in film boiling can persist as droplets settle to the bottom of a water pool, accumulate, and coalesce into a melt pool.

The course mixing phase of melt interactions with water can be interrupted. A pressure pulse passing through the system can cause the steam film surrounding a droplet to collapse. Much of the current

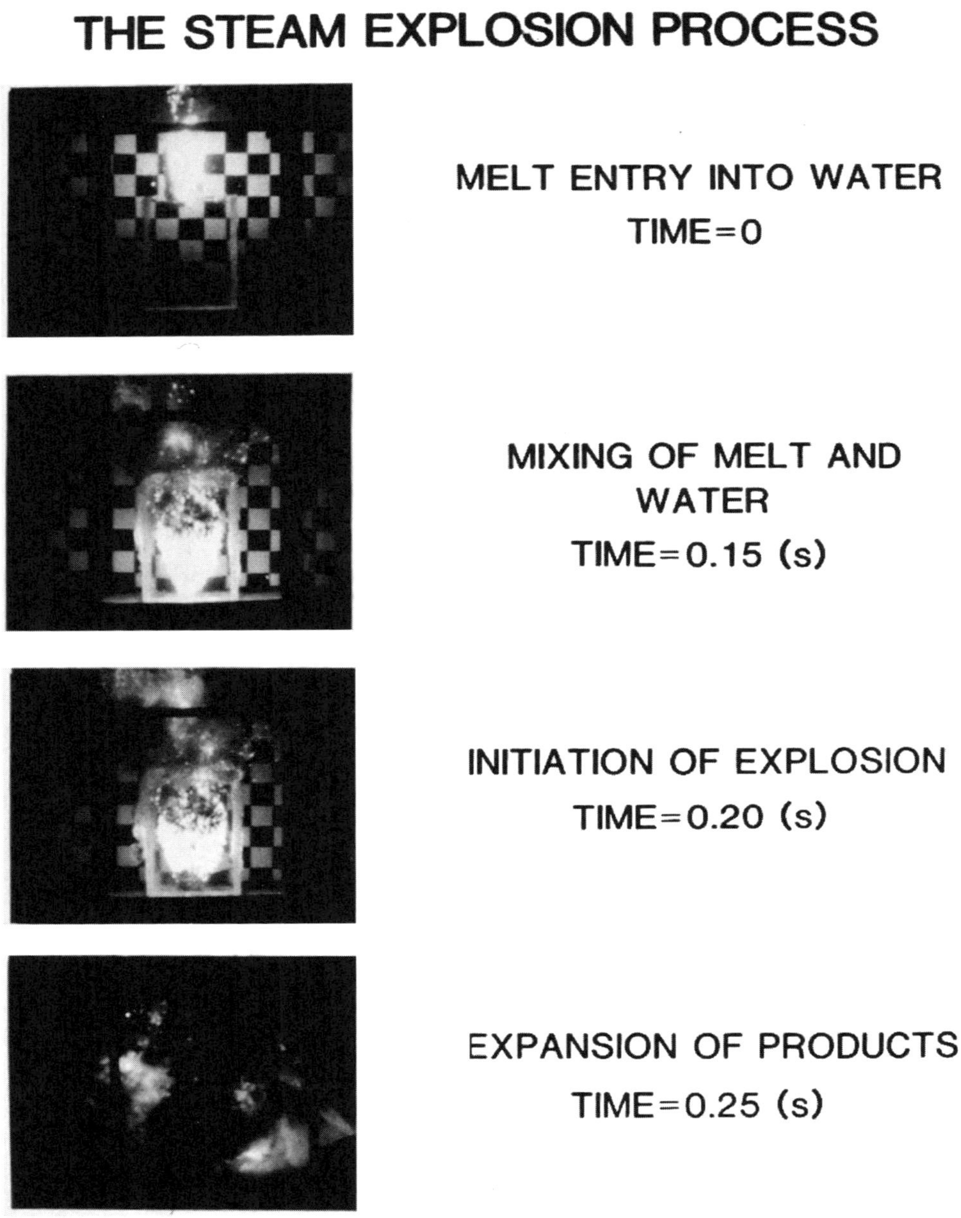

FIG. 7.9
MAJOR STAGES OF THE STEAM EXPLOSION PROCESS

understanding of what happens following collapse of the steam film comes from experiments with single molten droplets [40, 41] and models of these experiments [42]. Once the steam film collapses, there is intimate contact between the molten surface and liquid water. The resulting rapid heat transfer leads to rapid steam generation and causes melt to solidify at the surface. The solidified material is not structurally sound. It fragments and exposes more molten material to liquid water. This leads to further, rapid, steam production. The sudden generation of steam produces a pressure pulse that can trigger collapse of steam films on adjacent droplets. These adjacent droplets also quench, fragment and produce large amounts of steam and additional pressure pulses. Individual pressure pulses combine and accumulate at a propagation front to produce a shock wave. A plot of pressure against time recorded during a steam explosion experiment is shown in Fig. 7.10. Quite clearly, the sharp pressure rise is indicative of a shock wave.

The exact nature of the pressure pulse to trigger the energetic steam explosion process initially is not well understood [43]. Pressure pulses can be produced artificially to cause steam explosions once some level of coarse mixing of melt with water has been achieved. Magnitudes of the pressure pulses required for triggering are similar to pressure pulses expected to occur naturally during the course of severe accident progression by collapses of equipment and structures. Rarefaction as the settling coarse mixture of droplets approaches the bottom of the water pool often appears in experiments to be sufficient to initiate a steam explosion. Triggering of explosive interactions of molten material with water under pressurized conditions is more difficult because the pressurized steam film surrounding molten droplets is "stiffer." Stiffening the vapor blanket can also occur when steam surrounding the droplet reacts to produce hydrogen [44] which does not condense under conditions of interest.

Triggering indubitably introduces a stochastic character into the steam explosion process. One of the singular frustrations of the experimental investigations of steam explosion phenomena is that replicate experiments with boundary conditions as nearly identical as possible will yield quite different results. In addition to this undiagnosed variability, there is variability in the probability of steam explosions with the composition of the melt material. Molten aluminum, aluminum oxide, and molten iron appear to be quite likely to produce steam explosions when mixed with water. On the other hand, molten uranium dioxide, despite a much higher melting temperature, does not readily undergo explosive interactions with water unless some external pressure pulse is applied to initiate the interaction. This variability in explosiveness with melt composition may well have to do with formation of a solidified crust at the melt-vapor interface during the coarse mixing phase of melt interactions with water. Hot solid materials even when finely divided prior to addition to water do not cause explosive interactions. Slow heat transfer and consequent slow steam generation when water interacts with hot solids may not lead to pressure pulses adequate to propagate the vapor film collapse throughout the mixture. A crust at the interface of melt and vapor of sufficient thickness to maintain integrity during vapor collapse may also retard heat transfer and steam generation enough to retard the propagation of the vapor

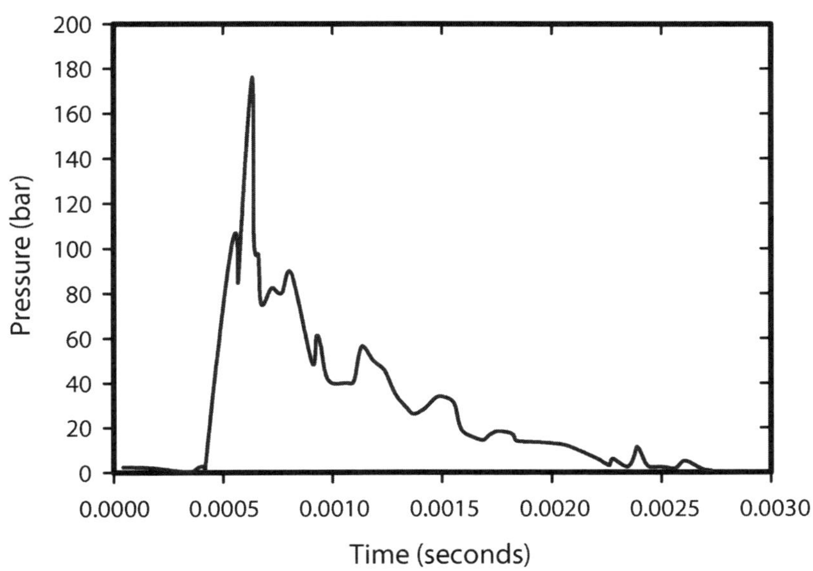

FIG. 7.10
PRESSURE AS A FUNCTION OF TIME DURING A STEAM EXPLOSION EXPERIMENT

collapse process. Current thinking is that crusts on the order of 2×10^{-4} m thick may be sufficient to prevent steam explosions. Of course, thick crusts are most likely to form on low thermal conductivity, high melting point materials such as uranium dioxide.

The development and propagation of the shock wave during steam explosions remains a topic of research. Early models were based on an analogy to shock fronts produced in the detonation of gas mixtures [45]. The idealized configuration assumed a steady shock wave moving through a homogeneous mixture. Mechanical and thermal equilibrium were assumed to be reached in a narrow zone behind the shock wave. Such modeling predicted shock pressures and propagation velocities well in excess of those observed in experiments. Critical assumptions made in the model included rapid droplet fragmentation and complete mixing of the hot and cold fluids in the reaction zone.

Current modeling of the propagation phase of steam explosions is still based roughly on an analogy to gas detonations, but some of the original assumptions have been relaxed. The essential challenge faced by modelers is that there is no equivalent to Schlieren photography for imaging the shock front propagating through the melt–water mixture as can be done for combustible gases. Details of both the mechanisms and the rates of droplet fragmentation and steam production remain topics of discussion.

A vast number of experiments to investigate steam explosion for purposes of nuclear reactor safety has been done. Corradini et al. [38] provide an extensive list of tests and conditions. Melt masses have varied from very small for single droplet experiments to in excess of 100 kg. The testing has been largely heuristic in nature. Variables such as melt composition, melt–water volume ratio, water depth, water subcooling and the like have been studied. The stochastic nature of the steam explosion process has greatly complicated development of an informative data base.

An important issue for the analysis of steam explosion consequences is the efficiency with which thermal energy in the high temperature melt is converted into mechanical work. There is, of course, a thermodynamic limit known as the Hicks-Menzies limit [46] of about 29%. Measured conversion efficiencies are always much lower than this thermodynamic limit. Some typical values for experiments of the type shown in Fig. 7.8 are plotted in Fig. 7.11 against the ratio of the water mass divided by melt mass. Results shown in this figure illustrate also the variability of data obtained in steam explosion experiments. Conversion efficiencies shown in the figure are

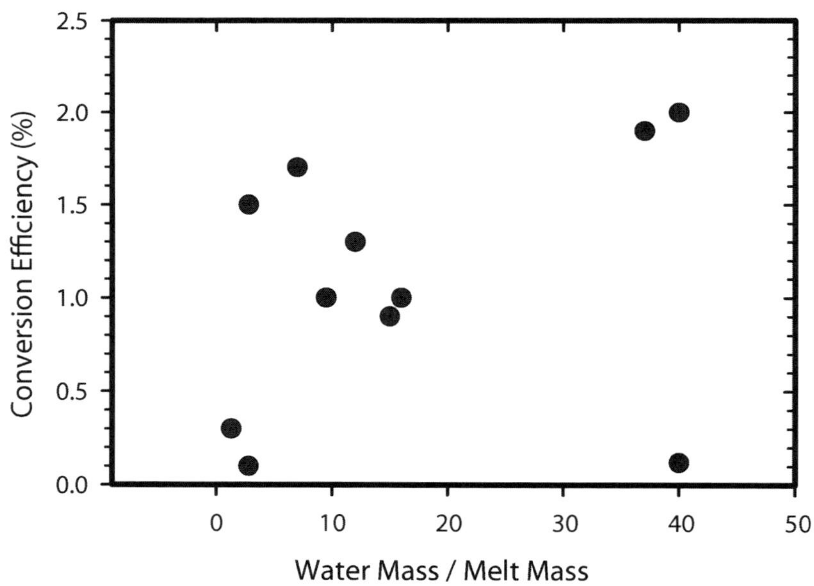

FIG. 7.11
THERMAL TO MECHANICAL EFFICIENCIES DETERMINED IN STEAM EXPLOSION EXPERIMENTS VARYING THE MASS RATIO OF COOLANT AND MELT

less than two percent. The largest conversion efficiency reported in experiments pertinent to reactor accident analyses is about 12% in a test with a single droplet. It does appear that conversion efficiency can be increased by system pressurization though this may well inhibit triggering of steam explosions.

Though the efficiencies with which thermal energy is converted to mechanical energy in steam explosions are low, there is a large amount of thermal energy available in reactor core melts. Consequently, much attention has been paid to identifying the limits on the total mechanical energy that could be released during steam explosion processes. Limits to the extent of coarse mixing have been examined [36, 47]. The reliability of triggering by rarefaction at the water pool base so that only melt suspended in water can participate in an explosion has been examined [43]. Computer models of the complicated three or four phase hydrodynamic and heat transfer situation are being developed to assess limitations on the phenomenon of steam explosions [48].

The likelihood that steam explosions would threaten the integrities of the reactor pressure vessel and the reactor containment was examined both in the aftermath of publication of the Reactor Safety Study and as part of the preparation of the NUREG-1150 study of severe reactor accidents in representative US nuclear power plants [49]. In the Reactor Safety Study, the alpha containment failure mode was assigned somewhat arbitrarily a probability of 1% given the occurrence of a severe accident initiating event. Expert judgments used for analyses in the preparation of the NUREG-1150 study assigned lower conditional probabilities when core degradation and melt relocation took place within a pressurized reactor pressure vessel. Subsequent expert opinion elicitations [50] led to a consensus that conditional probabilities for reactor pressure vessel rupture by steam explosions were exceptionally low even if the system was not pressurized. This same expert panel did note that small steam explosions are quite likely to occur. Though they are not likely to be either large enough or involve sufficiently efficient thermal to mechanical energy conversion to threaten structural integrity of either the reactor pressure vessel or the reactor containment, they could affect the continued progression of a severe accident.

The steam explosion process quenches and finely fragments the melt that interacts with water. The size distribution of debris produced by steam explosions is compared in Fig. 7.12 with the size distribution of debris produced by simple, nonexplosive quenching of a similar melt. More than 96% of the mass of debris from the simple quenching process has particles sizes in excess of 0.5 mm based on simple sieve analysis.

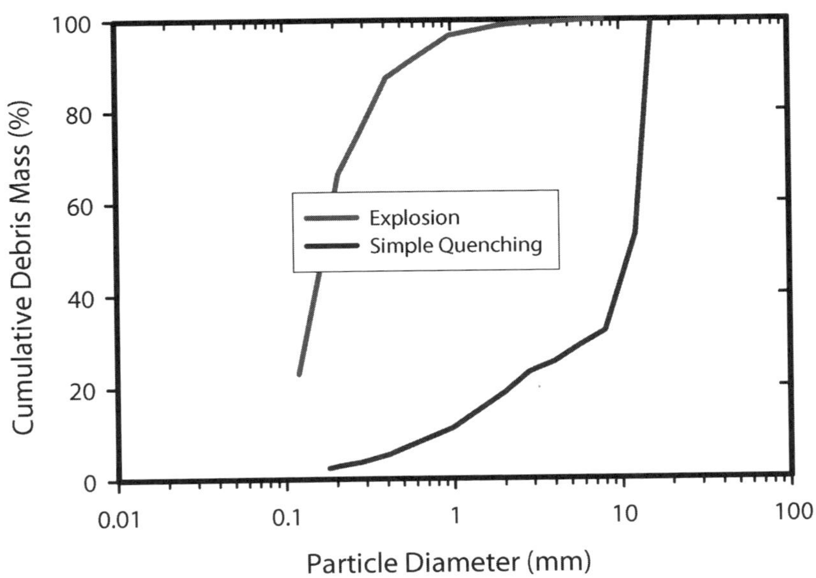

FIG. 7.12
SIZE DISTRIBUTION OF SOLID DEBRIS COLLECTED FOLLOWING A STEAM EXPLOSION AND FOLLOWING A SIMPLE, NONEXPLOSIVE MELT QUENCHING EVENT

Only 37% of the debris collected following a steam explosion has such a large particle size. Most of the debris from the steam explosion is less than 0.1 mm in size.

The debris produced by a steam explosion is fine enough that sufficient accumulations of the debris within areas of either the reactor pressure vessel or the reactor containment will not be coolable even if submerged in water [31]. Flooding limits the ability of liquid water to penetrate deeply into beds of fine particulate. Because water is unable to penetrate the bed, the bed will dry; the debris will reheat and eventually melt. The hot debris will then interact with the substrate which for many reactor containments will be concrete. The interactions of core debris with structural concrete are discussed in Section 7.5. Included in these discussions are the effects of water on core debris interacting with concrete.

A concern at the time steam explosions were identified as a potential hazard to containment integrity was that steam explosions could loft finely divided reactor core materials into the containment atmosphere. Air might chemically react with the debris particles. Such air reactions were expected to release large fractions of the radioactive inventory of ruthenium which had not been considered to be a major cause of reactor accident consequences. This contention has largely been refuted by more detailed analyses [51].

Steam explosions, even ones that do not threaten the structural integrity of either the reactor pressure vessel or the reactor containment but do affect the progression of the reactor accident, are not routinely considered in reactor accident analyses. Computer models of the steam explosion process are not now viewed as predictive in capability. There are, then, ongoing investigations of steam explosions within the nuclear community. Experimental studies are underway in Italy [52] and in Korea [53]. Because of the poorly resolved status of the technical issues of steam explosions, safety analyses rely heavily on expert opinion [50] and uncertainty analyses [37, 54].

7.4 HIGH PRESSURE MELT EXPULSION AND DIRECT CONTAINMENT HEATING

Direct containment heating is a threat to containment integrity that did not enter into reactor safety discussions until after the accident at Three Mile Island. Prior to that accident, severe reactor accidents involving extensive reactor fuel degradation were treated as extensions of design basis accidents. The most challenging of the design basis accidents is one initiated by rupture of the main coolant system. Consequently, it was usually assumed in the analysis of severe accidents that the reactor coolant system would be depressurized by the time molten core debris could penetrate the reactor vessel and cascade into the reactor containment.

The accident at Three Mile Island showed that accidents involving severe core degradation could be initiated by means other than a large coolant system rupture and that core degradation and relocation to the lower head of the reactor pressure vessel could take place in a highly pressurized reactor coolant system. This raised the question of what would happen if core debris penetrated the reactor vessel while the reactor coolant system was pressurized. Significant debates erupted over the modes of vessel failure. An important research program on lower vessel head failure was sponsored by the U.S. Nuclear Regulatory Commission and some members of the Organization for Economic Cooperation and Development (OECD) [55]. These tests showed that core debris in the lower plenum could lead to extensive creep of the reactor vessel. For reactor vessels with bottom head penetrations, the most likely mode of failure was through the penetration orifice once the weldment tore due to vessel creep.

Initial speculations were that core debris penetrating the reactor pressure vessel would accumulate in the reactor cavity. The blowdown of steam and hydrogen from the reactor vessel once debris was expelled could be expected to entrain and disperse debris out of the reactor cavity and across the reactor containment floors. There was the hope that this dispersal would reduce local accumulations of such debris so that it was easily quenched, if not by air cooling, then by spray waters [56].

This hopeful expectation of debris dispersal was dashed by the first realistic experiment done by Tarbell et al. [57]. These investigators prepared an 80 kg thermitic melt ($Fe + Al_2O_3$) in a vessel pressurized with nitrogen to about 80 bar and connected to a 1:20 scale model of the reactor cavity in a large dry containment. The cavity model included a wide, inclined opening for instrumentation tubes mimicking that found in Westinghouse and Babcock and Wilcox plants (see Fig. 7.13). Photographs taken from the motion picture record of the outdoor test are shown in Fig. 7.14. Debris was not just entrained during gas blowdown. It was fragmented into droplets and lofted over 38 m into the air above the reactor cavity. The lofted debris chemically reacted with the atmosphere. Substantial aerosol was formed during the melt ejection process.

Detailed examination of pressure and temperature records from the test showed that blowdown of gases in the pressure vessel began before melt was fully expelled. As expected [58, 59], streaming melt considerably enlarged the release pathway provided in the pressure vessel. Eventually, gases were able to punch through the melt layer and there was simultaneous expulsion of melt and gas from the vessel. Subsequent experiments [60] monitored by x-ray showed that the expelled melt was also disrupted by the effervescence of gas dissolved in the melt at high pressure (see Fig. 7.15).

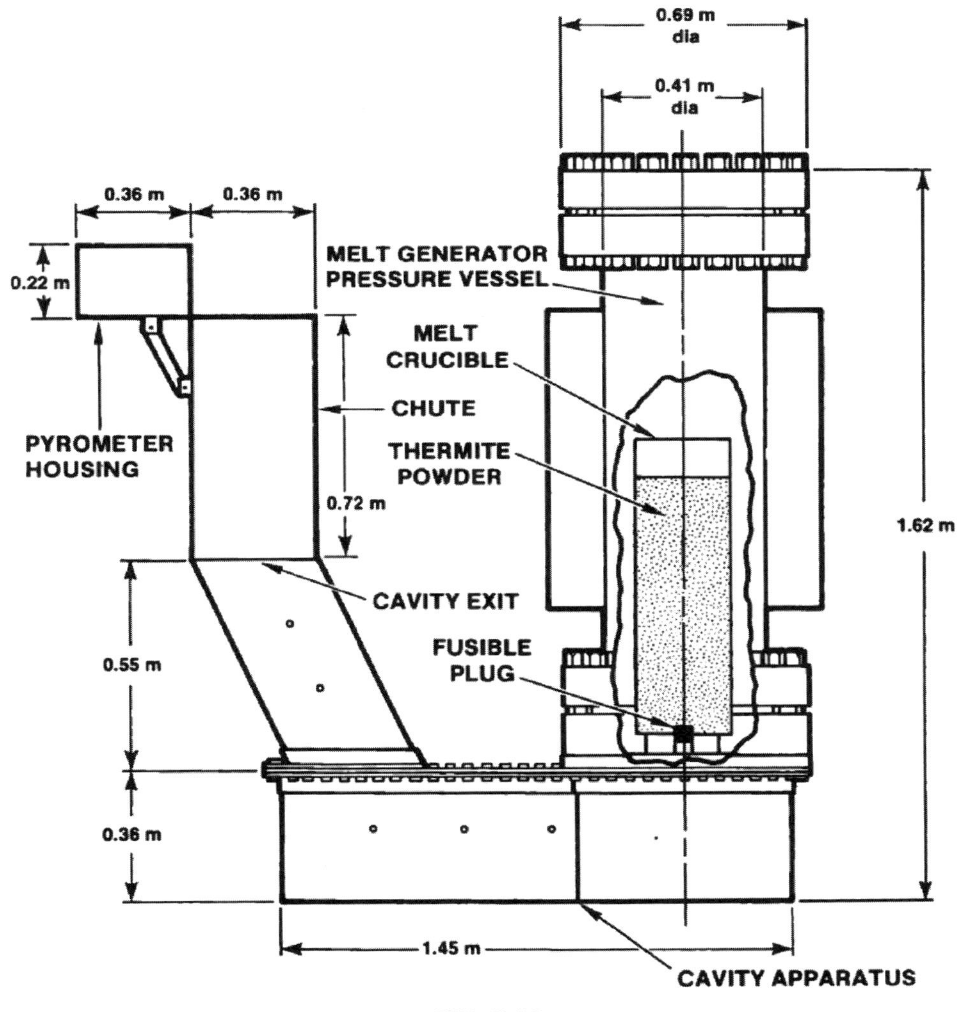

FIG. 7.13
PRESSURE VESSEL AND REACTOR CAVITY MODEL USED FOR THE FIRST TEST OF HIGH PRESSURE MELT EJECTION (HPME)

Time = 0.1 seconds

Melt just emerging from reactor cavity

Time = 0.5 seconds

Melt ejection from reactor cavity

Time = 1.0 seconds

Ejected melt reaches maximum

elevation, > 38 meters

Time = 1.5 seconds

Separation of melt particles and

aerosol cloud complete

FIG. 7.14
PHOTOGRAPH SEQUENCE OF FIRST TEST OF THE EXPULSION OF A HIGH TEMPERATURE MELT FROM A PRESSURIZED VESSEL INTO A SCALE MODEL OF THE REACTOR CAVITY OF A LARGE DRY CONTAINMENT FOR A PRESSURIZED WATER REACTOR [57]

This first test made it evident that expulsion of molten core debris from a pressurized reactor vessel could lead to rapid heat transfer and consequent pressurization of the containment atmosphere. The high pressure melt ejection threatened containment failure early in the progression of a reactor accident — a situation that could lead to substantial accident consequences. The balance of this section will discuss research into the high pressure melt ejection (HPME) and direct containment heating (DCH) process.

7.4.1 Experimental and Analytic Investigations of Direct Containment Heating

The threat of early containment failure due to DCH prompted rapid response within the reactor safety community. The U.S. Nuclear Regulatory Commission formed a working group [61]. Significant experimental investigations were undertaken using both high temperature melts and simulant materials [62]. At least three analytical efforts were initiated to model test results and predict the effects of DCH on reactor accidents.

Experimentation with high temperature melts used experimental configurations much like those used in the first test, but the apparatus was housed inside the Surtsey experimental vessel. Surtsey is a 103-m^3 vessel rated for both pressurized and elevated temperature operations. Use of this facilitate made it possible to quantify atmospheric heating and pressurization as a result of DCH. Tests were conducted eventually using

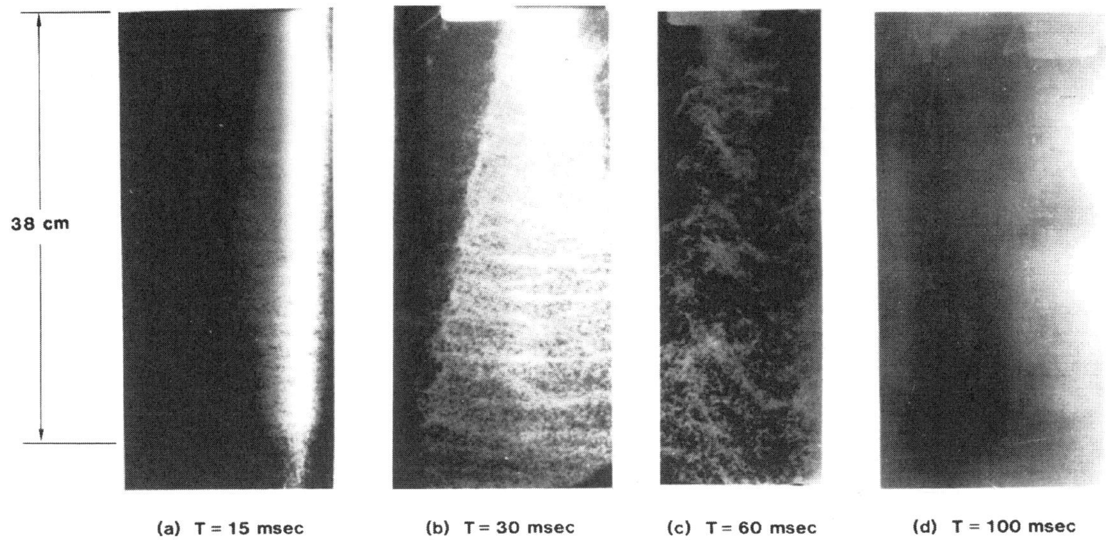

FIG. 7.15
SEQUENCE OF X-RAY PHOTOGRAPHS TAKEN OF A HIGH TEMPERATURE MELT BEING EXPELLED FROM A PRESSURIZED VESSEL [60]

the reactor cavity configuration for large, dry PWR containments, subatmospheric containments, ice condenser containments and the unique configurations for Combustion Engineering plants.

Very quickly and to no one's surprise, it was established that expelled particulate debris could serve as an ignition source for hydrogen that might accumulate in the containment atmosphere as a result of core degradation processes within the reactor coolant system early in an accident [63, 64, 65]. This meant further loads could be placed on containment. It was also established that pressurization of the reactor coolant system needed to be on the order of 10–25 bar to obtain safety significant direct containment heating [66]. Consequently, no high temperature tests were done with boiling water reactor containment configurations. There was a faith that the automatic depressurization system of boiling water reactors would assure that HPME and DCH would not be a concern.

The blowdown of gases from the reactor pressure vessel entrains molten core debris and breaks debris into droplets. A representative particle size distribution from a high temperature test is shown in Fig. 7.16. The median debris droplet size is about 1 mm. Such fine particles can of course transfer heat to the gas phase quite efficiently. Furthermore, the debris could include substantial amounts of readily oxidized metals. The metals could include zirconium from fuel cladding and structures that were not oxidized by steam during the core degradation phase of an accident. Molten stainless steels (Fe, Cr, and Ni) from in-vessel structures could also make up a substantial portion of the debris. Mild steel from the reactor pressure vessel was thought to be a minor constituent of the debris. These metals could be oxidized exothermically by the blowdown steam to form hydrogen. They could also be oxidized by reaction with oxygen in the containment atmosphere to produce even more chemical heat. At the temperature of the molten metals when expelled from the reactor pressure vessel, the rates of oxidation reaction are limited by mass transport of oxidant to the reactive metal surface. Because of the small droplet sizes, mass transport rates could be quite rapid.

An important set of high temperature tests were reported by Allen et al. [67, 68]. They examined the pressurization of the Surtsey facility as a function of the flight path available to the debris which was varied from 0.91 to 7.7 meters. Findings from these experiments suggested that structures and compartments above the opening to the reactor cavity would have an important bearing on the extent of direct containment heating. Experimental investigations of the direct heating phenomena thereafter included simulations

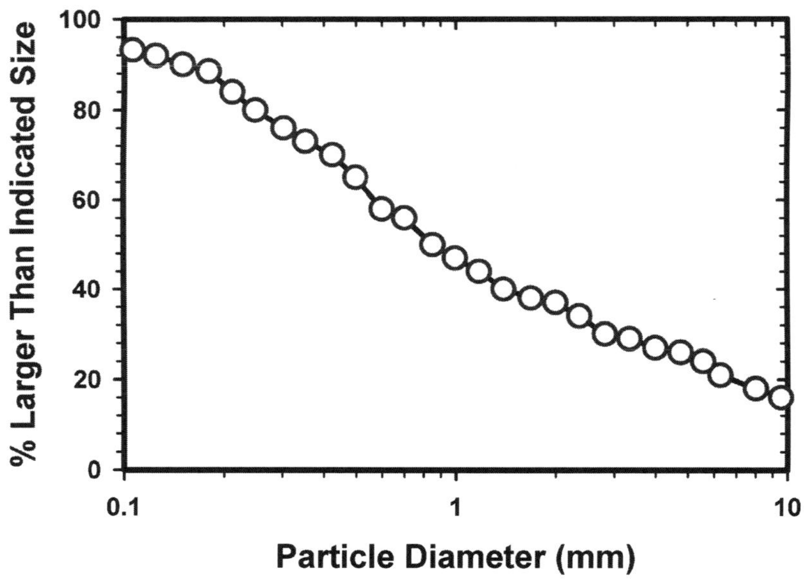

FIG. 7.16
COMPOSITE SIZE DISTRIBUTION OF DEBRIS COLLECTED AFTER HPME EXPERIMENTS WITH THERMITIC MELTS IN THE SURTSEY FACILITY

of compartments or structures above the reactor cavity opening. Experiments done with compartments and structures limited the flight paths of droplets and prevented much of the expelled debris from reaching the larger Surtsey volume. It was found that droplets that encountered structures above the reactor cavity opening were quickly removed from the flow stream out of the cavity and did participate significantly in the heating of the test facility atmosphere.

In all the high temperature tests, it was observed that large amounts of aerosol were produced. Photomicrographs of particles produced in a test are shown in Fig. 7.17. Typically, three size classes of particles were collected in tests [69, 70]. Very fine particles (<1 μm) composed of yet smaller 'primary particles' were found. These particles were thought to come from the condensation of vapors evolved from the high temperature melts and debris particles. These very fine particles were found to be enriched in volatile simulants of fission products incorporated into the high temperature melts. Larger particles (~5 μm) were spherical and were thought to be solidified melt droplets that had been produced by mechanical processes including the bursting of bubbles from effervescence of gases dissolved in high temperature melts. The third size of class of yet larger, irregular particles was thought to come from comminution of solidified melt. The spherical particles and the larger particles were not enriched in volatile melt constituents.

Modeling was being developed as the experimental investigations progressed. Early models were largely adiabatic in character and yielded predictions of pressures inconsistent with observations. Eventually, three models that considered heat and mass transport limitations emerged as useful for the analysis of reactor accidents:

- Two-cell Equilibrium Model (TCE) [71, 72]
- Convection Limited Containment Heating (CLCH) [73]
- CONTAIN DCH models [74, 75]

Modeling showed that the steam and hydrogen mixture expelled from the vessel along with core debris either by entrainment or by aspiration makes the biggest contribution to the pressurization by direct containment heating. Steam and hydrogen that is mixed with the debris passing through the cavity is termed 'coherent' in the parlance of the field and defined somewhat differently in various models. The contribution to containment pressurization comes about by heat transfer from the debris droplets to the gas and by exo-

FIG. 7.17
PHOTOMICROGRAPHS OF AEROSOL PARTICLES COLLECTED DURING DCH TESTS WITH HIGH TEMPERATURE MELTS

thermic reactions of the metallic constituents with either the steam or air entrained into the gas flow. Most of this heating of the gas occurs in the reactor cavity and subcompartments above the opening to the reactor cavity. The amount of heat transfer from the droplets to the gas is limited by thermal saturation of the gas. Similarly, the extent of chemical reaction is limited by the availability of oxidant as long as droplets do not successfully negotiate passage to the larger containment dome.

7.4.2 Resolution of the Direct Containment Heating Issue

Resolution of the DCH issue was focused on the question of whether high pressure melt ejection and direct containment heating could pose a significant threat to containment integrity early in a reactor accident immediately following melt penetration of the reactor vessel. A large number of experiments at various scales using both high temperature molten materials and low-temperature liquid were performed. Some of these experiments have been mentioned. Pilch [72] provides a more complete list. From these experiments and the associated modeling a large number of variables had been found to affect the containment pressurization as a result of HPME and DCH. Predominant among these were pressure in the reactor coolant system at the time of vessel rupture, mass of core debris expelled from the reactor vessel, the fraction of the core debris that was metallic and could react exothermically with steam and oxygen, the geometry of the reactor cavity and the vent pathways to the containment, and the amount of hydrogen that had accumulated in the containment prior to vessel penetration.

The strategy for resolution of the issue was to use the computer models calibrated by comparison to the many experiments done at various scales to predict pressurization at the much larger scale of reactor accidents [71, 76]. Multiple analyses were done considering a variety of accident scenarios where the reactor coolant system was thought likely to remain pressurized during core degradation such as small-break loss of coolant accidents and station blackout accidents. Replicate analyses were done sampling values of important variable such as molten core debris mass from hypothesized probability distributions. The analyses had to be done for major classes of pressurized water reactor containments (large dry, subatmospheric, ice condenser, and combustion engineering) because of the importance of geometry especially in the reactor cavity and the compartments above pathways out of the cavity. Combustion engineering plants do not have lower head penetrations and consequently do not have the wide inclined pathways found in other plants. The predominant escape route for debris expelled into the reactor cavities of these plants is upward around the reactor vessel into containment dome without intervening structures and compartments found in other kinds of reactor containments [77].

Unfortunately, some of the important variables in determining the containment pressurization could not be directly measured. This included the mass of core debris present in the lower plenum at the time of reactor pressure vessel failure and the metal content of this debris as well as the reactor coolant system pressure. Values of these important variables had to be derived instead from results of accident analysis codes. These models of reactor accidents have continued to evolve over the years, and it is not certain that distributions for variables important to DCH predicted today would be very similar to those that were used in the issue resolution process.

The distributions of pressurizations derived from the multiple analyses were then compared to the distributions of containment fragilities [78]. It was concluded that large, dry reactor containments and subatmospheric containments are very likely to survive pressurization by high pressure melt ejection and direct containment heating. Containments for Combustion Engineering plants are more vulnerable because of the cavity design and venting or the cavity to the containment dome. Conclusions were harder to reach for ice condensers because of the probability that the reactor cavity would be deeply flooded at the time a vessel failure and this condition had not been investigated in DCH experiments.

7.4.3 Ongoing Research

There is no longer ongoing research in the USA on direct containment heating. HPME and DCH are not routinely considered in accident analyses. Indeed, neither the TCE nor the CLCH computer codes that were so pivotal in the risk resolution of the direct containment heating issue is now readily available. To an important extent, this is because accident analyses predict that the reactor coolant system will not remain pressurized throughout the core degradation process until vessel failure though this certainly happened nearly to vessel failure in the Three Mile Island accident [79]. Procedures call for operator depressurization of the reactor coolant system to make available low pressure sources of water to arrest accidents and to take advantage of the greater cooling capability of low pressure water. Even if depressurization by the operators is not possible, many accident analysis computer codes predict that heat transfer from the degrading core to the primary pressure boundary will lead to a creep failure of that boundary prior to melt penetration of the reactor pressure vessel.

There is some continuing research on direct containment heating underway in Europe. Reactor cavity configurations are different in European reactors [80]. Many of these reactors have small pathways out of the cavity. It is more likely in these cases that expelled core debris will be forced up around the reactor vessel into the open spaces of containment. A new generation of computer codes for the modeling of experiments and the prediction of containment pressurization is emerging from this research [81].

7.5 INTERACTION OF CORE DEBRIS WITH CONCRETE

The previous sections of this chapter dealt with accident phenomena that posed the threat of containment failure early in the progression of a severe reactor accident. In this section, the longer term threats to containment integrity are discussed. If heat cannot be removed from the containment, the continuing energy release from the decay of radionuclide will result in containment failure either by overpressurization or over heating whether or not the core debris is quenched and cooled. The pressurization and heating of the containment is complicated if core debris outside the reactor coolant system can interact with structural concrete.

The potential for core debris interactions with concrete has been long recognized in reactor accident analyses. In the Reactor Safety Study [32], penetration of the basemat by core debris interactions with concrete was the default containment failure mode should containment survive other threats to its integrity. The possibility of core debris interactions with structural concrete gave rise to the term "China Syndrome" made famous by a motion picture released shortly before the accident at Three Mile Island. Indeed, core debris interactions did occur during the accident at Chernobyl, and there is current speculation that such interactions may have occurred in some of the units involved in the reactor accident at Fukushima Daiichi [82].

Early modeling of core debris interactions with concrete was based on very little experimental data. Guidance from the modeling came from experiences with concrete in fires. Consequently, the model in the Reactor Safety Study assumed that initial attack on the concrete was by spallation. Subsequent attack was expected to be by melting. Gases were assumed to bypass the melt because of the hydrostatic head provided by the molten core debris and there was not a major release of radioactive material to the containment as a result of the melt interactions with concrete. As will be discussed further below, these assumptions were disproved in experimental investigations especially those following the accident at Three Mile Island, and it was shown that threats posed to containment during core debris interactions with concrete are more than basemat penetration.

Before launching into discussions of core debris interactions with concrete, it is useful to review some of the salient features of core debris and concrete.

7.5.1 Nature of Ex-Vessel Core Debris

The process of core degradation results in core debris slumping into the lower plenum of a reactor pressure vessel. Such core debris was quenched by residual water in the accident at Three Mile Island. If the core debris in the lower plenum cannot be quenched and permanently cooled, it will eventually penetrate the reactor vessel and fall into the reactor cavity in pressurized water reactors or the pedestal region in the case of boiling water reactors. If the expelled debris is not dispersed by blowdown of steam and hydrogen from the reactor coolant system or does not encounter large volumes of water below the reactor pressure vessel, it will promptly interact with the concrete.

The core debris interacting with the concrete will consist of two major metallurgical phases. There will be an oxide phase composed predominantly of uranium dioxide fuel and the zirconium dioxide product of steam oxidation of cladding on the fuel. There will be a metal phase composed predominantly of stainless steel from in core structures and structures melted during the progression of the accident. There may be mild steel from the penetration of the reactor vessel. There will also be metallic zirconium that escaped steam oxidation during the core degradation within the reactor coolant system. The contribution of zirconium to the metallic phase of core debris is likely to be especially important in core melts in boiling water reactors because of the large mass of zirconium structures in addition to fuel cladding. Also, boron and carbon from control blades will be present in the metallic phase of core debris produced by boiling water reactors. The metal phase and oxide phases will be quite fluid when molten (viscosity < 0.1 Pa-s). the densities of the two phases will be sufficiently different that they will tend to stratify with the metallic phase being the more dense [83].

Both phases of the core debris will contain fission products and other radionuclides. Most of these radionuclides will be in the oxide phase. Some such as isotopes of ruthenium and molybdenum will partition preferentially into the metal phase. There will be, then, a continuing decay heat source in both phases.

If, on the other hand, core debris expelled from the reactor vessel, quenched and dispersed, it may be coolable if the particle sizes are sufficiently coarse. If the particles are small as they are following steam explosions or high pressure melt ejection it may not be possible to permanently cool the debris. Substantial accumulations of core debris will heat to sufficient temperatures that they too begin an attack on concrete.

In both modes of core debris attack on concrete, there will be continuing additions of core debris from the reactor vessel. Clad reactor fuel about the periphery of the core that is slow to degrade will continue to heat, chemically react and slump from the pressure vessel into the reactor cavity. The continued accumulation of core debris in the reactor cavity will be especially likely if air from the containment circulates into the reactor vessel and reacts with zirconium alloy cladding on the fuel.

7.5.2 Nature of Concrete

Structural concretes used in nuclear power plants are predominantly coarse aggregate, fine aggregate and Portland cement. The aggregates used are diverse and depend largely on local availability. The aggregates are broadly classed as either siliceous or calcareous. Siliceous aggregates include basalt which is a eutectic rock [84] that melts at relatively low temperatures and more siliceous materials such as granite and oolite. Calcareous aggregates are usually limestone ($CaCO_3$) though dolomite ($MgCa(CO_3)_2$) may be used. Fine aggregates are usually silica (SiO_2) sand though crushed limestone is also used.

Chemical compositions of three representative concretes are shown in Table 7.1. Of important note are the water and carbon dioxide contents of the concretes. These are the constituents that produce gas when concrete interacts with core debris. Figure 7.18 is a thermogram of a limestone concrete showing the residual weight of a concrete specimen as a function of temperature. The initial weight loss taking place around 100°C is vaporization of molecular water in the concrete which is sometimes called "gel water." The contribution of gel water depends on the ambient relative humidity and can be as high as 8 weight percent. The second weight loss event at about 450°C is also due to the vaporization of water, but in this case it is water bound in the concrete as hydroxides such as calcium hydroxide ($Ca(OH)_2$). The final weight loss shown in the figure at about 800°C is the very endothermic decomposition of calcium carbonate to produce calcium oxide and carbon dioxide.

Concrete is a heterogeneous material so melting can be complicated. Typically melting will begin in the cementious phase at about 1100°C. Melting may not be complete until temperatures in excess of 1600°C are reached especially when concretes rich in silica or calcium are used. Melting leads to exothermic reactions of silica with bases such as calcium oxide and alkali metal oxides. These irreversible reactions complicate determination of the enthalpy of melting.

7.5.3 Experimental Investigations of Core Debris Interactions with Concrete

There have been many investigations of high temperature melts with concrete. The first large scale investigations were undertaken by Powers [85] and by Powers and Arellano [86]. These investigations focused on the interactions of the metallic phase. A photograph taken from an experiment is shown in Fig. 7.19. The important safety features of melt interactions with concrete are illustrated in this photograph. The attack is predom-

TABLE 7.1
CHEMICAL COMPOSITIONS OF REPRESENTATIVE CONCRETES

	Composition (Weight Percent as Indicated Oxide)		
Oxide	**Basaltic Coarse Aggregate, Silica Fine Aggregate Concrete**	**Limestone Coarse and Fine Aggregate Concrete**	**Limestone Coarse Aggregate and Silica Fine Aggregate Concrete**
SiO_2	54.7	3.6	35.7
CaO	8.8	45.4	31.2
Al_2O_3	8.3	1.6	3.6
MgO	6.2	5.7	0.5
Fe_2O_3	6.3	1.2	1.4
K_2O	5.4	0.7	1.2
TiO_2	1.1	0.12	0.2
Na_2O	1.8	0.1	0.8
MnO	-	0.01	0.03
Cr_2O_3	-	0.004	0.014
H_2O	5.0	4.1	4.8
CO_2	1.5	35.7	22.0

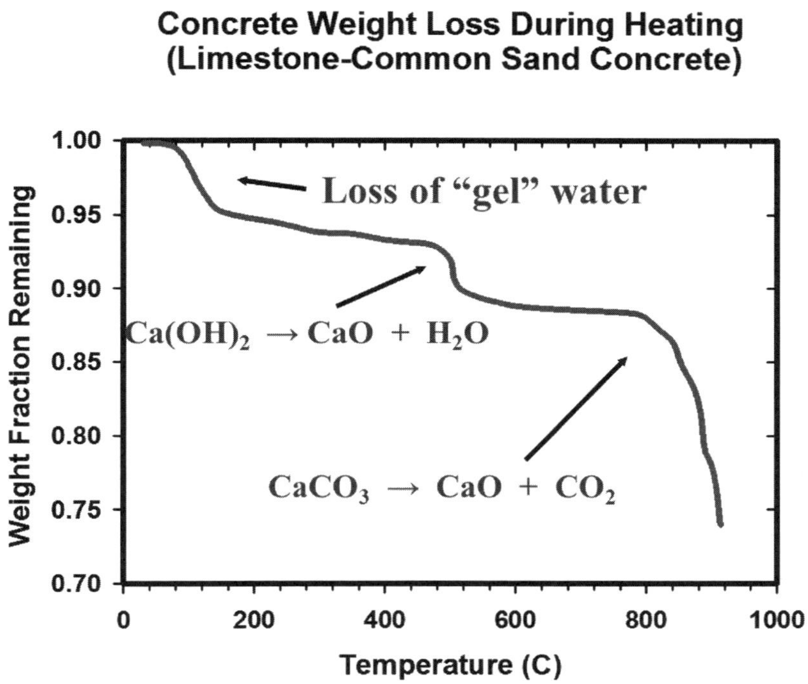

FIG. 7.18
THERMOGRAM OF CONCRETE COMPOSED OF LIMESTONE COARSE AGGREGATE AND SILICA FINE AGGREGATE HEATED AT 10°C/MIN IN FLOWING NITROGEN

inantly thermal and there is very little concrete spallation. A quasi-steady thermal profile marked by concrete decomposition events and melting passes into the concrete. Gases produced by the concrete decomposition sparge through the melt and chemically react to form combustible hydrogen and carbon monoxide:

$$Zr + 2\,H_2O \rightarrow ZrO_2 + 2\,H_2$$

$$Zr + 2\,CO_2 \rightarrow ZrO_2 + 2\,CO$$

$$2\,Cr + 3\,H_2O \rightarrow Cr_2O_3 + 3\,H_2$$

$$Fe + H_2O \rightarrow FeO + H_2$$

etc.

The exothermic reactions provide chemical heating that in the case of reactor accidents would supplement decay heating to sustain the attack on concrete.

Powers and Arrellano showed that radial attack on the concrete relative to axial attack was affected by gas generation [87]. By using x-rays to monitor the interaction in real time, they showed that melt contact with the concrete was interrupted episodically by gas bubble formation [88].

As shown in Fig. 7.19, there is vigorous generation of aerosol throughout the interaction of high temperature melts with concrete. Much of the aerosol comes from the more volatile constituents of concrete such as alkali metal oxides, iron oxides, and silicon monoxide. Many of the tests have included fission product simulants and have shown that these are released including some release of more refractory (less volatile) radionuclides such as cerium and lanthanum.

FIG. 7.19
PHOTOGRAPH TAKEN DURING INTERACTION OF 200 KG OF STAINLESS STEEL AT 1700°C WITH A LIMESTONE CONCRETE

Somewhat similar experiments with inductively sustained thermitic melts were undertaken in Germany at the Beta facility [89].

Copus et al. [90, 91] extended the investigations to sustained interactions of uranium dioxide melts using an embedded susceptor for induction heating. They showed that condensed products of concrete decomposition were readily dissolved into the oxide melt greatly expanding its volume. Later, Argonne National Laboratory (ANL) investigators introduced direct electrical heating to sustain thermitically prepared, predominantly oxidic melts in contact with concrete [92]. Lately, the VULCANO experiment series has been undertaken again using inductively sustained melt interactions [93].

The interactions of uncoolable, particulate core debris with concrete have been less extensively studied. Copus and Bradley [94] and Tarbell et al. [95] reported some tests that showed the interaction was consistent with lower temperature phases of tests with molten core debris. Interestingly, the solid core debris continued to reduce gaseous products of concrete decomposition to form combustible gases. Volatile fission products were observed to be released from the attack of solidified core debris on concrete.

7.5.4 Modeling Core Debris Interactions with Concrete

Major, integrated models of core debris interactions with concrete are the CORCON code [96] and the WECHSL code [97]. A similar model is included in the MAAP proprietary computer code for the systems analysis of severe reactor accidents. Because core melts are vigorously stirred by the sparging gases, keys to the modeling are accounting for chemical processes and modeling the heat transfer from the molten material to the concrete. In general, the melt chemistry is treated thermodynamically with rate effects controlled by mass transfer of oxidizing gases into and through the melt as bubble swarms. Bradley [98] has developed the most elaborate model of the heat transfer to the concrete. The model accounts for intermittent melt contact and a film of melting concrete at the interface. A comparison of the predictions of the CORCON computer code with concrete ablation observed in a test [99] of molten UO_2-ZrO_2 with concrete is shown in Fig. 7.20. Example predictions of concrete erosion and gas production for a reactor accident are shown in Figs. 7.21 and 7.22. Accident analyses show that typically concrete ablation by core debris is slow and many days would be required before typical basemats are penetrated. Core debris can erode and weaken pedestal supports for reactor vessels in boiling water reactors. Collapse of these supports and the vessel might rupture the drywell containment.

The aerosol production and radionuclide release during core debris interactions with concrete is modeled with the VANESA module [100] coupled with the CORCON computer code. The model considers aerosol production from the vaporization of melt constituents and the mechanical formation of aerosol by bubbles bursting at the melt surface. Vaporization is done kinetically assuming mass transfer in the liquid phase limits the vapor content in gas bubbles sparging through the melt.

7.5.5 Mitigation of Core Debris Interactions with Concrete

Combustible gas generation is the most important threat of early containment failure posed by core debris interactions with concrete. The magnitude of this threat depends very much on the plant configuration around the site of the interactions. In relatively open geometries where air from the containment can readily circulate over the core debris, combustible gases are burned as they emerge from the mixture of core debris and ablated concrete. In more secluded configurations with limited access to the air of containment, emerging gases

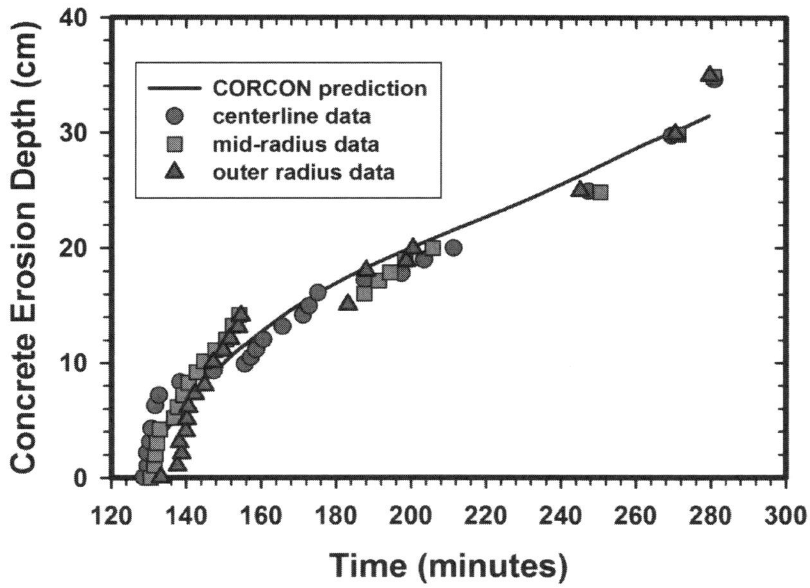

FIG. 7.20
COMPARISON OF CORCON PREDICTIONS [95] OF CONCRETE EROSION BY ABOUT 100 KG OF A MOLTEN MIXTURE OF UO_2 AND ZRO_2 TO DATA COLLECTED DURING THE SURC-2 TEST [90]

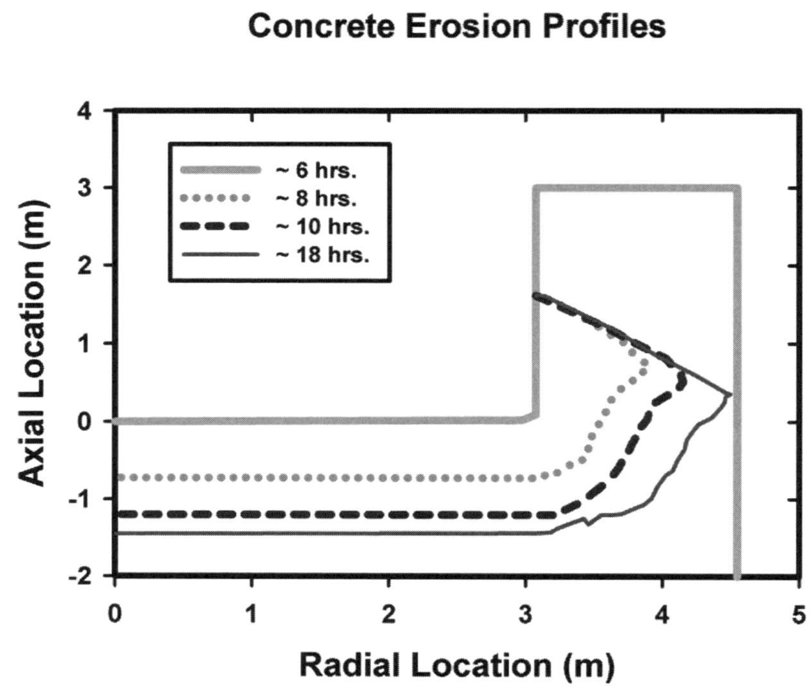

FIG. 7.21
CONCRETE EROSION PROFILES PREDICTED BY THE CORCON COMPUTER CODE FOR A REACTOR ACCIDENT IN A BOILING WATER REACTOR

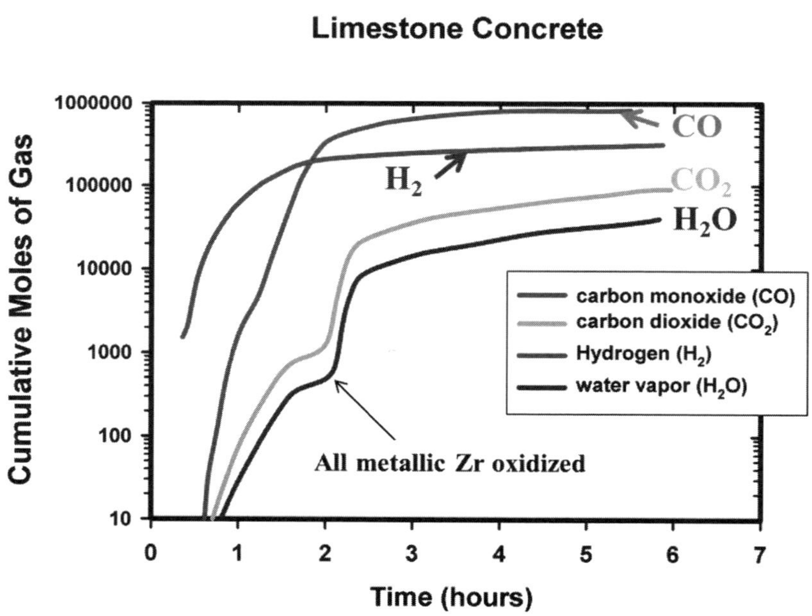

FIG. 7.22
GAS PRODUCTION DURING CORE DEBRIS INTERACTIONS WITH CONCRETE AS PREDICTED WITH THE CORCON COMPUTER CODE

cool so they do not self ignite (see Section 7.2) when they encounter air. They can accumulate and augment the threat of global deflagrations or even detonations in containment.

An interest has developed, then, in engineered systems or accident management measures to mitigate core debris interactions with concrete. Two mitigation approaches have attracted the most attention. Use of water to quench the melt and replacing the concrete with materials better suited to retain concrete ("core catchers"). Several tests have been done in which water pools were sustained over melts interacting with concrete [101, 102]. In all these tests, addition of water simply led to formation of a gas permeable crust between the water pool and the core debris. No evidence of explosive melt–water interactions were detected for core debris similar to that expected from a power reactor. Core debris simulating a reactor with aluminum clad fuel did produce steam explosions [99]. Addition of water did sharply attenuate the aerosol generation (see Section 7.6). On the other hand, the water pool substantially cools the gases emerging from the core debris-concrete mixture to the point that they are unlikely to spontaneously ignite even if abundant air is available. That is, a water pool augments the probability that core debris interactions with concrete would lead to the accumulation of combustible gases in the reactor containment.

Concern developed that tests of water effects on core debris interactions with concrete were done at too small a scale to observe realistic fragmentation of melt upon water addition. A set of very large tests were done at Argonne National Laboratory [Farmer et al., 2006]. Crust formation rather than melt fragmentation was also observed in these tests. The tests did reveal some enhanced heat transfer from core debris early in the course of the interaction with concrete [103, 104]. This enhanced heat transfer was attenuated as more concrete was incorporated into the molten core debris and the crust separating the core debris from the concrete.

The concept of core catchers was first investigated in connection with sodium-cooled fast reactors where core melting is considered in the design basis. The first consideration for light water reactors arose in connection with the floating nuclear power plant [105].

Two varieties of core catchers have been tested. Passive core catchers simply replace some of the basemat concrete with a material that interacts in a more benign way with core debris. Refractory materials such as magnesium oxide, fire brick (used in the Fast Flux Test Facility — a sodium-cooled fast reactor) or high alumina cement have been proposed. Sacrificial materials such as borax or lead absorb heat by melting and dissolve and dilute the core debris [106]. Powers [107] has reported experiments on high temperature melt interactions with both types of materials. Tarbell et al. [108] has reported on particle beds to dilute and quench core debris. An essential problem with passive core catchers is that eventually core debris will erode even the refractory materials. Once the core catcher material is eroded, core debris interactions with concrete are again possible.

Active core catchers spread core debris over a sacrificial layer of material. The underside of this active layer is then cooled and water may be added on the top of core debris. Alsmeyer et al. [109] has reported experiments with this design. Such a core catcher is included in the EPR reactor design.

7.6 AEROSOL BEHAVIOR IN REACTOR CONTAINMENTS

Reactor containments exist to provide a physical barrier to limit the accidental release of radioactivity to the environment. A test of containment adequacy is prescribed in the US code of federal regulations (10 CFR Part 100). Licensees must show that the containment can adequately attenuate the release of radioactivity given an inventory of radionuclides suspended in the containment consistent with a reactor accident involving substantial core melting. Most of the currently operating reactors in the USA were licensed assuming that the inventory of radioactivity suspended in the containment as a result of an accident involving substantial core melting was 100% of the core inventory of noble gases and 25% of the iodine. Most of the iodine (91%) is assumed to be molecular iodine (I_2). The rest is in the form of volatile organic iodide (4%) and iodine aerosol (5%) [110, 111]. This prescription was based on studies of furnace heating of irradiated reactor fuels [112]. The source term is mostly gaseous material with rather little in the form of aerosol.

In the aftermath of the accident at Three Mile Island, the US Nuclear Regulatory Commission launched a massive investigation of core melting accidents in nuclear power reactors. This work culminated in the publication of risk assessments for five representative nuclear power plants [113]. Based on these risk assessments and the underlying mechanistic analyses of accident progression and radionuclide releases to the reactor containment, revisions to the inventory of suspended radioactivity were proposed for evaluating containment adequacy [114] and incorporated with modifications into regulatory guidance [115]. Separate, revised inventories were defined for boiling water reactors and pressurized water reactors. The releases to the containment were prescribed for two periods of time called "gap release" and "in-vessel release." The prescriptions included the noble gases. They also included radionuclides from seven other chemical classes. The chemical form of halogen release was taken to be 95% CsI, 4.85% I_2, and 0.15% volatile organic iodide. Releases of the other radionuclides were taken to be in the form of aerosol. It was further noted that additional aerosol composed of things other than radionuclides would be released to the reactor containment.

Clearly, application of the revised source term for containment evaluation requires the capability to predict the physical behavior of aerosols in the reactor containments. The US Nuclear Regulatory Commission has invested heavily to develop the technology for making such predictions. The state of the art in this field has recently been reviewed [116].

7.6.1 Aerosol Formation and Growth

A small fraction of the aerosol produced in reactor accidents comes from the comminution of bulk materials (see Sections 7.4 and 7.5). Most of the aerosol comes from vapors produced during the high temperature degradations and interactions of reactor accidents. The vapors flow from the sites of generation to cooler regions where the vapors become sufficiently supersaturated that because of density fluctuations stable particle nuclei can form [117]. These nuclei grow by vapor condensation until supersaturation is relieved. The resulting very small particles (diameters ~ 0.01 μm) are often called 'primary' particles. These primary particles coagulate to form larger particles. An example aerosol particle obtained in experimental investigations of direct containment heating discussed in Section 7.4 is shown in Fig. 7.23. Primary particles making up the overall aerosol particle are resolved in this photograph.

Typically, primary particles are composed of a single chemical species or in some cases solid solutions of two or more chemical species. Composite particles formed by agglomeration of primary particles are chemical mixtures of all the species that have formed aerosol particles. There are, then, not distinct aerosols of iodine and aerosols of other radionuclide species. Agglomeration tends to homogenize the chemical composition of the aerosol.

Agglomeration occurs when particles can cross streamlines of gas flow and come into contact. Primary particles are so small that Brownian diffusion is the most important mechanism for causing the particles to deviate from streamlines of flow. Once the particles become larger, other forces such as gravity and differential velocities caused by turbulence in the atmosphere can lead to additional particle agglomeration [118].

Prediction of aerosol agglomeration becomes challenging when additions of vapor and primary particles are made to an "aging" aerosol. These problems were overcome by the advent of the "sectional" method for analysis developed by Gelbard [119]. Today all mechanistic models of aerosol behavior in reactor containment are based on the sectional method.

Aerosol agglomeration proceeds at a rate proportional to the square of the number density of particles. The agglomeration process, of course, reduces the particle number density. The number density is also reduced by natural particle deposition and by engineered safety features. These removal mechanisms are discussed in the following sections.

7.6.2 Natural Particle Removal Processes

The predominant natural mechanism for the removal of aerosol particles from a containment atmosphere is gravitational settling. That is, the particles agglomerate to the sizes such that the gravitational force on

FIG. 7.23
AEROSOL PARTICLE COLLECTED IN A DCH EXPERIMENT. THE OVERALL PARTICLE IS SOMEWHAT LARGER THAN 1 μM AND IS COMPOSED OF MUCH SMALLER "PRIMARY" PARTICLES

particles is greater than the natural convection force or forced convection force for keeping the particles suspended. Gravitational settling has to be considered in all types of containments. Gravitational settling is only rapid for large particles.

Gravitational settling can be enhanced if water vapor will condense on the aerosol particles. Indeed some materials vaporized in the course of an accident such as cesium iodide and cesium hydroxide are quite water soluble and thus they are hygroscopic. That is, they will absorb water vapor from the atmosphere even if the relative humidity is less than 100%. Enhanced deposition of aerosol particles as a result of hygroscopicity has been demonstrated in tests done with simple hygroscopic particles such as CsI or CsOH particles [120]. Aerosols expected in reactor accidents will be composed of agglomerated mixtures of both hygroscopic and nonhygroscopic particles. The latter will typically outweigh substantially the hygroscopic component. It is not at all evident how these composite particles will be affected by whatever water absorption occurs.

Particles in regions that favor the condensation of steam can also deposit by diffusiophoresis. That is, the flux of condensing steam can sweep particles into surfaces where will stick due to the surface tension of water. Diffusiophoresis is a particularly important mechanism of aerosol removal in the AP1000 reactor design where the containment shell is cooled by an external source of water. Diffusiophoresis can also be a significant mechanism for the removal of aerosol from gas flows through the ice beds of ice condenser containments. Diffusiophoretic deposition of aerosol particles is to first order independent of aerosol particle size.

Many other aerosol deposition mechanisms have been identified including inertial deposition, turbulent deposition, thermophoresis and diffusive deposition. These mechanisms tend to be far more important in the piping system of a reactor coolant system. In the reactor containment, these other mechanisms typically make only incremental changes to aerosol removal rates calculated considering gravitational settling and where applicable diffusiophoresis.

Many tests have been conducted to validate the ability of modern computer codes to predict the aerosol depletion by natural processes within large volumes like reactor containments. Many of these are discussed in proceedings of specialists' meetings [121]. An example comparison of code predictions and experimental data is shown in Fig. 7.24.

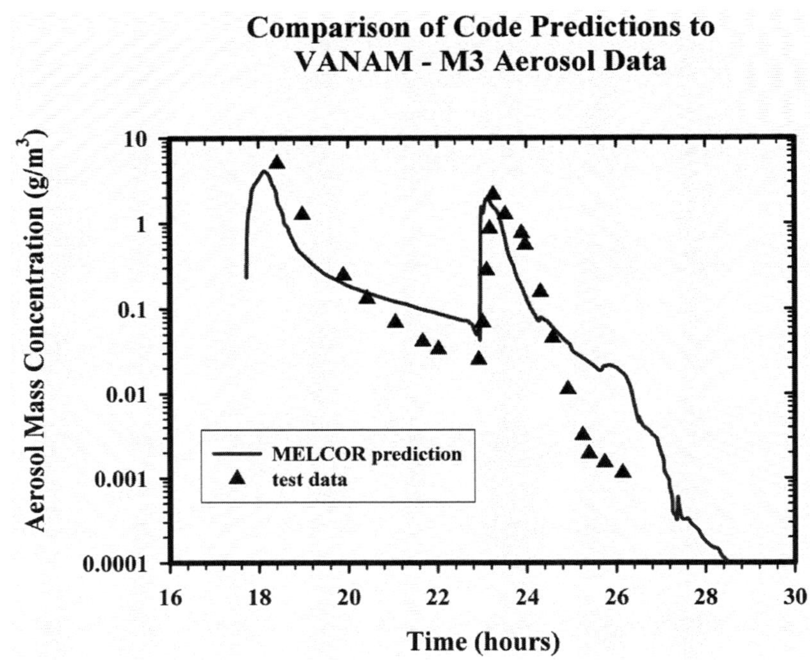

FIG. 7.24
COMPARISON OF AEROSOL BEHAVIOR PREDICTIONS WITH THE MELCOR COMPUTER CODE AND RESULTS OF THE VANAM-M3 INTEGRAL TEST [115]

7.6.3 Effects of Engineered Safety Features

Engineered safety features in nuclear power plant containments such as containment sprays or drywell sprays, fan coolers, and suppression pools were installed to limit containment pressurization during design basis accidents when large quantities of steam are expelled into the containment from the reactor coolant system. Some of these engineered safety features also work quite well at removing aerosol particles from the containment atmosphere. Especially noteworthy are containment sprays in pressurized water reactors and steam suppression pools in boiling water reactors.

Containment sprays act much like rain to cleanse the air. Falling droplets can impact the largest aerosol particles. Smaller particles will follow streamlines of flow around the droplet. But, aerosol particles are finite in size so particles whose center of mass is along a streamline very near the droplet surface will contact the droplet and be captured by surface tension. Very small particles can avoid both impaction and interception by falling droplets, but they will diffuse to the droplet surface. There is then a minimum in the efficiency of spray droplet capture of aerosol particles at a size often called the "maximum penetration size" which is often found to be about 0.2 μm [122]. As a consequence of this minimum in efficiency, containment sprays not only reduce the amount of suspended particle mass, they also shift the size distribution of the remaining aerosol so that it narrows around the maximum penetration size. Results of some example calculations are shown in Fig. 7.25.

Gases blowing through steam suppression pools are also very effectively cleansed of aerosol contamination [123, 124]. Much of the decontamination occurs during formation of the bubbles at orifices in sparging systems. Once a bubble detaches and rises through the pool, gases within the bubble circulate. Larger aerosol particles cannot follow streamlines of the gas flow and impact the bubble surface. Small particles can diffuse to the bubble surface. Intermediate particles that diffuse slowly but can follow streamlines of flow are intercepted when streamlines are close to the bubble surface. Particles are also removed by bubble oscillations when par-

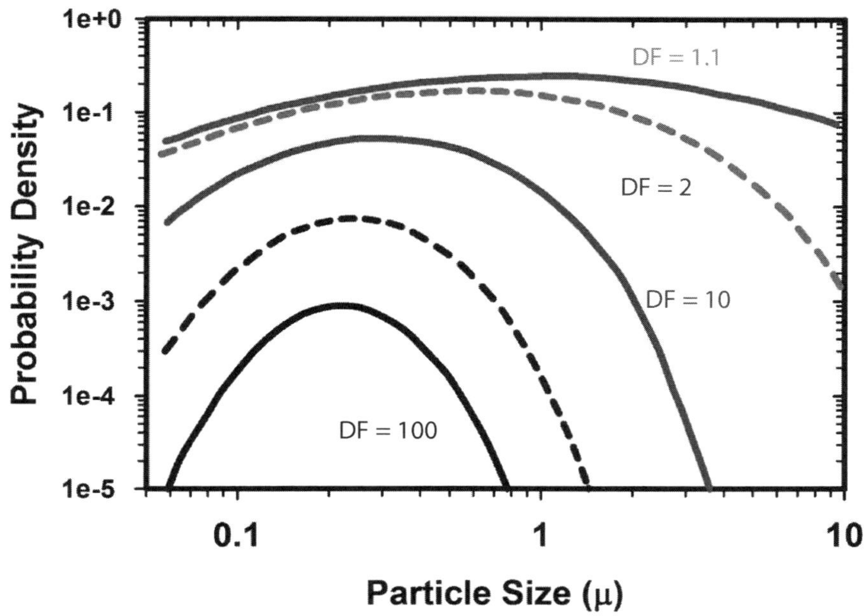

FIG. 7.25
EVOLUTION OF THE PARTICLE SIZE DISTRIBUTION OF AN AEROSOL SUBJECTED TO THE ACTIONS OF A REACTOR CONTAINMENT SPRAY

ticles have too much inertia to respond to the oscillations. Very substantial decontaminations can be achieved by water pools of depths over about one meter. Mechanisms for particle removal do depend on the particle size, so, as with sprays, there is a size that is removed with minimal efficiency. Suppression pools then both attenuate aerosol concentration and shift the size distribution toward the maximum penetration size. A validated computer code is available for calculating the decontamination by suppression pools [125].

7.6.4 Aerosol Leakage Out of Containment

No containment can perfectly retain radionuclide gases and aerosols even when intact. There is always some leakage. The potential for aerosol leakage through cracks and holes in containment has been reviewed recently [1].

A persistent issue is whether aerosol passing through small openings will eventually plug the opening at least for continued particle passage if not for aerosol penetration. A map of flow regimes showing where passage plugging could be expected and where it would not be anticipated is shown in Fig. 7.26. This map is cast in terms of the ratio of the particle settling velocity, $V_{settling}$, to the gas flow velocity, $V_{sampling}$, and the particle Stokes number defined by:

$$\text{Stokes number} = \frac{\rho_{part} D_{part}^2 V}{9_{gas} d_{crack}}$$

where:

$$D_{part} = \text{particle diameter}$$

$$d_{crack} = \text{crack or passage diameter}$$

$$V = \text{gas velocity} = V_{sampling}$$

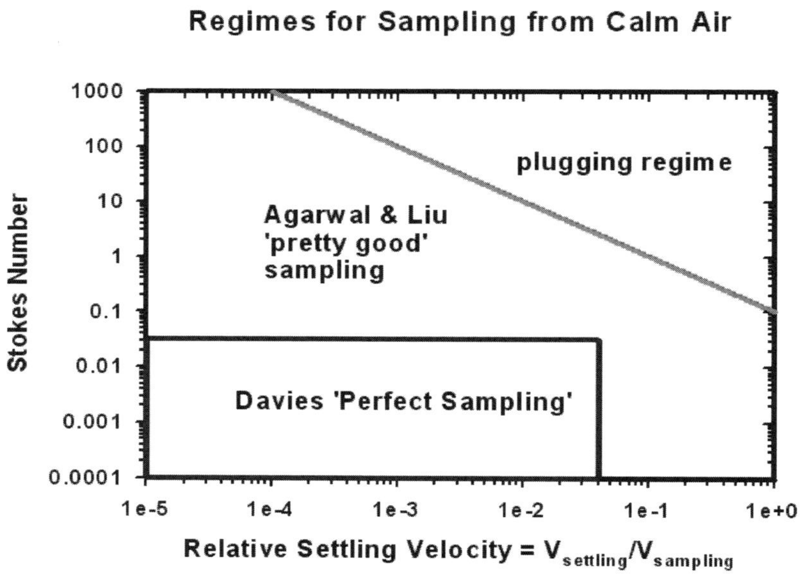

FIG. 7.26
MAP SHOWING REGIMES WHERE PARTICLE PLUGGING OF A LEAK PATH WOULD BE EXPECTED

ρ_{part} = density of material making up patricle

9_{gas} = gas viscosity

Typical conditions for leak paths in containments that are pressurized during accidents fall in the so-called pretty good sampling regime, and plugging is not expected.

7.6.5 Filtered Vents

Around the world, many reactors are equipped with filtration systems so that containments can be vented before they pressurize. Reactors in France have sand bed filters and reactors in Sweden have been equipped with gravel bed filters and venture scrubber filters [116]. Such filter systems can efficiently remove aerosol particles from the vented gases if the aerosol size distribution has not been altered by the action of other processes such as containment sprays or suppression pools. Particles remaining in the gases after the actions of such processes have size distributions that will maximize their abilities to pass through these external filtration systems.

External filtration systems must be equipped with the capability to handle decay heat from collected radionuclides. Water is usually used to remove the decay heat. An issue with water is the behavior of radioisotopes of iodine that can partition from the aqueous phase back into the gas phase. This issue is discussed in the next section of the chapter.

7.7 GASEOUS IODINE IN CONTAINMENT

Though most of the iodine injected into the reactor containment atmosphere under accident conditions is now thought to be in the form of iodine particulate, some are assumed to be in gaseous form. Furthermore, particulate iodine is expected to accumulate eventually in the containment sumps as a dissolved species. Iodine is unusual among radionuclides in the containment because it can undergo transformations in aqueous solution to form molecular iodine that will partition back into the containment atmosphere where it can leak or be released to the environment. Pertinent solution equilibrium for thermal conditions are:

$$2\,I^- + IO_3^- \leftrightarrow 3\,OI^-$$

$$HOI \leftrightarrow OI^- + H^+$$

$$I^- + HOI + H^+ \leftrightarrow I_2(aq) + H_2O$$

$$I^- + I_2(aq) \leftrightarrow I_3^-$$

$$I_2(aq) \leftrightarrow I_2(gas)$$

These equilibria are very strongly dependent on the solution pH as is shown in Fig. 7.27. As the solution becomes acidified, molecular iodine makes a larger contribution to the speciation of iodine in solution. Acidification is, of course, expected in reactor accidents because of the formation of nitric acid by air radiolysis [127] and the formation of both hydrochloric acid and sulfuric acid by radiolysis of cable insulation [128]. Acidification also facilitates the reaction of dissolved iodine with trace organic species to form volatile organic iodides such as methyl iodide, CH_3I, which can readily partition from the liquid phase back into the containment atmosphere [129]. Radiolysis of the aqueous solution appears to accentuate the formation of volatile iodine products at lower pH.

The behavior of gaseous iodine remains an issue of safety now much as it was when plants were licensed and the containments were evaluated with an accident source term that had iodine predominantly in gaseous form. Iodine continues to be a subject of ongoing research study [130–135]. An empirical approach to the suppression of iodine partitioning adopted by licensees is to fit sumps and reactor cavities in power plants with buffers such as trisodium phosphate or base such as sodium hydroxide to maintain levels of pH in excess of seven. Integral testing of this empirical approach showed it was not sufficient.

The Phébus FPT-1 test [136] involved the simulation of a severe reactor accident by degradation of irradiated fuel rods. Effluent radionuclides were allowed to flow through a model of the reactor coolant system and into a model containment. As expected iodine reaching the containment was in both particulate and gaseous form. The concentration of gaseous iodine is shown as a function of time in Fig. 7.28. Gaseous iodine concentrations fell rapidly once fuel degradation ceased. But, the concentration arrested at a low but very constant value. This arrest in the gaseous iodine concentration was shown to be independent of pH in the sump. Close analysis of the test results showed that gaseous iodine behavior did not follow characteristics expected of iodine partitioning from the sump waters.

Current thinking is that gaseous iodine is interacting with paint. Also particulate iodine dissolved in water surfaces also interacts with the paint. Irradiation of the paint releases both molecular iodine and volatile organic iodides. These vapors are then oxidized to iodine oxides perhaps by radiolytically generated ozone in the containment atmosphere:

$$I_2(gas) + 2\,O_3 \rightarrow I_2O_4 + 2\,O_2$$

$$I_2(gas) + 3\,O_3 \rightarrow I_2O_5 + 2\,O_2$$

$$I_2O_5 + H_2O \rightarrow 2\,HIO_3$$

Research in the atmospheric sciences field has shown that these iodine oxides nucleate to form very fine (~0.1 μm) aerosol particles [137]. Such fine particles will sediment very slowly. They are more likely to diffuse or deposit on surfaces, redissolve and again interact with the paint.

The results of the Phébus FPT-1 test, which have been replicated, present a different paradigm for the behavior of iodine in reactor containments following accidents than that considered in the past. The technical issues of this new aspect of iodine behavior are being investigated in international collaborative research programs underway in Canada and France.

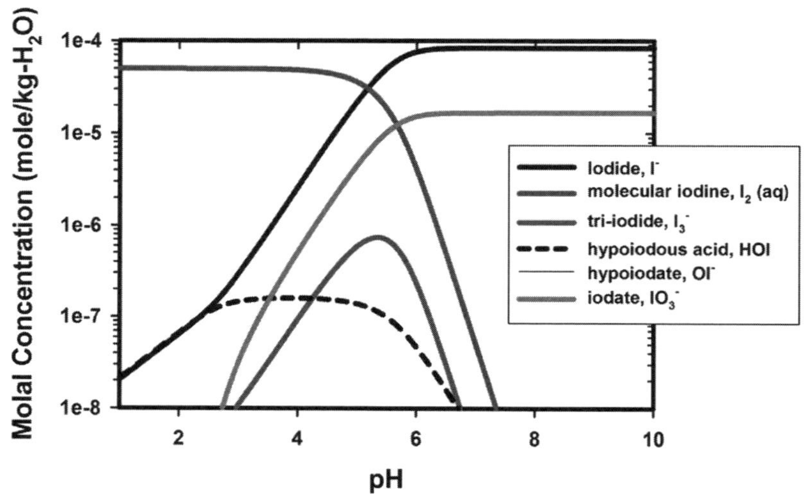

FIG. 7.27
SPECIATION OF IODINE DISSOLVED IN WATER AS A FUNCTION OF SOLUTION PH

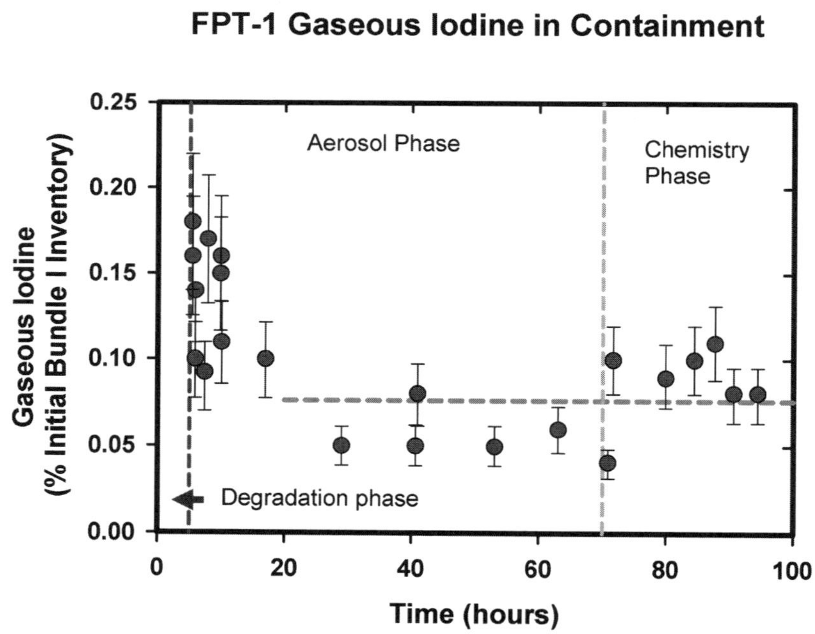

FIG. 7.28
GASEOUS IODINE CONCENTRATION IN THE PHÉBUS CONTAINMENT MODEL AS A FUNCTION OF TIME DURING THE FPT-1 TEST

7.8 CONSIDERATION OF SEVERE ACCIDENTS IN REGULATORY FRAMEWORK

Sections 7.2 to 7.7 of this chapter describe the phenomena associated with severe accidents. They also describe the state of the knowledge associated with mitigating and managing these phenomena, so that the loads on containment system can be reduced or eliminated. This section describes the regulatory actions taken by the NRC to minimize the adverse effects of severe accidents on the plant structures, systems, and components (SSCs).

7.8.1 Operating Reactors

On August 8, 1985, NRC issued a Policy Statement on severe accidents regarding future Designs and existing Plants (50 Federal Register Notice 32138) that introduced the Commission's plan to address severe accident issues for existing commercial nuclear power plants. In the following years, the Commission developed an approach to implement this plan for the existing plants and issued a Generic Letter 88-20 [139] that communicated this plan to all licensees. Each nuclear power plant licensee was requested to perform a plant examination that would look for vulnerabilities to severe accidents and cost-effective safety improvements that could reduce or eliminate the potential vulnerabilities. The specific objectives for these Individual Plant Examinations (IPEs) for each licensee consisted of (1) develop an overall appreciation of severe accident behavior; (2) understand the most likely severe accident sequences that could occur at its plant; (3) gain a more quantitative understanding of the overall probability of core damage and radioactive material releases; and (4) if necessary, reduce the overall probability of core damage and radioactive material release by appropriate modifications to procedures and hardware that would help prevent or mitigate severe accidents. Upon completion of the examination, the licensee was required to submit a report to NRC describing the results and conclusions of the examination.

In order for the licensees to systematically organize the submittals, the NRC (and its contractors) provided administrative and technical guidance through the following documents (partial listing).

- NUREG 1335 [140]: "IPE Submittal Guidance Document" established format and content guidelines for the licensees' submittals. The appendices to this document contain: (1) an approach to the back-end portion of a PRA, (2) references to PRAs performed or reviewed by the NRC, (3) NRC responses to questions and comments raised at the IPE workshop, and (4) staff review guidance.
- NUREG/CR-5132 [141]: "Severe Accident Insights Report," This report describes the conditions and events that nuclear power plant personnel may encounter during the latter stages of a severe core damage accident and what the consequences might be of actions they may take during these latter stages. The report also describes what can be expected of the performance of the key barriers to fission product release (primarily containment systems), what decisions the operating staff may face during the course of a severe accident, and what could result from these decisions based on our current state of knowledge of severe accident phenomena.
- NUREG/CR-4920, Volumes 1–5 [142]: "Assessment of Severe Accident Prevention and Mitigation Features," This series of reports describes plant features and operator actions found to be important in either preventing or mitigating severe accidents in LWRs with five different types of containments.
- NUREG/CR-2300 [143]: "PRA Procedures Guide," This report is a guide to the performance of probabilistic risk assessments (PRAs) for nuclear power plants.
- NUREG/CR-3485 [144]: "PRA Review Manual," This report describes an approach for reviewing a Level 1 type PRA (a PRA that carries the accident analysis up to the point of calculating the probability of core damage or core melt).

The NRC, Office of Research, evaluated the licensees' submittals for IPE and documented the results of its evaluations in NUREG-1560 [2]. The report provides perspectives gained by reviewing 75 IPE submittals pertaining to 108 nuclear power plant units (existing in 1997). The NRC review was focused on gaining perspectives in three major areas: (1) improvements made to individual plants as a result of the IPE program, (2) plant specific design and operational features and modeling assumptions that significantly affect the estimates of core damage frequency (CDF) and containment performance, and (3) strengths and weaknesses of the models and methods used in the IPEs. In particular, these results are assessed in relation to the design and operational characteristics of the various reactor and containment types, and by comparing the IPEs to PRA characteristics. In addition to the IPE program, the NRC also implemented a program, "Individual Plant Examination for External Events" (IPEEE). This program is discussed in Chapter 8 of this book.

7.8.2 Advanced Reactors

The NRC originally issued 10 CFR Part 52 on April 18, 1989. This rule provided for issuing early site permits (ESPs), standard design certifications (DCs), and combined operating licenses (COLs) for nuclear power reactors. The rule stated the review procedures and licensing requirements for applicants of these new licenses and certifications. The rule was intended to achieve the early resolution of licensing issues, as well as to enhance the safety and reliability of nuclear power plants. With regard to severe accidents, 10 CFR Part 52 codified some parts of the guidance in the Severe Accident Policy Statement and Standardization Policy Statement.

In 2007, the NRC published a revision to 10 CFR Part 52 and 10 CFR Part 50. The revision to 10 CFR Part 52 included the requirement for a COL applicant to conduct a plant-specific PRA, and to provide a description of the plant-specific probabilistic risk assessment (PRA) and its results within its final safety analysis report (FSAR). The revision to 10 CFR Part 50 included the requirement for the COL holder to maintain and upgrade the PRA periodically throughout the life of plant. The NRC has previously issued guidance for addressing PRA and severe accidents in new plant licensing, including the following:

- NRC Policy Statement [145] "Severe Reactor Accidents Regarding Future Designs and Existing Plants," 50 FR 32138, August 8, 1985.
- NRC Policy Statement [146], "Safety Goals for the Operations of Nuclear Power Plants," 51 FR 28044, August 4, 1986.
- NRC Policy Statement [147], "Nuclear Power Plant Standardization," 52 FR 34844, September 15, 1987.
- NRC Policy Statement [148], "Regulation of Advanced Nuclear Power Plants," 59 FR 35461, July 12, 1994.
- NRC Policy Statement [149], "The Use of Probabilistic Risk Assessment Methods in Nuclear Regulatory Activities," 60 FR 42622, August 16, 1995.
- SECY-90-016 [150], "Evolutionary Light-Water Reactor (LWR) Certification Issues and Their Relationship to Current Regulatory Requirements," January 12, 1990, and the related staff requirement memoranda (SRM) dated June 26, 1990.
- SECY-93-087 [151], "Policy, Technical, and Licensing Issues Pertaining to Evolutionary and Advanced Light-Water Reactor Designs," April 2, 1993, and the related SRM dated July 21, 1993.
- SECY-96-128 [152], "Policy and Key Technical Issues Pertaining to the Westinghouse AP600 Standardized Passive Reactor Design," June 12, 1996, and the related SRM dated January 15, 1997.
- SECY-97-044 [153], "Policy and Key Technical Issues Pertaining to the Westinghouse AP600. Standardized Passive Reactor Design," February 18, 1997, and the related SRM dated June 30, 1997.

The first five documents provide guidance regarding the appropriate course of action to address severe accidents and the use of PRA. The SRMs relating to SECY-90-016, SECY-93-087, SECY-96-128, and SECY-97-044 provide Commission-approved guidance for implementing features in new designs to prevent severe accidents and to mitigate their effects, should they occur. Regulatory Guide 2.206 [154] and Standard Review Plan (SRP) Chapter 19 [155] provide additional guidance for implementing the NRC's requirements.

Today, with continued appreciation in understanding of reactor safety, especially with respect to severe accident managements, advanced technologies based on evolutionary and passive designs with enhanced safety features provide better defense-in-depth and safety margins in accident preventions and mitigations than the operating reactors. Table 7.2 (courtesy: Jim Xu, author of Chapter 9 of this book) shows some of the severe accident design features seen in various design certification applications in the U. S. Also included in the table are comparisons of various risk metrics associated with different designs. The calculated core damage (CDF) frequencies for Generation III/III+ (the advanced reactors) are typically in a range of 10^{-6} to 10^{-8} per year (internal events at full power) and corresponding large release frequencies (LRF) are between 10^{-7} and 10^{-8} per year. These results are not directly comparable to core damage frequencies and large early release frequencies predicted for existing reactors since the advanced reactors have not been built and operated

TABLE 7.2
SEVERE ACCIDENT DESIGN FEATURES AND RISK METRIC COMPARISON*

Feature	ABWR	ESBWR	AP1000	U.S. EPR	U.S. APWR
CDF/yr LRF/yr CCFP (internal events at power)	2E-7 <1E-8 <0.1	2E-8 1E-9 0.08	2E-7 2E-8 0.08	3E-7 2E-8 0.08	1E-6 1E-7 0.1
Core melt stabilization (CMSS) BiMAC or core catcher	core catcher passive core debris flooding	BiMAC, passive cooling of molten debris	IRWST flooding for cooling core debris in-vessel	core catcher passive cooling of molten debris	core debris trap
Reactor cavity flooding capability	AC-independent with alternate thermally actuated flooder system	GDCS with firewater as back-up	manual actuation	SAHRS and CMSS	dedicated cavity flooding pathway
Containment overpressure protection system (vent)	filtered vent	filtered vent		containment opening reserved for filter vent	
Containment combustible gas control	inerted containment	inerted containment	manual igniters, 2 PARs for LOCAs	PARs	glow type igniters
In-containment refueling water storage tanks			included	included	included
RCS severe accident de-pressurization	normal ADS	normal ADS	ADS valves	2 trains, manually actuated	1 train manually actuated
Reactor cavity concrete floor	sacrificial layer	sacrificial layer	large floor area with thick concrete layer	sacrificial layer	large floor area

*Information taken from respective application FSAR or DCD.

for sufficient periods to compile suitable inventories of plant-specific data essential for reliable probabilistic risk assessments. It must also be emphasized that the core damage frequencies often quoted for advanced light water reactors account for only so-called 'internal events' and do not include risks associated with site-specific external events which are likely to be similar for advanced designs and current plant designs. For severe accident mitigations, new containment features are included such as: in-containment refueling water storage tanks (IRWST), core melt stabilization systems, containment overpressure protection systems, and containment combustible gas controls, etc. These features seen in standard designs provide much better and measured safety performance than previous generations of reactors. In addition, all standard designs ensure that containments can withstand external impacts from commercial aircrafts in compliance with NRC regulation 10 CFR 50.150.

REFERENCES

[1] R. Gauntt et al., 2012, "Fukushima Daiichi Accident Study (Status as of April 2012)," SAND2012-6173, Sandia National Laboratories, Albuquerque, NM.

[2] Office of Nuclear Regulatory Research, 1997, "Individual Plant Examination Program: Perspectives on Reactor Safety and Plant Performance. Part 1. Final Summary Report," NUREG-1560, U.S. Nuclear Regulatory Commission, Washington, D.C.

[3] M. Berman, J. Larkins, and L. Thompson, 1982, "Hydrogen and Reactor Safety — Proceedings of the Second International Conference on the Impact of Hydrogen on Water Reactor Safety," NUREG/CP-0038; EPRI RP 1932-35; SAND82-2456.

[4] G. Schanz, 2003, "Recommendations and Supporting Information on the Choice of Zirconium Oxidation Models in Severe Accident Codes," FZKA 6857; SAM-COLOSS-P043, Forschungszentrum Karlsruhe, Karlsruhe, Germany.

[5] D. A. Powers, 1981, "A Review of Steam Oxidation of Steels — the Forgotten Source of Hydrogen," Proceedings of the Workshop on the Impact of Hydrogen on Water Reactor Safety, NUREG/CR-2017, SAND81-0661, Sandia National Laboratories, Albuquerque, NM.

[6] M. P. Sherman, M. Berman, J. C. Cummings, G. W. Perkins, D. A. Powers, P. O. Bieniarz, and O. R. Green, 1980, "The Behavior of Hydrogen During Accidents in Light Water Reactors," NUREG/CR-1561, SAND80-1495, Sandia National Laboratories, Albuquerque, NM.

[7] J. E. Shepherd and O. B. Crump, 1982, "Hydrogen-Steam Jet Flame Experiments," Proc. 2nd Workshop on the Impact of Hydrogen on Water Reactor Safety, Albuquerque, NM.

[8] G. T. Kalghatyi, 1981, "Blow-out Stability of Diffusion Flames: Part I. In Still Air," Comb. Sci. Tech., *26*, 233–239.

[9] N. Cohen, 1992, "Flammability and Explosion Limits of H_2 and H_2/CO: A Literature Review," SMC-TR-93-19, The Aerospace Corporation, El Segundo, CA.

[10] J. E. Hustad and O. K. Sønju, 1988, "Experimental Studies of Lower Flammability Limits of Gases and Mixtures at Elevated Temperature," Comb. Flame, *71*, 283–294.

[11] I. Wierzba and V. Kilchyk, 2001, "Flammability limits of hydrogen–carbon monoxide mixtures at moderately elevated temperatures," Int. J. Hydrogen Energy, *26*, 639–643.

[12] M. Berman, 1981, "Light Water Reactor Safety Research Program Quarterly Report January–March, 1981," NUREG/CR-2163 (1 of 4), SAND81-1216 (1 of 4), Sandia National Laboratories, Albuquerque, NM.

[13] M. Berman and J. C. Cummings, 1984, "Hydrogen Behavior in Light-Water Reactors," Nuclear Safety, *25*, 53–74.

[14] W. E. Lowry, B. R. Bowman, and B. W. Davis, 1982, "Final Results of the Hydrogen Igniter Experimental Program," NUREG/CR-2486, UCRL-53036, Lawrence Livermore Laboratory, Livermore, CA.

[15] J. H. S. Lee, 1984, "Dynamic parameters of gaseous detonations," Ann. Rev. Fluid Mech., *16*, 331–336.

[16] A. L. Camp, J. C. Cummings, M. P. Sherman, C. F. Kupiec, R. J. Healy, J. S. Caplan, J. R. Sandhop, and J. H. Saunders, 1983, "Light Water Reactor Hydrogen Manual," NUREG.CR-2726; SAND82-1137, Sandia National Laboratories, Albuquerque, NM.

[17] J. M. McGlaun, S. L. Thampson, L. N. Kmetyk, and M. G. Elrick, 1990, "A Brief Description of the Three-Dimensional Shock Wave Physics Code CTH," SAND89-0607, Sandia National Laboratories, Albuquerque, NM.

[18] R. K. Byers, 1982, "CSQ Calculation of H_2 Detonations in the Zion and Sequoyah Nuclear Plants," SAND81-2216, Sandia National Laboratories, Albuquerque, NM.

[19] W. Breitung et al., 2000, "Flame Acceleration and Deflagration-to-Detonation Transition in Nuclear Safety," NEA/CSNI/R(2000)7, OECD Nuclear Energy Agency.

[20] E. S. Oran and V. N. Gamezo, 2007, "Origins of the degragration–to-detonation transition in gas phase combustion," Combustion and Flame, *148*, 4–47.

[21] S. B. Dorofeev, V. P. Sidorov, A. E. Dvoinishnikov, and W. Breitung, 1996, "Deflagration to Detonation Transition in Large Confined Volume of Lean Hydrogen–Air Mixtures," Comb. Flame, *104*, 95–110.

[22] R. Knystautas, J. H. Lee, I. Moen, and H. G. G. Wagner, 1979, "Direct Initiation of Spherical Detonation by a Hot Turbulent Gas Jet," Symposium (International) on Combustion, *17*, 1235–1245.

[23] M. P. Sherman and M. Berman, 1988, "The possibility of local detonations during degraded-core accidents," Nuclear Tech., *81*, 63–77.

[24] G. Ciccarelli and S. Dorofeev, 2008, "Flame acceleration and transition to detonation in ducts," Progress in Energy and Combustion Science, *34*, 449–550.

[25] Committee on Hydrogen Combustion, 1987, "Technical Aspects of Hydrogen Control and Combustion in Severe Light-water Reactor Accidents," National Academy Press, Washington, D.C.

[26] F. Morfin, J.-C. Sabroux and A. Renouprez, 2004, "Catalytic combustion of hydrogen for mitigating hydrogen risk in case of a severe accident in a nuclear power plant: study of catalysts poisoning in a representative atmosphere," Applied Catalysis B: Environmental, *47*, 47–58.

[27] L. L. Hegedus and R. W. McCabe, 1981, "Catalyst Poisoning," Catal. Rev.-Sci. Eng., *23*, 377–476.

[28] E. L. Tolman, P. Kuan, and J. M. Broughton, 1988, "TMI-2 accident scenario update," Nucl. Eng. Design, *108*, 45–54, and references therein.

[29] N. Reinke, T. Drath, T. V. Berlepsch, H. E. Unger, and M. K. Kuch, 2006, "Formation, characterization and cooling of debris: Scenario discussion with emphasis on TMI-2," Nucl. Eng. Design, *236*, 1955–1964.

[30] L. Sepold, P. Hofmann, W. Leiling, A. Miassoedov, D. Piel, L. Schmidt, and M. Steinbrück, 2001, "Reflooding experiments with LWR-type fuel rod simulators in the QUENCH facility," Nucl. Eng. Design, *204*, 205–220.

[31] R. J. Lipinski, 1984, "A coolability model for post-accident nuclear reactor debris," Nucl. Tech., *65*, 53–66.

[32] US Nuclear Regulatory Commission, 1975, "Reactor Safety Study An Assessment of Accident Risks in U.S. Commercial Nuclear Power Plants," WASH-1400; NUREG 75/014.

[33] S. A. Colgate and T. Sigurgeirsson, 1973, "Dynamic mixing of water and lava," Nature, *244*, 552–555.

[34] G. Long, 1957, "Explosions of molten metal in water — causes and prevention," Metal Progress, *71*, 107–112.

[35] L. D. Buxton and L. S. Nelson, 1977, "Steam Explosions," Chapter 6 in "Core Meltdown Experimental Review," W. B. Murfin, editor, NUREG-00205, SAND 74-0382, Sandia National Laboratories, Albuquerque, NM.

[36] M. K. Denham, A. P. Tyler, and D. F. Fletcher, 1994, "Experiments on the mixing of molten uranium dioxide with water and initial comparisons with CHYMES code calculations," Nucl. Eng. Des., *146*, 97–108.

[37] K. Moriyama, S. Takagi, K. Muramatsu, H. Nakamura, and Y. Maruyama, 2006, "Evaluation of Containment Failure Probability by Ex-Vessel Steam Explosion in Japanese LWR Plants," Nuclear Science and Tech., *43*, 774–784.

[38] M. L. Corradini, B. J. Kim, and M. D. Oh, 1988, "Vapor Explosions in Light Water Reactors: A Review of Theory and Modeling," Progress in Nuclear Energy, 22, 1–117.

[39] G. Ciccarelli, D. L. Frost and C. Zrafonitis, 1991, "Dynamics of explosive interactions between molten tin and water in stratified geometry," Prog. Astronaut. Aeronaut., *134*, 307–325.

[40] L. S. Nelson and P. M. Duda, 1982, "Steam explosion experiments with single drops of iron oxide melted with a CO2 laser," High Temperature — High Pressure, *14*, 259–281.

[41] L. S. Nelson and P. M. Duda, 1988, "Photographic Evidence for the Mechanism of Fragmentation of a Single Drop of Melt in Triggered Steam Explosion Experiments," J. Non-Equilibrium Thermodyn., *13*, 27–55.

[42] B. Kim and M. L. Corradini, 1988, "Modelling of Small-Scale Single Droplet Fuel-Coolant interactions," Nuclear Sci. Eng., *98*, 16–28.

[43] D. F. Fletcher, 1994, "A Review of the Available Information on the Triggering Stage of a Steam Explosion," Nuclear Safety, *35*, 36–57.

[44] R. Akiyoshi, S. Nishio, and I. Tanasawa, 1990, "A study on the effect of non-condensable gas in the vapor film on vapor explosion," Int. J. Heat Mass Trans., *33*, 603–609.

[45] D. F. Fletcher and R. P. Anderson, 1990, "A review of pressure-induced propagation models of the vapour explosion process," Prog. Nucl. Energy, *23*, 137–179.

[46] E. P. Hicks and D. C. Menzies, 1965, "Theoretical Studies on the Fast Reactor Maximum Accident," ANL-7120, Argonne National Laboratory.

[47] M. L. Corradini and G. A. Moses, 1985, "Limits to fuel coolant mixing," Nucl. Sci. Eng., *90*, 19–27.

[48] R. Meignen and D. Magallon, 2005, "Comparative Review of FCI computer Models Used in the OECD-SERENA Program," Proceedings 2005 International Congress on Advances in Nuclear Power Plants, American Nuclear Society, Paper 5087.

[49] US Nuclear Regulatory Commission, 1990, "Severe Accident Risks: An Assessment for Five U.S. nuclear Power Plants," NUREG-1150.

[50] S. Basu and T. Ginsberg, 1996, "A Reassessment of the potential for an Alpha-Mode containment Failure and a Review of Current Understanding of Broader Fuel-Coolant Interactions Issues — Second Steam Explosion Review Group Workshop," NUREG-1524, U.S. Nuclear Regulatory Commission.

[51] D. A. Powers, 1985, "A Re-examination of the Steam Explosion Source Term" paper IAEA-SM-291-34, IAEA Symposium on the Reactor Accident Source Term, Columbus, OH, November 1985.

[52] I. Huhtiniem, D. Magallon and H. Hohmann, 1999, "Results of recent KROTOS FCI tests: alumina versus corium melts," Nuclear Eng. Design, *189*, 379–389.

[53] J. H. Kim, I. K. Park, B. T. Min, S. W. Hong, Y. S. Shin, J. H. Song, and H. A. Kim, 2004, "The influence of Variation in Water Depth and Melt Composition on a Spontaneous Steam Explosion in the TROI Experiments," Proceedings, 2004 International Congress on Advances in Nuclear Power Plants, American Nuclear Society, pp. 1210–1219.

[54] B. D. Turland, D. F. Fletcher, K. I. Hodges and G. J. Attwood, 1995, "Quantification of the probability of containment failure caused by an in-vessel steam explosion for the Sizewell B PWR," Nuclear Eng. Design, *155*, 445–458.

[55] T. Y. Chu, M. M. Pilch, J. H. Bentz, J. S. Ludwigsen, W.-Y. Lu, and L. L. Humphries, 1998, "Lower Head Failure Experiments and Analyses," NUREG/CR-5582; SAND98-2047.

[56] ZPSS, 1981, "Zion Probabilistic Safety Study," Commonwealth Edison Co.

[57] W. W. Tarbell, J. E. Brockmann, M. Pilch, J. W. Ross, M. S. Oliver, D. A. Lucero, T. E. Kerley, F. E. Arellano, and R. D. Gomez, 1987, "Results from the DCH-1 Experiment," NUREG/CR-4871, SAND86-2483, Sandia National Laboratories, Albuquerque, NM.

[58] D. A. Powers, 1984, "Erosion of Steel Structures by High Temperature Melts," Nucl. Sci. Eng., *88*, 357.

[59] B. W. Morris and R. V. MacBeth, 1988, "Experimental modeling and computer simulation of core debris dispersal," IAEA-SM-296/70, Int. Sump Severe Accidents in Nuclear Power Plants.

[60] W. Frid, 1988, "Behavior of a Corium jet in High Pressure melt Ejection from a Pressure Vessel," SAND85-1726, Sandia National Laboratories, Albuquerque, NM.

[61] U.S. NRC, 1985, "Estimates of Early Containment Loads from Core Melt Accidents," NUREG-1079, U.S. Nuclear Regulatory Commission, Washington, D.C.

[62] M. M. Pilch, M. D. Allen, and D. C. Williams, 1997, "Heat Transfer During Direct Containment Heating," Advances in Heat Transfer, *29*, 215–344.

[63] M. D. Allen, M. Pilch, R. O. Griffith, D. C. Williams, and R. T. Nichols, 1992a, "The Third Integral Effects Test (IET-3) in the Surtsey Test Facility," SAND92-0166, Sandia National Laboratories, Albuquerque, NM.

[64] M. D. Allen, T. K. Blanchat, M. Pilch, and R. T. Nichols, 1992b, "Results of an Experiment in a Zion-Like Geometry to Investigate the Effect of Water on Containment Basement Floor on Direct Containment Heating (DCH) in the Surtsey Test Facility: The IET-4 Test," SAND92-1241, Sandia National Laboratories, Albuquerque, NM.

[65] M. D. Allen, T. K. Blanchat, M. Pilch, and R. T. Nichols, 1992c, "An Integral Effects Test to Investigate the Effects of Condensate Levels of Water and Preexisting Hydrogen on Direct Containment heating in the Surtsey Test Facility: The IET-7 Experiment," SAND92-2021, Sandia National Laboratories, Albuquerque, NM.

[66] B. W. Spencer, M. Bengis, S. Baronowsky, and J. J. Sienicki, 1982, "Sweepout Thresholds in Reactor Cavity Interactions," ANL/LWR/SAF 82-1, Argonne National Laboratory, Argonne, IL.

[67] M. D. Allen, M. Pilch, R. T. Nichols, and R. O. Griffith, 1991, "Experiments to Investigate the Effect of Flight Path on Direct Containment Heating (DCH) in the Surtsey Test Facility" The Limited Flight Path (LFP) Tests," NUREG/CR-5728; SAND91-1105, Sandia National Laboratories, Albuquerque, NM.

[68] M. D. Allen, M. M. Pilch, R. O. Griffith, and R. T. Nichols, 1992, "Experiments of investigate the effect of flight path on direct containment heating," Nuclear Technology, *100*, 52–69.

[69] J. E. Brockmann and W. W. Tarbell, 1984, "Aerosol source term in high-pressure melt ejection." Nucl. Sci. Eng., *88*, 342–356.

[70] J. E. Brockmann, 1987, "Ex-Vessel Releases: Aerosol Source Terms in Reactor Accidents," Progress in Nuclear Energy, *19*, 7–68.

[71] M. M. Pilch, H. Yan, and T. G. Theofanous, 1994, "The Probability of Containment Failure by Direct Containment Heating in Zion," NUREG/CR-6075; SAND93-1535, Sandia National Laboratories, Albuquerque, NM.

[72] M. M. Pilch, 1996, "A two-cell equilibrium model for predicting direct containment heating," Nuclear Engineering and Design, *164*, 61–94.

[73] H. Yan and T. G. Theofanous, 1993, "The prediction of direct containment heating," ANS Pro. Natl. Heat Transfer Conf.

[74] D. C. Williams, K. D. Bergeron, D. E. Carroll, R. D. Gasser, J. L. Tills, and K. E. Washington, 1987, "Containment Loads Due to Direct Containment heating and Associated hydrogen Behavior: Analysis and Calculations with the CONTAIN Code," NUREG/CR-4896, SAND87-0633, Sandia National Laboratories, Albuquerque, NM.

[75] K. E. Washington and D. C. Williams, 1995, "Direct Containment heat Models in the CONTAIN Code," SAND94-1073, Sandia National Laboratories, Albuquerque, NM.

[76] M. M. Pilch, H. Yan, and T. G. Theofanous, 1996, "The probability of containment failure by direct containment heating in Zion," Nuclear Engineering and Design, *164*, 1–36.

[77] R. E. Schneider and R. R. Sherry, 1993, "Application of a two-cell adiabatic model for direct containment heat to ABB C-E System 80+ ALWR," Trans. ANS, *69*, 509–570.

[78] E. W. Klamerus, M. P. Bohn, and D. A. Wesley, 1996, "Containment performance of prototypical reactor containments subjected to severe accident conditions," SAND96-2445, NUREG/CR-6433, CTF/8708346, Sandia National Laboratories, Albuquerque, NM.

[79] Office of Nuclear Regulatory Research, 2012, "State-of-the-Art Reactor Consequence Analysis Project Volume 2: Surry Integrated Analysis," NUREG/CR-7110 Vol. 2, Sandia National Laboratories, Albuquerque, NM.

[80] L. Meyer, G. Albrecht, C. Caroli, and I. Ivanov, 2009, "Direct containment heating integral effects tests in geometries of European nuclear power plants," Nuclear Engineering and Design, *239*, 2070–2084.

[81] R. Meignen, S. Mikasser, C. Spengler, and A. Bretault, 2007, "Synthesis of Analytical Activities for Direct Containment Heating," *Proceedings of the Second European Review Meeting on Severe Accident Research*, ERMSAR-2007.

[82] R. Gauntt et al., 2012, "Fukushima Daiichi Accident Study (Status as of April 2012)," SAND2012-6173, Sandia National Laboratories, Albuquerque, NM.

[83] D. A. Powers, 1987, "Chemical Phenomena and FP Behavior during Core Debris/Concrete Interactions," *Proceedings CSNI Specialist Meeting on Core Debris Concrete Interaction*, EPRI NP-5054-SR, Electric Power Research Institute, Palo Alto, CA.

[84] S. A. Morse, 1980, Basalts and Phase Diagrams — An Introduction to the Quantitative Use of Phase Diagrams in Igneous Petrology, Springer-Verlag.

[85] D. A. Powers, 1978, "Sustained Molten Steel/Concrete Interactions Tests: A Preliminary Report on the Feasibility of Experimental Techniques," SAND77-1423, Sandia National Laboratories, Albuquerque, NM.

[86] D. A. Powers and F. E. Arellano, 1982a, "Large-Scale Transient Tests of the Interaction of Molten Steel with Concrete," NUREG/CR-2282, SAND81-1753, Sandia National Laboratories, Albuquerque, NM.

[87] D. A. Powers, 1980, "Influence of Gas Generation on Melt/Concrete Interactions," Thermodynamics of Nuclear Materials, Volume II, International Atomic Energy Agency, Vienna Austria, paper IAEA-SM-236/58.

[88] D. A. Powers and F. E. Arellano, 1982b, "Direct Observation of Melt Behavior During High-Temperature Melt/Concrete Interactions," NUREG.CR-2283, SAND81-1754, Sandia National Laboratories, Albuquerque, NM.

[89] H. Alsmeyer, 1987, "BETA experiments in verification of the Wechsl code: experimental results on melt-concrete interaction." Nucl. Eng. Design, *103*, 115–125.

[90] E. R. Copus, R. E. Blose, J. E. Brockmann, R. B. Simpson, and D. A. Lucero, 1992a, "Core concrete interactions using molten urania and zirconium on a limestone basemat. SURC-1 experiment," NUREG/CR-5443, SAND90-0087, Sandia National Laboratories, Albuquerque, NM.

[91] E. R. Copus, R. E. Blose, J. E. Brockmann, R. B. Simpson, and D. A. Lucero, 1992b, "Core concrete interactions using molten urania and zirconium on a basaltic basemat. SURC-2 experiment," NUREG/CR-5564, SAND90-1022, Sandia National Laboratories, Albuquerque, NM.

[92] D. H. Thompson and J. K. Fink, 1991, "ACE MCCI Test L6 Test Data Report, Vol. 1 and 2," ACE-TR-C26 v.1 and v.2, Argonne National Laboratory, Argonne, IL.

[93] C. Journeau, J. M. Bonnet, L. Ferry, J. F. Haquet and P. Piluso, 2009, "Interaction of Concretes with Oxide + Metal Corium: The Vulcano VBS Series," ICAPP 2009, Int. Conf. Advances Nuclear Power Plants.

[94] E. R. Copus and D. R. Bradley, 1986, "Interaction of Hot Solid Core Debris with Concrete," NUREG/CR-4558, SAND85-1739, Sandia National Laboratories, Albuquerque, NM.

[95] W. W. Tarbell, R. E. Blose, J. W. Ross, and D. W. Gilbert, 1987, "Sustained Concrete Attack by Low-Temperature, Fragmented Core Debris," NUREG/CR-3024, SAND82-2476, Sandia National Laboratories, Albuquerque, NM.

[96] D. R. Bradley, D. R. Gardner, J. E. Brockmann, and R. O. Griffith, 1993, "CORCON-MOD3: An Integrated Computer Model for Analysis of Molten Core-Concrete Interactions," NUREG/CR-5843, SAND92-0167, Sandia National Laboratories, Albuquerque, NM.

[97] M. Reimann and S. Stiefel, 1989, "The WECHSL-Mod2 Code: A Computer Program for the Interaction of a Core Melt with Concrete including the Long Term Behavior — Model Description and User's Manual," KfK 4477, Kernforschungszentrum Karlsruhe, Karlsruhe, Germany.

[98] D. R. Bradley, 1988, "Modeling of heat transfer between core debris and concrete," *ANS Proceedings of the 1988 National Heat Transfer Conference.*

[99] E. R. Copus et al., 1989, "Core-Concrete Interactions Using Molten Aluminum on a Siliceous Basemat: The SLR-1 Experiment," SAND89-2166, Sandia National Laboratories, Albuquerque, NM.

[100] D. A. Powers, J. E. Brockmann, and A. W. Shiver, 1986, "VANESA: A Mechanistic Model of Radionuclide Release and Aerosol Generation During Core Debris Interactions with Concrete," NUREG/CR-4308; SAND85-1370, Sandia National Laboratories, Albuquerque, NM.

[101] R. E. Blose, J. E. Gronager, A. J. Suo-Antilla and J. E. Brockmann, 1987, "SWISS: Sustained Heated Metallic Melt/Concrete Interactions with Overlying Water Pools," NUREG/CR-4727, SAND85-1546, Sandia National Laboratories, Albuquerque, NM.

[102] R. E. Blose, D. A. Powers, E. R. Copus, J. E. Brockmann, R. B. Simpson, and D. A. Lucero, 1993, "Core-Concrete Interactions with Overlying Water Pools — The WETCOR-1 Test," NUREG/CR-5907, SAND92-1563, Sandia National Laboratories, Albuquerque, NM.

[103] M. T. Farmer, D. J. Kilsdonk, R. W. Aeschlimann, 2009, "Corium Coolability Under Ex-vessel Accident Conditions," Nucl. Eng. Tech., *41*, 575–609.

[104] M. T. Farmer, S. Lomperski, D. J. Kilsdonk, R. W. Aeschlimann, and S. Basu, 2006, "OECD MCCI Project 2-D Core Concrete Interaction (CCI) Tests: Final Report," OECD/MCCI-2005-TR05.

[105] T. Hunter, 1977, "Core-Meltdown Considerations for Offshore Siting," Chapter 12, Core Meltdown Experimental Review, W.B. Murfin, editor, NUREG-0205; SAND74-0382, Sandia Laboratories, Albuquerque, NM.

[106] M. Dalle Donne, S. Dorner and G. Schumacher, 1976, "Preliminary Design of a Borax Internal Core-Catcher for a Gas-cooled Fast Reactor," KfK-2352, EUR-5505e, Kernforschungszentrum Karlsruhe, Karlsruhe Germany.

[107] D. A. Powers, 1979, "A Survey of Melt Interactions with Core Retention Materials," Proc. Int. Mtg. Fast Reactor Safety Technology, vol. 1, pp. 379ff.

[108] W. W. Tarbell, R. E. Blose, and F. E. Arellano, 1984, "Molten Thermite Teeming into an Iron Oxide Particle Bed," NUREG/CR-3023, SAND82-2475, Sandia National Laboratories, Albuquerque, NM.

[109] H. Alsmeyer, T. Cron, B. Flurer, G. Messemer, A. Miassoedov, S. Schmidt-Stiefel, and T. Wenz, 2006, "The Comet-L3 Experiment on long-term melt–concrete interaction and cooling by surface flooding," FZKA 7244, Forschungszentrum Karlsruhe, Karlsruhe, Germany.

[110] US AEC, 1974a, "Assumptions Used for Evaluating the Potential Radiological Consequences of a Loss of Coolant Accident for Boiling Water Reactors," Regulatory Guide 1.3, Atomic Energy Commission, Washington, DC.

[111] US AEC, 1974b, "Assumptions Used for Evaluating the Potential Radiological Consequences of a Loss of Coolant Accident for Pressurized Water Reactors," Regulatory Guide 1.4, Atomic Energy Commission, Washington, DC.

[112] J. J. DiNunno, F. D. Anderson, R. E. Baker, and R. L. Waterfield, 1962, "Calculation of Distance Factor for Power and Test Reactor Sites," TID-14844, U.S. Atomic Energy Commission, Washington, DC.

[113] USNRC, 1990, "Severe Accident Risks: An Assessment for Five U.S. Nuclear Power Plants," NUREG-1150, U.S. Nuclear Regulatory Commission, Washington, DC.

[114] L. Soffer, S. B. Burson, C. M. Ferrell, R. Y. Lee, and J. N. Ridgely, 1995, "Accident Source Terms for Light-Water Nuclear Power Plants," NUREG-1465, U.S. Nuclear Regulatory Commission, Washington, DC.

[115] Office of Nuclear Regulatory Research, 2000, "Alternative Radiological Source Terms for Evaluating Design Basis Accidents at Nuclear Power Plants," Regulatory Guide 1.183, U.S. Nuclear Regulatory Commission, Washington, DC.

[116] H.-J. Allelein, A. Auvinen, J. Ball, S. Güntay, L. E. Herranz, A. Hidaka, A. V. Jones, M. Kissane, D. A. Powers, and G. Weber, 2009, "State-of-the-Art Report on Nuclear Aerosols," NEA/CSNI/R(2009)5, OECD Nuclear Energy Agency, Issy-les-Moulineaux, France.

[117] H. Reiss, J. I. Katz, and E. R. Cohen, 1968, "Translation-rotation Paradox in the Theory of Nucleation," J. Chem. Phys., *48*, 5553–5560.

[118] M. M. R. Williams and S. K. Loyalka, 1991, "Aerosol Science Theory and Practice with Special Applications to the Nuclear Industry," Pergamon Press.
[119] F. Gelbard, 1982, "MAEROS User Manual," NUREG/CR-1391, SAND80-0822, Sandia National Laboratories, Albuquerque, NM.
[120] H. G. Scheibel, G. Poss, and D. Weber, 1992, "KAEVER: An experiment for an improved understanding of aerosol depletion processes in a reactor containment," J. Aerosol Sci., *23*, Supplement 1, 209–212.
[121] OECD, 1998, "Third OECD Specialist Meeting on Nuclear Aerosols in Reactor Safety," GRS-166, NEA/CSNI/R(98)4, Gesellschaft für Reaktorsicherheit mbH, Cologne, Germany.
[122] D. A. Powers and S. B. Burson, 1993, "A Simplified Model of Aerosol Removal by Containment Sprays," NUREG/CR-5966, SAND92-2689, Sandia National Laboratories, Albuquerque, NM.
[123] D. A. Powers, 1997, "A Simplified Model of Decontamination by BWR Steam Suppression Pools," NUREG/CR-6153, Sandia National Laboratories, Albuquerque, NM.
[124] D. A. Powers, 1993, "Source Term Attenuation by Water in the Mark I Boiling Water Reactor Drywell," NUREG/CR-5978, SAND92-2688, Sandia national Laboratories, Albuquerque, NM.
[125] P. C. Owczrski and K. W. Burk, 1991, "SPARC-90: A Code for Calculating Fission Product Capture in Suppression Pools, NUREG/CR-5765, Pacific Northwest Laboratory, Richland, WA.
[126] D. A. Powers, 2009, "Aerosol Penetration of Leak Pathways — An Examination of the Available Data and Models," SAND2009-1701, Sandia National Laboratories, Albuquerque, NM.
[127] H. Karasawa, E. Ibe, S. Uchida, Y. Etoh and T. Yasuda, 1991, "Radiation Induced Decomposition of Nitrogen," Rad. Phys. Chem., *37*, 193–197.
[128] E. C. Beahm, R. A. Lorenz and C. F. Weber, 1992, "Iodine Evolution and pH Control," NUREG/CR-5950, ORNL/TM-12242, Oak Ridge National Laboratory, Oak Ridge, TN.
[129] R. J. Hunter-Smith, P. W. Balls and P. S. Liss, 1983, "Henry's Law constants and the sir-sea exchange of various low molecular weight halocarbon gases," Tellus, *35B*, 170–176.
[130] A. M. Deane and P. E. Potter, editors, "Proceedings of the specialists' workshop on iodine chemistry in reactor safety," AERE R 11974, United Kingdom Atomic Authority, Oxfordshire, UK, 1986.
[131] A. C. Vikis, editor, "Proceedings of the Second CSNI Workshop on Iodine Chemistry in Reactor Safety," CSNI-149, AECL-9923, Atomic Energy of Canada Limited, Toronto, CA, 1989.
[132] K. Ishigure, M. Saeki, K. Soda, and J. Sugimoto, editors, "Proceedings of the Third Workshop on Iodine Chemistry in Reactor Safety," NEA/CSNI/R(91)15, JAERI-M 92-102, Japan Atomic energy Research Institute, Tokai-mura, Naka-gun, Ibaraki-ken, Japan, 1992.
[133] S. Güntay, editor, "Proceedings of the Fourth CSNI Workshop on the Chemistry of Iodine in Reactor Safety," NEA/CSNI/R(96)6, PSI Report Nr. 97-02, Paul Scherrer Institut, Villigen, Switzerland, 1996.
[134] NEA, "Iodine Aspects of Severe Accident management," NEA/CSNI/R(99)7, OECD Nuclear Energy Agency, 1999.
[135] B. Clément, L. Cantrel, G. Ducros, F. Funke, L. Herranz, A. Rydl, G. Weber, and C. Wren, 2007, "State of the art report on iodine chemistry," OECD/CSNI Report NEA/CSNI/R 1.
[136] N. Girault, S. Dickinson, F. Funke, A. Auvinen, L. Herranz, and E. Krausmann, 2006, "Iodine behavior under LWR accident conditions: Lessons learnt from analysis of the first two Phebus FP tests," Nucl. Eng. Design, *236*, 1293–1308.
[137] S. Pechtl, E. R. Lovejoy, J. B. Burkholder, and R. von Glasow, 2006, "Modeling the possible role of iodine oxides in atmospheric new particle formation," Atmos. Chem. Phys., *6*, 505–523.
[138] D. A. Powers, 1978, "Sustained Molton Steel/Concrete Interaction Tests." A Preliminary Report on the Feasibility of Experimental Techniques, Sandia National Laboratories, Albuquerque NM.
[139] NRC Generic Letter 88-20: "Individual Plant Examination for Severe Accident Vulnerabilities." — U.S. Nuclear Regulatory Commission, 1988.
[140] NUREG 1335: "IPE Submittal Guidance Document," August 1989.
[141] NUREG/CR-5132: "Severe Accident Insights Report," prepared by Brookhaven National Laboratory, for the NRC, April 1988.

[142] NUREG/CR-4920, Volumes 1–5: "Assessment of Severe Accident Prevention and Mitigation Features," prepared by Brookhaven National Laboratory, for the NRC, July 1988.
[143] NUREG/CR-2300: "PRA Procedures Guide," published by American Nuclear Society (ANS) and Institution of Electrical and Electronic Engineers (IEEE), January 1983.
[144] NUREG/CR-3485: "PRA Review Manual," prepared by Brookhaven National Laboratory, for the NRC, September 1985.
[145] NRC Policy Statement: "Severe Reactor Accidents Regarding Future Designs and Existing Plants," 50 FR 32138, August 8, 1985.
[146] NRC Policy Statement: "Safety Goals for the Operations of Nuclear Power Plants," 51 FR 28044, August 4, 1986.
[147] NRC Policy Statement: "Nuclear Power Plant Standardization," 52 FR 34844, September 15, 1987.
[148] NRC Policy Statement: "Regulation of Advanced Nuclear Power Plants," 59 FR 35461, July 12, 1994.
[149] NRC Policy Statement: "The Use of Probabilistic Risk Assessment Methods in Nuclear Regulatory Activities," 60 FR 42622, August 16, 1995.
[150] SECY-90-016: "Evolutionary Light-Water Reactor (LWR) Certification Issues and Their Relationship to Current Regulatory Requirements," January 12, 1990, and the related staff requirement memoranda (SRM) dated June 26, 1990.
[151] SECY-93-087: "Policy, Technical, and Licensing Issues Pertaining to Evolutionary and Advanced Light-Water Reactor Designs," April 2, 1993, and the related SRM dated July 21, 1993.
[152] SECY-96-128: "Policy and Key Technical Issues Pertaining to the Westinghouse AP600 Standardized Passive Reactor Design," June 12, 1996, and the related SRM dated January 15, 1997.
[153] SECY-97-044: "Policy and Key Technical Issues Pertaining to the Westinghouse AP600 Standardized Passive Reactor Design," February 18, 1997, and the related SRM dated June 30, 1997.
[154] Regulatory Guide 1.206: "Combined License Applications for Nuclear Power Plants," U.S. Nuclear Regulatory Commission, Washington DC 20555, June 2007.
[155] NUREG-0800, Chapter 19, "Probabilistic Risk Assessment and Severe Accident Evaluation for Nuclear Power Plants," U.S. Nuclear Regulatory Commission, Washington DC 20555, Draft Revision 3, September 2012.

CHAPTER

8

Design Basis and Beyond Design Basis Considerations of Natural Phenomena

Nilesh Chokshi and Goutam Bagchi

Containment structures are designed to remain functional when subjected to the loading from the natural events, such as earthquakes, flood water heights and associated effects, and high winds. Traditionally, the design bases for these natural hazards, particularly for seismic and flooding, have been evaluated using deterministic methods considering maximum historical events and accounting for uncertainties in characterizing such events. Our understandings of the natural hazard phenomena and our experiences from actual events have shown that the design bases are not bounding events and could be exceeded, albeit with low likelihoods. The evolution of probabilistic hazard methods has given us an ability to characterize hazard in terms of frequency of exceedance. These methods, in conjunction with the risk assessment methods, allow us to evaluate performance of a plant when subjected to design bases or beyond design bases natural events.

This chapter provides an overview of how the design bases were determined for the natural phenomena, and how they were incorporated in the containment design. The discussion then focuses on the current understanding of the natural phenomena and how the methods of probabilistic hazard and risk analyses have evolved. The methods for capacity and fragility evaluations for containments are treated in some detail with examples and insights from past evaluations, experiences, and test programs. Finally, some of the recent initiatives that have followed the occurrence of the Fukushima event are described.

This chapter is by no means a comprehensive and full treatment of containment design or performance with respect to the natural phenomena hazards. Also, the intent is not to provide a comprehensive bibliography, instead, to provide broad insights and discussions of some of the key aspects of the topic. The contents of the chapter reflect authors experiences related to the NRC activities and practices in the United States. Many other countries and organizations have also had activities related to containments and containment performance.

Keywords: Natural phenomena, containment design, seismic, flooding, high-winds, containment fragility, probabilistic hazard analysis, probabilistic risk-assessment methods, seismic margin analysis, design basis, beyond design basis

8.1 INTRODUCTION

Natural phenomena hazards pose a common cause threat that could affect the entire site and large areas surrounding the site and all of the units located at a site compromising accident prevention and mitigation measures including multiple defense-in-depth barriers. The containment is a final barrier preventing uncontrolled release of radioactive materials to the outside environment. Section 8.2 provides a summary of how the design bases were derived for currently operating plants, governing regulations and associated guidance and some of the key design concepts. The containment structures are designed to withstand the natural phenomena loads in combination with the accidental pressure loads to further assure adequate safety performance. For the natural phenomena like earthquakes and high winds, the containment structures are designed to essentially remain elastic and in the linear range of the material properties, providing substantial margins to withstand loads beyond design basis. As our knowledge of the natural hazards phenomena has evolved and recent experiences have indicated, the design bases events are not necessarily bounding events and there is some, albeit small, likelihood of exceeding the design bases events. Section 8.3 provides more details of the recent understanding of the hazards, advances in the state-of-the-art, and insights from some of the recent experiences. Section 8.3 examines why beyond design basis considerations are needed. As discussed earlier, a containment structure has significant margin (capacity) to withstand events beyond the design basis events. Section 8.4 discusses methods to estimate available margins beyond the design basis events. The probabilistic risk assessments (PRA) provide a framework to look at the integrated performance of a plant when subjected to various accident initiators, including natural phenomena hazards. Section 8.4 describes the PRA concepts as they relate to the natural phenomena and the functional and integrity aspects of containment structures. Section 8.5 summarizes various initiatives that are being implemented in the United States in aftermath of the Fukushima event.

Insights described in this chapter reflect authors' experiences at the NRC and their involvement in the NRC's domestic and international activities. As such, the practices in the United States are primarily discussed. Many other countries and organizations have also had activities related to containments and containment performance.

8.2 SUMMARY OF DESIGN BASIS FOR NATURAL PHENOMENA

The containment structure is designed to provide protection against uncontrolled release of radioactive materials by providing an essentially leak tight barrier against internal pressure and temperature due to an accident. For many of the earlier power plants, the containment structure was designed for internal pressure due to the postulated design basis accident. After the Three Mile Island accident in 1979, severe accident management strategies were developed for existing plants by being able to take advantage of the inherent margins in containment structure capacity.

Containment structures are designed to remain functional when subjected to the loading from the natural events. Natural events induced loads consist of normal and extreme wind loads, external surface and ground water flooding loads, and dynamic load induced by ground motion from earthquakes. Effects of vibratory motion from earthquakes pose important design challenges because it is a common cause that has the potential for causing subsoil liquefaction, soil and structure interaction, interaction of adjacent structures below and above ground, and inertia and anchor motion loads on supported equipment and system. The ground motion can combine with simultaneous flooding and high waves due to seiche in adjacent water bodies.

Both concrete and steel containment structures are designed in accordance with the provisions of the American Society of Mechanical Engineers (ASME) Boiler and Pressure Vessel Code [1]. Steel containments are designed to Subsection NE of ASME B&PV Code. Steel containment design is discussed in detail in Chapter 3 of this book.

Containment structures are designed by applying the design basis natural events loading in combination with other loads. Loads that are to be combined for design purposes are in accordance with the Section III, Division 2, of the American Society of Mechanical Engineers (ASME) Boiler and Pressure Vessel Code [1]. The containment structure is designed to 1.5 times the design basis accident pressure, P_a; combination of P_a and safe shutdown earthquake (SSE); combination of 1.25 times P_a and 1.25 times the operating basis earthquake (OBE) along with dead and live loads, including any effects of temperature. The resulting design using the Code allowable stresses and strains contribute to the robustness of the containment structures. Since earthquake ground motion results from stress release within the earth's crust, large earthquakes in the order of SSE frequently produce additional sequences of vibratory ground motion until the stress regime within the crust comes to equilibrium. The additional sequence of earthquake motion is termed aftershock. Consequently, the design of steel containments and containment penetration attachments of steel require evaluation of fatigue due to earthquake motion cycles. Seismic design of containment and all Category I structures is carried out on the basis of combined loading of SSE and design basis accidental pressures using elastic stress limits; therefore, the effects of an aftershock need not be explicitly considered in the structural design. Consequently, it is necessary and sufficient to design seismic Category I structures for only the SSE loading.

Effects of both internal and external flooding are considered in the containment design. Some containment structures are designed for post-LOCA internal flooding effects, and the load combination includes the effects of OBE. External flooding is a site related condition and the evaluation is also considered on a site-specific basis.

Traditionally, the containment and other Seismic category I structures were designed for only tornado and tornado-associated missile effects. However, as discussed later, for the coastal sites, particularly in the southeastern United States, extreme wind loads and wind borne missiles can be caused by either hurricanes or tornadoes. The containment structure is designed for the governing effect of missile impact and extreme loads. When the eye of a tornado is on the containment structure, a negative pressure is imposed on the outer shell of a structure. The containment structure is designed for the applicable negative pressure from a tornado. As discussed in Regulatory Guide 1.221 [2], hurricane wind speed in southeastern United States can be equal to or greater than tornado wind speed. Consequently, effects of hurricane wind and wind bourn missiles would need to be evaluated.

8.2.1 Key Regulations Related to Containment Design

The containment structure is the most important structure in a nuclear power plant since it provides the overall barrier to release of radioactive material to the atmosphere. U.S. commercial nuclear power plants have been designed with containment structures; see Chapter 1 for additional discussion. The defense in depth philosophy in the design of containment structures is embedded in the "Title 10 of the Federal Code of Regulation (CFR), Part 50," [3]. The following is a list of key regulations related to containment design and natural hazard phenomena.

Appendix A to Part 50 — General Design Criteria (GDC) for Nuclear Power Plants

GDC 2, Design Bases for Protection against Natural Phenomena
GDC 16, Containment Design
GDC 50, Containment Design Basis
GDC 51, Fracture Prevention of Containment Pressure Boundary
GDC 52, Capability for Containment Leakage. Rate Testing
GDC 53, Provisions for Containment Testing and Inspection

Appendix J to Part 50 — Primary Reactor Containment Leakage Testing for Water-Cooled Power Reactors

Chapter 4 provides additional details on containment leakage tests and their frequencies. In addition to above, another important aspect of the design of containment structures is provided in GDC 4: Environmental and dynamic effects design bases, which states that "Structures, systems, and components important to safety shall be designed to accommodate the effects of and to be compatible with the environmental conditions associated with normal operation, maintenance, testing, and postulated accidents, including loss-of-coolant accidents. These structures, systems, and components shall be appropriately protected against dynamic effects, including the effects of missiles, pipe whipping, and discharging fluids, that may result from equipment failures and from events and conditions outside the nuclear power unit. However, dynamic effects associated with postulated pipe ruptures in nuclear power units may be excluded from the design basis when analyses reviewed and approved by the Commission demonstrate that the probability of fluid system piping rupture is extremely low under conditions consistent with the design basis for the piping."

In dealing with the natural hazard, the GDC 2 specifies: (1) appropriate consideration of the most severe of the natural phenomena that have been historically reported for the site and surrounding area, with sufficient margin for the limited accuracy, quantity, and period of time in which the historical data have been accumulated, (2) appropriate combinations of the effects of normal and accident conditions with the effects of the natural phenomena, and (3) the importance of the safety functions to be performed.

The following Section 8.2.2 describes the application of these regulations to develop natural hazards design basis for currently operating reactors. Section 8.3 covers recent developments in determining natural hazards design bases in more detail.

8.2.2 Seismic

In dealing with the design basis for seismic considerations, the application of GDC 2 led to the use of safe shutdown earthquake (SSE) that is based on maximum historical events with sufficient margin to account for uncertainties in our historical knowledge. The SSE is the design basis earthquake that is characterized by the maximum magnitude plus a margin based on seismic knowledge of the site. The ground motion is specified in terms of a peak ground acceleration (PGA) value associated with SSE that is anchored to a standard response spectrum shape, such as the shape of Regulatory Guide 1.60, "Design Response Spectra for Seismic Design of Nuclear Power Plants" [4]. Figure 8.1 shows an example of a SSE spectrum for an operating reactor. Eventually, the siting regulation was revised in 1997 in 10 CFR 100.23, Geologic and seismic siting criteria [5]. The revised criteria required that uncertainties inherent in such estimates be addressed through an appropriate analysis, such as a probabilistic seismic hazard analysis or suitable sensitivity analyses. Therefore, the SSE ground motion spectra are currently developed using seismic probabilistic hazard analysis.

Regulatory guide 1.208, "A Performance-Based Approach to Define the Site-Specific Earthquake Ground Motion," [6] provides a step by step procedure for conducting a probabilistic seismic hazard analysis, and deriving a site-specific ground motion that is consistent with a performance goal of not exceeding significant inelastic deformation of structures, systems and components at a frequency of 10^{-5} per year. Chapter 3 of NUREG 0800, "Review of Safety Analysis Reports for Nuclear Power Plants," [7] provides seismic analysis and design criteria for structures, systems and components.

8.2.3 Flooding

Design basis flooding at a site is determined by evaluating all potential causes that produce the maximum flood elevation. The flood hazard elements are: local intense precipitation, probable maximum precipitation in the site drainage area, stream flooding, surges due to hurricanes, seiche, tsunami, dam failure effects, landslide induced flooding, ice blockage, and the worst flooding at a site that may result from a reasonable combination of individual flooding mechanisms.

Hurricanes and tsunamis are complex natural phenomena and they can pose challenges to plant safety due to concomitant effects. Hurricanes can cause flooding due to precipitation and surge with concomitant loss of off-site power. Similarly, tsunami waves can inundate a site and create impact due to floating debris, loss of off-site power, etc. A tsunami induced flooding and seismic ground motion that precedes it, like that occurred in March 11, 2011 event in Japan requires considerations of both seismic and tsunami effects on the plant and emergency response.

Regulatory Guide 1.59, "Design Basis Floods for Nuclear Power Plants" dated 8. 1977 [8] has provisions for flooding evaluations at nuclear power plant sites. RG 1.59 is currently being revised. RG 1.102, "Flood Protection for Nuclear Power Plants," dated September 1976 [9] has provisions for flood protections including those for design of protective structures for plants that need exterior flood barriers to protect plant safety.

8.2.4 High Winds

This natural hazard group can be subdivided into two categories; one, the design basis severe wind and two, the extreme wind, such as those due to tornados and hurricanes. Severe winds are based on an annual frequency of 10^{-2}. Severe wind loads on structures are generally enveloped by seismic loads. Tornado or hurricane wind loads are generated by extreme winds with an annual frequency of 10^{-7}. Extreme wind load needs to be combined with effects of wind bourn missiles. Regulatory Guide 1.76, "Design-Basis Tornado and Tornado Missiles for Nuclear Power Plants," Dated April 1974, Rev 0 [10] has been used to determine tornado wind speeds and associated spectrum of missiles consistent with applicable wind velocity ranges. RG 1.76 subdivided the continental U. S. into three regions for ranges of wind speed. Recent developments in this area are further discussed in Section 8.3.2.3.

8.2.5 Other Natural Hazards

Other natural hazards that can affect safety are listed as follows: avalanche, biological events, drought, forest fire, hurricane, ice cover, landslide, lightning, river diversion, sandstorm, severe wind, snow, frost, hail, high temperatures, low water, low temperatures, precipitation, tornado, volcanic activity and water spouts These hazards are very site specific and many of these hazards can be eliminated based on location, site elevation, geological and tectonic setting, and other factors. However, those that cannot be eliminated should be considered in the design.

8.3 DESIGN BASIS AND BEYOND DESIGN BASIS EVENTS

8.3.1 Historical Perspective

As discussed in the preceding section, the design bases for the currently operating plants were established in 60s, 70s, and 80s, using the knowledge and state-of-the-art technology available then. Historically, for seismic and flooding, the design bases were established using deterministic approaches taking into account the historical events and making implicit adjustments for the uncertainties in our knowledge. Our understanding of the natural phenomena continue to evolve through advances in the state-of-the-art, new technology, and from the actual experiences. Some of the major advances that have occurred include development of the probabilistic hazard assessment methods, better instrumentation and investigative tools, investigations of actual events, enhanced modeling, and paleo-seismology and paleo-flood evaluations. Along with the enhanced understanding, the recent seismic events have shown that the design basis ground motions used in the past could have significantly different characteristics when compared to current day predictions of the ground motions at a site. Although, the design bases have a very low probability of exceedance, the recent experiences have shown that they can be exceeded. The following sections discuss in more detail evolution of the hazard understanding and observations from the recent experiences. The emergence of probabilistic risk assessment (PRA) technology also provide a framework to assess the potential risk from the natural phenomena events and establish risk-informed/performance based approaches to establish design bases and design methods. This framework is also discussed in Section 8.3.4.

Collectively, the above elements highlight the need for considering events beyond a design basis event and periodical assessment of a design basis. In fact, the regulatory framework in the US has processes which allow for evaluation of new knowledge. The US industry in 90s under a program called, Individual Plant Examination for External Events (IPEEE), examined to some extent, a plant's capability to withstand events larger than the design basis events (this is further discussed in Section 8.4). The NRC Near-Term Task Force (NTTF) in its report [11] has recommended re-evaluation of the seismic and flooding hazards using the present day licensing methods for the new reactor applications for the currently operating reactors. The NTTF recommendations are aimed at improvement of overall plant safety against important external hazards that can severely affect how a plant is to be maintained in a safe shutdown condition when subjected to initiating events that are beyond the design basis and pose common cause challenges to plant protection. The containment structure is the last barrier to uncontrolled release of radioactivity; consequently, a systematic examination of the integrity of the containment barrier is necessary under the overall NTTF initiative. The NTTF has also recommended periodical assessment of the natural hazards. These aspects are discussed in more detail in Section 8.5.

8.3.2 Evolution of Hazard Understanding

8.3.2.1 Seismic. As discussed in Section 8.2, seismic design basis for a nuclear power plant (NPP) involves determination of the vibratory ground motion or SSE at the site for which the Seismic Category I structures, systems, and components, including the containment structure and systems are to be designed. The SSE is defined in terms of response spectra. For most operating plants, these spectra were derived using estimates of peak ground acceleration (PGA) at the site and anchoring a standard spectral shape to the PGA; for example, RG 1.60 [4]. Figure 8.1 shows a comparison between SSE spectrum anchored to PGA of 0.2 g and a ground motion response spectrum (GMRS) derived for a recent combined license (COL) application using probabilistic seismic hazard analysis (PSHA) and risk-informed/performance-based approach (RG 1.208 [6] and ASCE/SEI Standard 43-05-[12]).

There are several things to be noted here. The shape of the SSE spectrum in the Fig. 8.1 has been derived from using the western US recorded earthquakes that were available at the time of development of the Regulatory Guide 1.60 in early seventies. This spectrum is rich in the lower frequencies when compared to the GMRS. The GMRS reflects the recent understanding of seismic sources and ground motion attenuations in the Central and Eastern US (east of the Rockies). The NPP SSCs, in general, have their fundamental

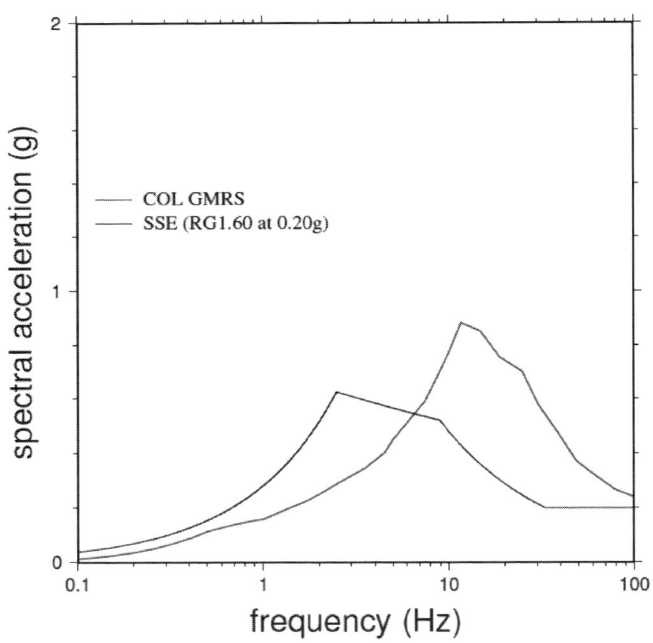

FIG. 8.1
COMPARISON OF DESIGN SPECTRA: PAST APPROACH VS. CURRENT APPROACH

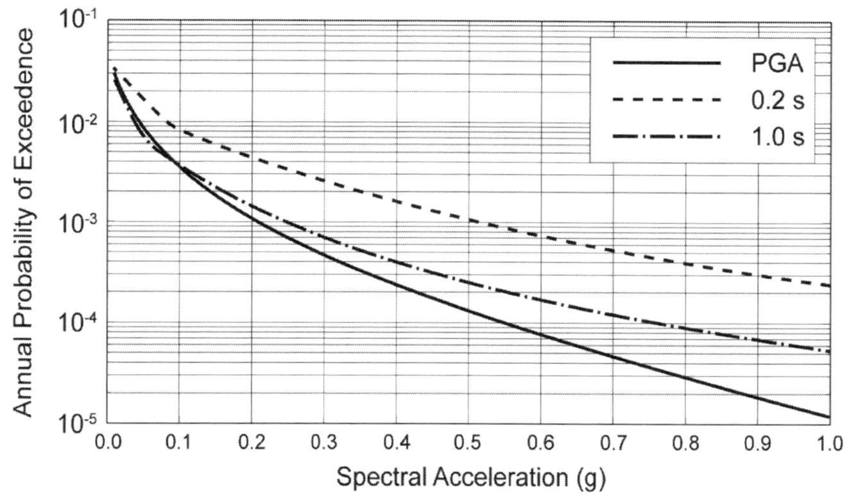

FIG. 8.2
EXAMPLES OF SEISMIC HAZARD CURVES FOR A SITE IN TERMS OF SPECTRAL ACCELERATIONS AT DIFFERENT RESPONSE PERIODS

frequencies below 10 Hz, and therefore, the higher ground motion estimated in the high frequency range is of less concern, except for chatter prone equipment. Modern solid state control instruments and components are generally not chatter prone. In fact, because of the richness in the low frequency ground motions, the existing NPP SSCs have significant margin beyond the design basis ground motions.

The advent of PSHA came with the seminal publication by Prof. Allin Cornell in 1968 [13]. This publication has allowed explicit consideration of the uncertainties and frequencies of exceedance in estimates of ground motions of various levels at a given site. Figure 8.2 of typical seismic hazard curves show likelihood of occurrence of ground motion of various levels at a site, from the application of a PSHA. NUREG/CR-6372, "Recommendations for Probabilistic Seismic Hazard Analysis: Guidance on Uncertainty and Use of Experts," U.S. NRC (1997) [14], NUREG-2117, "Practical Implementation Guidelines for SSHAC Level 3 and 4 Hazard Studies, USNRC (2012) [15], Reiter [16], and Regulatory Guide 1.208 [6] are some of the references that discuss concepts and guidance for conducting a PSHA.

Advances such as paleo-liquefaction studies, more accurate dating techniques, and additional strong motion data from recent events have significantly enhanced our ability to estimate parameters, such as timing and size of pre-historical events and recurrence frequencies and magnitude distance variability of ground motion that are input to both deterministic and probabilistic hazard assessments.

By using the results from the PSHA, studies have shown that the mean probability of exceeding the deterministic seismic design basis (SSE) of the currently operating plants roughly ranges from 10^{-3}/year to 10^{-6}/year. Thus, the deterministic design basis is neither hazard nor risk consistent based on current knowledge. The use of PSHA in conjunction with the risk-informed/performance-based considerations allows establishment of risk-consistent seismic design basis, [6, 12, 17]. The GMRS shown in Fig. 8.1 is derived using such concepts.

8.3.2.2 Flooding. As discussed in Section 8.2.2, the estimate of design basis flood involves several flood causing mechanisms with different approaches. The last revision of Regulatory Guide 1.59 that was published by the U.S. NRC in 1977 [8] has been used to derive the flooding design basis for many of the operating plants. Since that time, flood estimation techniques have significantly advanced. The availability of more accurate datasets and advent of new analysis methodologies such as geographical information systems have facilitated rapid processing of large, spatially distributed datasets in the estimation of input for newer classes of hydrologic, hydraulic, and hydrodynamic models. The recently published report (NUREG/CR-7046) [18] has outlined relevant information that can be used in the estimation of design-basis floods at nuclear power plant sites. As noted in the NUREG/CR-7046 [18], the availability of more than 30 years of additional data since the last revision of Regulatory Guide 1.59 [8] also provides a longer and more robust validation dataset. Advances in computing speeds and availability of specialized software also reduce the time required for analysis, allowing the examination of multiple scenarios and thereby helping to reduce uncertainty in estimates of design-basis flood hazard. The recent experiences with the evaluation of early site permit (ESP) and combined license (COL) applications have demonstrated that using the current modeling techniques and databases the estimates of design basis could be significantly different than the earlier estimations. Thus, similar to the seismic hazard, the new knowledge and understanding of the flooding phenomena also indicate that there is a need to consider beyond design basis events in plant evaluations.

The design-basis floods are primarily derived from the deterministic procedures. Probabilistic flood hazard analysis (PFHA) is only possible for some of the flood causing mechanisms and is to be further developed before they gain wider acceptance. The NUREG/CR-7046 [18] provides additional insights into the current status of the PFHA.

As noted in NUREG/CR-7046 [18], the recent debate has focused on the projected climate change in the latter parts of the twenty-first century and what these projections mean to the estimation of design-basis flood hazards at sites where nuclear power plants may be expected to be in operation during the same period.

The strategy for providing protection against a design basis flood is significantly different than that used to protect against other natural hazards like seismic and wind. For seismic design basis, the seismic category I SSCs are designed or qualified by analysis or testing resulting in substantial margins. Protections against floods involve passive and active barriers, temporary measures, and other operational procedures as discussed in [9]. As noted in the NTTF report [11], the Task Force concluded that flooding risks are of concern due to a "cliff-edge" effect, in that the safety consequences of a flooding event may increase sharply with only a small increase in the flooding level.

8.3.2.3 High Winds. As discussed in Section 8.2, the extreme wind loads in the past were characterized by the tornado wind load and missile effects associated with the annual frequency of 10^{-7}. The following Fig. 8.3 represents the tornado wind speeds and some of the other associated parameters that were used in the past designs (RG 1.76, Revision 0 [10]).

The recent advances have led to decrease in the estimates of the tornado wind speeds over time. The following Fig. 8.4 and the accompanying Table 8.1 show the recent estimates of the tornado wind speeds and other associated parameters postulated in Revision 1 of Regulatory Guide 1.76 [19].

As a result of downward revision of tornado wind speeds, the effects of hurricanes which were considered bounded by the design basis tornados can no longer be considered bounded. Both the hurricane wind speed and missile effects now need to be considered for certain sites. The USNRC has recently issued Regulatory Guide 1.221, "Design-Basis Hurricane and Hurricane Missiles for Nuclear Power Plants," October, 2011 [2]. Consequently; effects of hurricane wind and wind bourn missiles would need to be evaluated. Table 8.2 below is excerpted from RG 1.221 and it shows the hurricane missile spectrum. In southeastern Florida hurricane missiles may need to be evaluated in addition to the tornado missile effects.

The protection against high winds is generally provided by the structures that house the safety related components and systems. The structures are designed for both wind induced external and internal pressure loads and providing adequate thickness with margins to prevent undesirable penetration of missiles or spalling of concrete walls.

Because of the very low probability of the design basis extreme wind as initiating events, these are generally not considered further in risk assessments. However, wind effects in combination with associated flooding (for example, during a storm surge event) need to be included in the risk evaluations.

RG 1.76, Revision 0

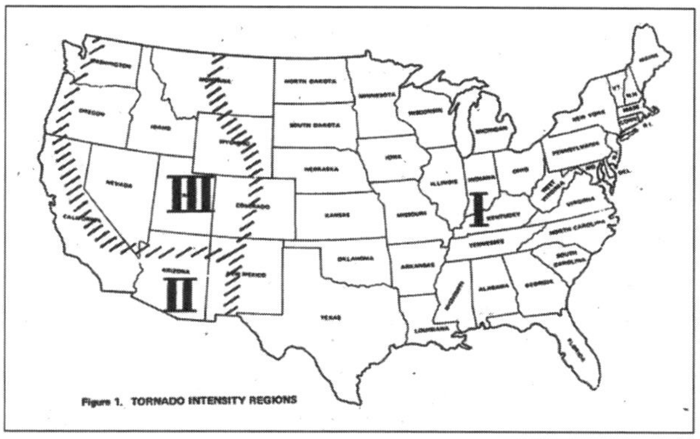

Figure 1. TORNADO INTENSITY REGIONS

Tornado Region	Max Wind Speed (mph)	Pressure Drop (psi)	Rate of Pressure Drop (psi/sec)
I	360	3.0	2.0
II	300	2.25	1.2
III	240	1.5	0.6

14

FIG. 8.3
TORNADO PARAMETERS USED IN PAST DESIGNS

Revision 1 to RG 1.76

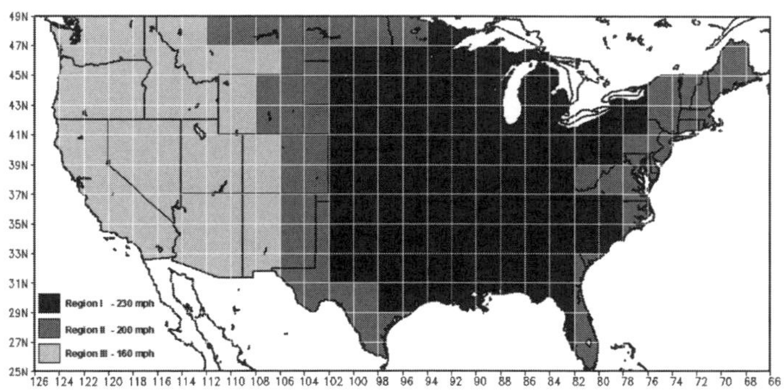

Tornado Region	Max Wind Speed (mph)	Pressure Drop (psi)	Rate of Pressure Drop (psi/sec)
I	230	1.2	0.5
II	200	0.9	0.4
III	160	0.6	0.2

25

FIG. 8.4
REVISED TORNADO WIND SPEEDS AND OTHER ASSOCIATED PARAMETERS

TABLE 8.1
DESIGN-BASIS TORNADO MISSILE SPECTRUM AND MAXIMUM HORIZONTAL SPEEDS

Missile Type		Schedule 40 Pipe	Automobile	Solid Steel Sphere
Dimensions		0.168 m dia × 4 .58 m long (6.62 5 in. dia × 1 5 ft long)	Region I and II 5 m × 2 m × 1.3 m (16.4 ft x 6.6 ft x 4.3 ft) Region III 4.5 × 1.7 × 1.5 m (14.9 ft x 5.6 ft x 4.9 ft)	2.54 cm dia (1 in. dia)
Mass		130 kg 287 lb)	Region I and II 1810 kg (4000 lb) Region III 1178 kg (2595 lb)	0.0669 kg (0.147 lb)
C_DA/m		0.0043 m^2/kg (0.0212 ft^2/lb)	Region I and II 0.0070 m^2/kg (0.0343 ft^2 /lb) Region III 0.0095 m^2/kg (0.0464 ft^2/lb)	0.0034 m^2/kg (0.0166 ft^2/lb)
max V_{Mh}	Region I	41 m/s (135 ft/s)	41 m/s (135 ft/s)	8 m/s (26 ft/s)
	Region II	34 m/s (112 ft/s)	34 m/s (112 ft/s)	7 m/s (23 ft/s)
	Region III	24 m/s (79 ft/s)	24 m/s (79 ft/s)	6 m/s (20 ft/s)

TABLE 8.2
HURRICANE MISSILE SPECTRA [2]

Missile Type	Dimensions	Mass
Automobile	5 m × 2 m × 1.3 m (16.4 × 6.6 × 4.3 ft)	1,810 kg (4,000 lb)
Schedule 40 Pipe	0.168 m dia × 4.58 m long (6.625 in. dia × 15 ft long)	130 kg (287 lb)
Solid Steel Sphere	25.4 mm (1 in.) diameter	0.0669 kg (0.147 lb)

8.3.3 Recent Experiences Related to Natural Phenomena Hazards

8.3.3.1 Seismic Experience. Since 2007, three earthquake events have occurred resulting in ground motions larger than the design basis ground motions at the affected plants. These three events are:

(1) The Niigataken Chuets-oki (NCO) earthquake of July 16, 2007, with epicenter 16 km from Kashiwazaki-Kariwa (KK) NPP site, Magnitude 6.8 (JMA Magnitude)

(2) Tohoku Pacific earthquake of March 11, 2011 affecting various sites including Fukushima-Daiichi, Magnitude 9.0 Mw

(3) August 23, 2011, Mineral, Virginia earthquake, epicenter about 11 miles from North Anna Power Station, Magnitude 5.8

Figure 8.5 shows observed maximum accelerations at the base mat of Reactor Buildings of all seven KK units with the corresponding design basis values. As seen in the figure, all seven units recorded significant but different motions and the observed motions significantly exceeded the design basis in some cases.

Observed Peak Ground Acceleration at Lowest Floor of the Reactor Building

(gal)

	North-South Component	East-West Component	Vertical Component
Unit #1	311(274)	680(273)	408(235)
Unit #2	304(167)	606(167)	282(235)
Unit #3	308(192)	384(193)	311(235)
Unit #4	310(193)	492(194)	337(235)
Unit #5	277(249)	442(254)	205(235)
Unit #6	271(263)	322(263)	488(235)
Unit #7	267(263)	356(263)	355(235)

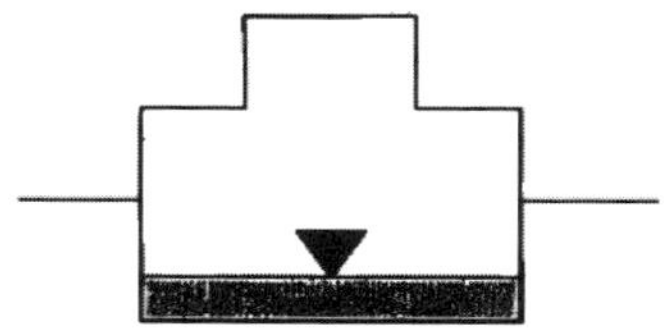

Numbers shown in ( ) are the design values.

FIG. 8.5
OBSERVED AND DESIGN BASIS MOTIONS AT KK UNITS

Figure 8.6 shows similar comparison of observed motion and design motion for Fukushima Daiichi units during the Tohoku Pacific earthquake. The highest peak acceleration measured at Fukushima Daiichi was 0.561 g in the horizontal direction and 0.308 g in the vertical direction at Unit 2. In three instances, at Units 2, 3, and 5, the observed motion exceeded the design basis. These exceedances were much smaller compared to the exceedances at the KK units.

North Anna earthquake:

At 1:51 p.m. EDT on August 23, a magnitude 5.8 earthquake occurred near Mineral, Virginia, close to the North Anna Power Station. The earthquake caused the reactor plants to automatically shut down and resulted in a loss of off-site power. The plant declared an Alert, the second lowest of the four emergency classification levels used by U.S. nuclear plants. No damage was reported to systems required to maintain the station in a safe condition. Several aftershocks felt in the region later that day did not affect the station.

	North–South Component	East–West Component	Vertical Component
Unit 1	460 (487)	447 (489)	258 (412)
Unit 2	348 (441)	550 (438)	302 (420)
Unit 3	322 (449)	507 (441)	231 (429)
Unit 4	281 (447)	319 (445)	200 (422)
Unit 5	311 (452)	548 (452)	256 (427)
Unit 6	298 (445)	444 (448)	244 (415)

FIG. 8.6
OBSERVED AND DESIGN BASIS MOTIONS (IN GAL) AT FUKUSHIMA DAIICHI UNITS (NUMBER SHOWN IN PARENTHESES ARE DESIGN VALUES)

FIG. 8.7
NORTH ANNA NUCLEAR POWER PLANT

As designed, the plant's four emergency diesel generators started and supplied power to important electrical equipment while off-site power was unavailable. One of the four emergency diesel generators was taken off-line to repair a coolant leak, and another generator available at the station was started to replace it until the off-site power was restored. Later that evening, off-site power to the plant was restored, eliminating the need to rely on the back-up generators. The North Anna Power Station has two Safe Shutdown Earthquake ground motions, one for structures, systems, and components located on top of rock, which is anchored at 0.12 g, and the other for structures, systems, and components located on top of soil, which is anchored at 0.18 g. The plant has two corresponding Operating Basis Earthquake ground motion spectra, anchored at 0.06 g for rock and 0.09 g for soil. At several frequencies, the spectral and peak ground accelerations as a result of the August 23, 2011 earthquake were greater than those used for the Operating Basis and Design Basis Earthquakes. Figures 8.8 and 8.9 below show recorded motions at the basemat of Unit 1 containment building. The observed horizontal spectrum (North-South component) and vertical spectrum show exceedances when compared to the corresponding SSE spectra.

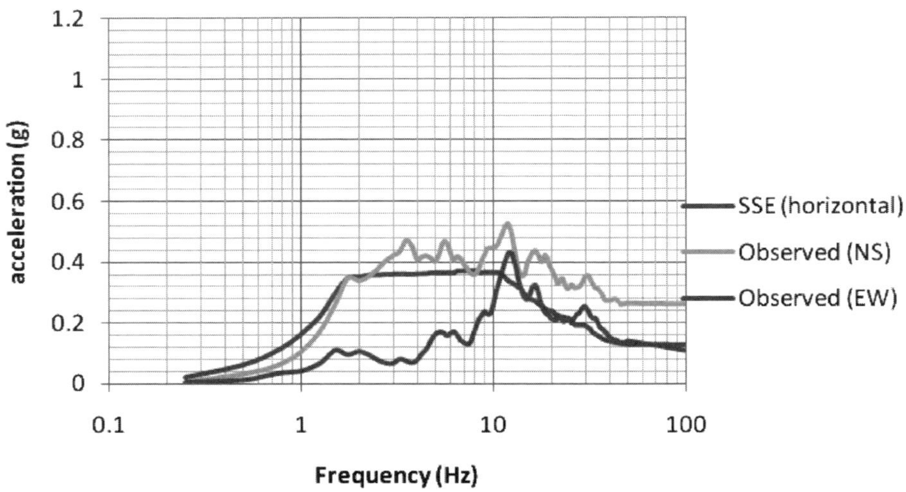

FIG. 8.8
UNIT 1 BASEMAT SPECTRUM COMPARISON (HORIZONTAL) BETWEEN THE DESIGNED AND THE OBSERVED

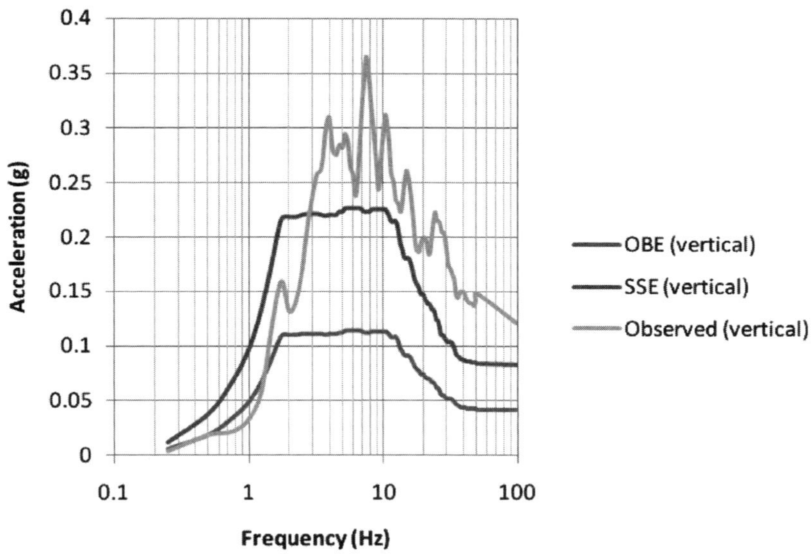

FIG. 8.9
UNIT 1 BASEMAT SPECTRUM COMPARISON (VERTICAL) BETWEEN THE DESIGNED AND THE OBSERVED

Several salient observations can be made from these three events:

(1) Common-cause effects of an earthquake, in-terms of affecting all of the units at a site, are clearly evident. Also, the observed damage of some of the non-safety related components at KK units highlighted a need for consideration of potential for significant correlation of earthquake responses and capacities of SSCs so that seismic risk is more realistically assessed. This is particularly an important issue in considering site risk for a site with multiple units.

(2) The recorded motions at units from the same event are significantly different, thus highlighting spatial variations and important effects of site topography, layering, and subsurface conditions under each unit that affect local site response.

(3) Although, the observed motions were higher, both at KK and North Anna Nuclear Power Plant, there was no observable damage to safety-related SSCs and these SSCs were able to perform their safety functions. The information available to date also indicates that there is no evidence of significant seismic induced damage to the safety related SSCs at the Fukushima units. This observation confirms the existence of margins and conservatisms associated with the seismic design processes.

(4) The occurrence of events that challenge or exceed a design basis could lead to extended shutdowns. Existence of procedures to deal with the post-earthquake inspections and conducting necessary plant evaluations can reduce the down time.

(5) Existence of functioning seismic instrumentation is also very important to fully understand the ground motions and other earthquake parameters, and to evaluate the plant responses.

8.3.3.2 Flood Experience. Blayais Nuclear Power Plant is located on the Gironde estuary near Blaye, South Western France. The 1999 flood took place on the evening of December 27, 1999. A combination of tide and high winds led to the sea walls of the Blayais Nuclear Power Plant in France being overwhelmed. The event resulted in the loss of the plant's off-site power supply and, it knocked out several safety-related systems, resulting in a Level 2 event on the International Nuclear Event Scale. The incident illustrated the potential for flooding to damage multiple items of equipment throughout a plant, weaknesses in safety measures, systems and procedures, and led to fundamental changes to the evaluation of flood risk at nuclear power plants and the precautionary measures to be taken.

The M9 Fukushima earthquake is one of the best documented and recorded events. The ground motion at the Daiichi station was slightly higher or equal to the design ground motion, see Section 8.3.3.1. The tsunami sequence of the event is well described in the, "Special Report on the Nuclear Accident at the Fukushima Daiichi Nuclear Power Station," by the Institute of Nuclear Power Operations (INPO), Report INPO 11-005 November 2011 [20]. The following excerpts are from the INPO report: "Three minutes after the earthquake, the Japan Meteorological Association issued a major tsunami warning, indicating the potential for a tsunami at least 3 meters high. Station workers were notified of the warning and evacuated to higher ground."

Forty-one minutes after the earthquake, at 1527, the first of a series of seven tsunamis arrived at the site. The maximum tsunami height impacting the site was estimated to be 46 to 49 ft (14 to 15 m). This exceeded the design basis tsunami height of 18.7 ft (5.7 m) and was above the site grade levels of 32.8 feet (10 meters) at units 1 to 4. All AC power was lost to units 1 to 4 by 1541 when a tsunami overwhelmed the site and flooded some of the emergency diesel generators and switchgear rooms. The seawater intake structure was severely damaged and was rendered nonfunctional. All DC power was lost on units 1 and 2, while some DC power from batteries remained available on unit 3. Four of the five emergency diesel generators on units 5 and 6 were inoperable after the tsunami. One air-cooled emergency diesel generator on unit 6 continued to function and supplied electrical power to unit 6, and later to unit 5, to maintain cooling to the reactor and spent fuel pool. The experience from this event indicates that there is significant uncertainty associated with tsunami hazard characterization. It is also important to consider and plan for beyond design basis flooding.

At the Fort Calhoun Nuclear Plant in Nebraska, flood seeped into Turbine Building early Sunday, June 26, 2011 following the collapse of an 8-foot-tall, water-filled temporary berm that was protecting the plant. Either floodwaters from the Missouri River or groundwater seeped into several of the peripheral buildings at Fort Calhoun, but all of the areas containing radioactive material or crucial safety gear remained dry. Fort Calhoun has at least nine backup power sources in place, including six different power lines and two diesel generators, which were just tested Sunday after the failure of the water-filled berm. In this incident there was no immediate nuclear safety issue. However, experiences such as these highlight the need for consideration of beyond design basis flooding and careful planning for layers of defense in depth protection.

FIG. 8.10
FT. CALHOUN NUCLEAR POWER PLANT FLOODING

8.3.4 Risk-Informed Considerations

Since publication of WASH-1400, "Reactor Safety study: An Assessment of Accident Risks in U.S. Commercial Nuclear Power Plants," [21], the NRC and its licensees have advanced significantly in their knowledge of PRA. PRA considers nuclear safety in a more comprehensive way by examining a broad spectrum of initiating events (circumstances that put a facility in an off-normal condition, such as a reactor trip or "scram" at a nuclear power plant). As a result, PRA analysts ask the additional question of how likely it is that something will go wrong. Analysts then explore the frequency and consequences of various scenarios, giving a measure of risk. The PRA framework provides an integrated response of a plant to an initiating event, such as an occurrence of an earthquake; by considering capacities of SSCs, operator actions, and non-seismic failures. Thus, this approach allows evaluation of what are the most significant risk contributors and what enhancements are most beneficial. This is an integrated evaluation of plant response to an initiating event, as opposed to the design process which is compartmentalized by various engineering disciplines looking at a single structure, system, or a component at a time.

In 1995, the U.S. Nuclear Regulatory Commission (NRC) issued a Policy Statement [22] on the use of probabilistic risk analysis (PRA), encouraging its use in all regulatory matters. The Policy Statement states that "...the use of PRA technology should be increased to the extent supported by the state-of-the-art in PRA methods and data and in a manner that complements the NRC's deterministic approach." The increased use of the PRA framework requires knowledge of probabilistic hazard for a full spectrum of events, below and above the design basis event as discussed in Section 8.3.2.1 (Fig. 8.2) for seismic events. The use of probabilistic hazard within a PRA type framework allows risk-informed/performance-based approaches to be incorporated in the regulatory framework. As discussed in Section 8.3.2.1, this has already been done to establish the site-specific seismic design basis for new reactors. The NRC in SECY paper 93-087, "Policy, Technical, And Licensing Issues Pertaining to Evolutionary and Advanced Light-Water Reactor (ALWR) Designs," [23] has explicitly required consideration of beyond design basis seismic events for new reactors. This is achieved by use of the seismic PRA (SPRA) framework, and this is further discussed in Section 8.4.5.

8.4 METHODS FOR BEYOND DESIGN BASIS EVALUATIONS

8.4.1 Historical Perspective — Evolution of Methods

As noted above, there has been significant advancement in the PRA technology since the publication of WASH-1400. The external event methodology, particularly the SPRA methodology, has been applied in many applications. The following paragraph related to the evolution of the seismic PRA (SPRA) has been taken from [24].

"The methodology for performing SPRA has been in existence for over three decades, having arisen in the late 1970s from the ground-breaking NRC sponsored work with a major research program called the "Seismic Safety Margins Research Program (SSMRP)" run by Lawrence Livermore National Laboratory and involving more than a dozen experts nationally [25]. The methodology has evolved and matured, like the rest of PRA over this 3-decade period, and has by now been applied to study several dozen nuclear power plants all around the world. SPRA has been used to support risk informed decision-making to upgrade the safety of existing plants, to help prioritize which proposed backfits are most urgent, to help regulatory agencies like the US NRC, Department of Energy (DOE), and international agencies like the International Atomic Energy Agency (IAEA) to develop regulations and regulatory guidance related to seismic risk, to support the prioritization of safety research projects, and to develop insights into the overall seismic risk from an individual plant and from the whole fleet of plants. What emerges from the large number of SPRA studies is that typically the seismic part of the overall reactor risk is a significant contributor to the overall risk, sometimes dominant, almost always important, although sometimes negligible. One of the reasons for this and a unique and important aspect of the seismic risk is that earthquakes are one of the only initiators that affect the entire site, all of the plant structures, systems, and components (SSCs), and also the surrounding areas."

In the US, SPRAs have been performed for several NPPS beginning in late seventies and early eighties (for example, Oyster Creek [26], Zion [27], IPPS [28], Limerick [29], Millstone [30], and Seabrook [31]). These have included several Level 3 PRAs including results related to seismic induced core damage sequences, seismic related containment performance, and seismic contribution to consequences. Since then, most of the PRAs have been Level 1 studies with consideration of containment performance that affects the large-early release frequencies (LERF).

Methodology for other external hazards, such as flood and wind, is conceptually same as for the seismic initiators, but are less developed and not widely applied. Screening approaches are applied to many of the external events. Some of the earlier studies have included flood and wind initiators in a limited sense. These hazards are further discussed in Section 8.4.6.

At this point it is also necessary to mention the "Seismic Margin Analysis Methodology" (SMA) and its relationship to SPRA methodology. The SMA approaches, as outlined in the standard methodology references [32] and [33] are derived from insights and results of SPRA studies conducted in 1980s. The SMA methodology uses the system insights to focus on initiating events and systems which were found risk significant in those SPRAs. It concentrates on selected accident scenarios. Another simplification is in the determination of seismic capacities using more deterministic approaches. Perhaps the most appealing aspect of the SMA methodology is that it does not make use of the seismic hazard curves with their uncertainties, but instead requires selecting a large enough earthquake for the margin assessment. Therefore, the plant capacity is evaluated at a seismic ground motion level that is chosen to be higher than the design motion. Use of a prescribed deterministic level altogether avoids the issue of large uncertainties in the PSHA. Results from SMA studies are primarily related to the plant's seismic capacity, and therefore risk insights are limited. It is necessary to understand the limitations of this methodology. The application of the SMA methodology is limited to light water reactors. The SMA methods do not provide risk results such as core-damage frequency and large-early release frequency.

In conjunction with the Advance Light Water Reactors program, a PRA-based margin approach has also been used in the evaluation of new reactors as discussed in Section 8.4.5.

In early to mid-1990s, the industry performed beyond design basis evaluations for external hazards under the IPEEE program. The objectives of the IPEEE for each licensee were to:

(1) Develop an appreciation of severe accident behavior;
(2) Understand the most likely severe accident sequences that could occur at its plant under full-power operating conditions;
(3) Gain a qualitative understanding of the overall likelihood of core damage and fission product releases; and
(4) If necessary, reduce the overall likelihood of core damage and fission product releases by modifying, where appropriate, hardware and procedures that would help prevent or mitigate severe accidents.

Several PRAs, SMAs, and other evaluations were conducted for some of the external hazards. NUREG/CR-1472 [34] summarizes insights from these studies. Pertinent insights are included in Section 8.4.4.

The ASME/ANS Standard [35] includes requirements for SPRA and other external hazards, and for the SMA methodology.

8.4.2 Seismic Probabilistic Risk Assessment (SPRA) Methods

8.4.2.1 General. There are three distinct analyses that together comprise seismic PRA, the seismic hazard analysis, the fragility analysis, and the systems-accident analysis. These are brought together in the quantification analysis to produce the bottom-line analysis of core-damage frequency or frequency of large releases. As stated in the ASME-ANS Standard [35]:

(1) Seismic hazard analysis is used to develop frequencies of occurrence of different levels of ground motion (e.g., peak ground acceleration [PGA]) at the site;

(2) Seismic-fragility evaluation is used to estimate the conditional probability of failure of important structures and equipment whose failure may lead to unacceptable damage to the plant (e.g., core damage); a plant walkdown is an important activity in conducting this task;

(3) Systems/accident sequence analysis is used to model the various combinations of structural and equipment failures that could initiate and propagate a seismic core damage sequence;

(4) Risk quantification assembles the results of the seismic hazard, fragility, and systems analyses to estimate the frequencies of core damage and plant damage states. A so-called "level-2" or "level-3" seismic PRA assesses the impact of seismic events on the containment and consequence analyses, and then integrates these results with the core damage analysis to obtain estimates of seismic risk in terms of effects on public health (e.g., early deaths and latent cancer fatalities).

This process is shown schematically in Fig. 8.11 [36].

Next, in the following separate sections, current status of each of these parts of SPRA is discussed.

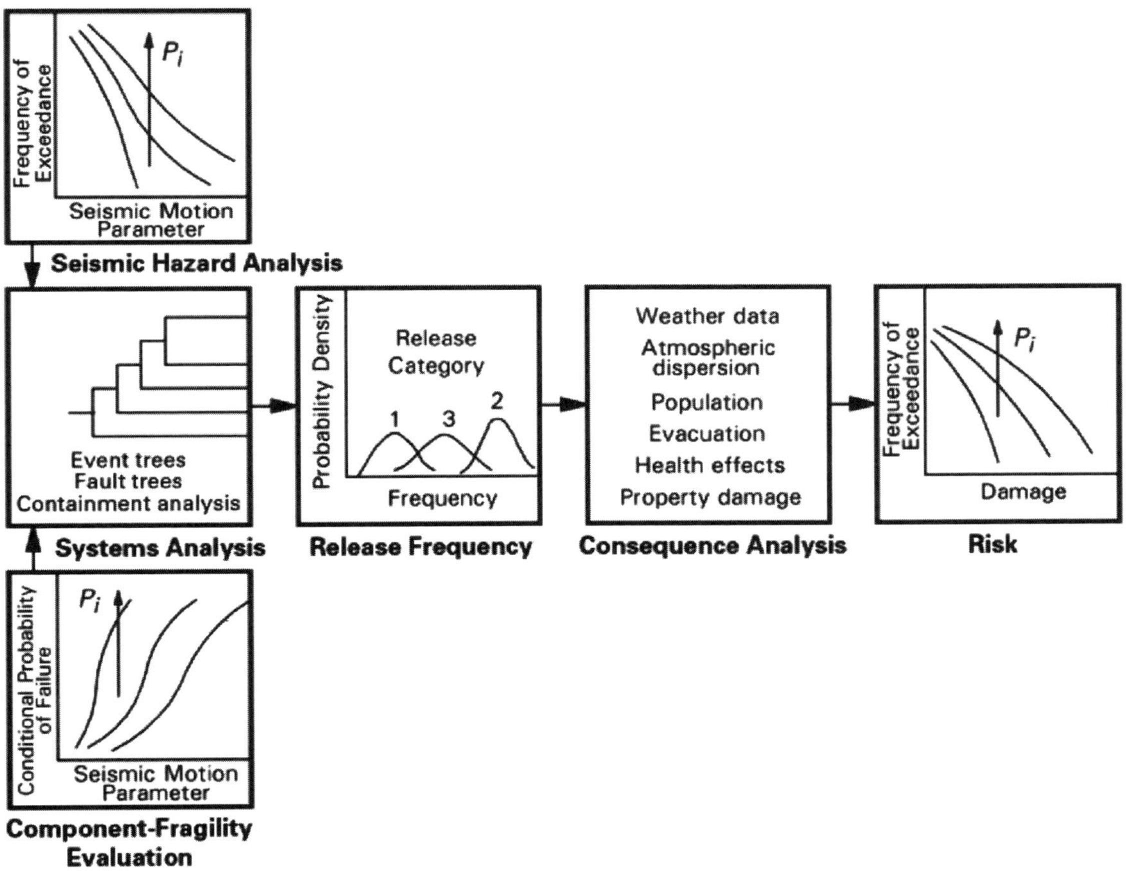

FIG. 8.11
SCHEMATIC OVERVIEW OF A SEISMIC PRA

8.4.2.2 Probabilistic Seismic Hazard Analysis. The objective of probabilistic seismic hazard analysis (PSHA) is to develop frequencies of occurrence of different levels of ground motion (e.g., peak ground acceleration [PGA] or a spectral acceleration) at a given site. Examples of hazard curves were shown in Fig. 8.2.

A useful description of PSHA is given in Appendix 5A of the ASME-ANS PRA Combined Standard [35] as follows:

Seismic hazard is usually expressed in terms of the frequency distribution of the peak value of a ground motion parameter (e.g., PGA) during a specified time interval. The different steps of this analysis are as follows:

(1) identification of the sources of earthquakes, such as faults and seismotectonic provinces;

(2) evaluation of the earthquake history of the region to assess the frequencies of occurrence of earthquakes of different magnitudes or epicentral intensities;

(3) development of attenuation relationships to estimate the intensity of earthquake induced ground motion (e.g., PGA) at the site;

(4) integration of the above information to estimate the frequency of exceedance for selected ground motion parameters.

The hazard estimate depends on uncertain estimates of attenuation, upper-bound magnitudes, and the geometry of the postulated seismic sources. Such uncertainties are included in the hazard analysis by assigning probabilities to alternative hypotheses about these parameters. A probability distribution for the frequency of occurrence is thereby developed.

Today, the standard methodology for PSHA that is used in the US and in some worldwide applications is the so-called "SSHAC methodology" named after the "Senior Seismic Hazard Analysis Committee" and its 1997 report [14]. This work was supported by DOE, NRC, and EPRI, and led to a bringing together into one place of all of the advances up to that time. A recent NRC publication [15] has provided additional guidance on PSHA methods.

8.4.2.3 Fragility Analysis. The objective of a seismic fragility analysis is to estimate the conditional probability of failure of a specific structure or equipment item, given an occurrence of earthquake ground motion. As discussed in Appendix 5A of the ASME-ANS PRA Combined Standard [35]:

> "This capacity is defined either in terms of average spectral acceleration value or PGA [peak ground acceleration] value for which the seismic response of a given component located at a specified point in the structure exceeds the component's resistance capacity, resulting in its failure. Although the average spectral acceleration is preferable, PGA has been used in many seismic PRAs and is acceptable provided that the uncertainties in the spectral shape are not too large. The ground acceleration capacity of the component is estimated using information on plant design bases, responses calculated at the design analysis stage, as-built dimensions, and material properties. Because there are many variables in the estimation of this ground acceleration capacity, component fragility is described by a family of fragility curves; a probability value is assigned to each curve to reflect the uncertainty in the fragility estimation. This family of fragility curves may be described by three parameters: the median acceleration capacity A_m, and logarithmic standard deviations, β_R and β_U, for randomness and uncertainty."

Figure 8.12 shows a typical family of fragility curves for a component including median, 5% confidence and 95% confidence, and mean fragility curves.

As noted in [24], the work of developing seismic fragilities is a specialized activity of engineers with specialized training and experience. Excellent general guidance exists on seismic fragility analysis. In recent years,

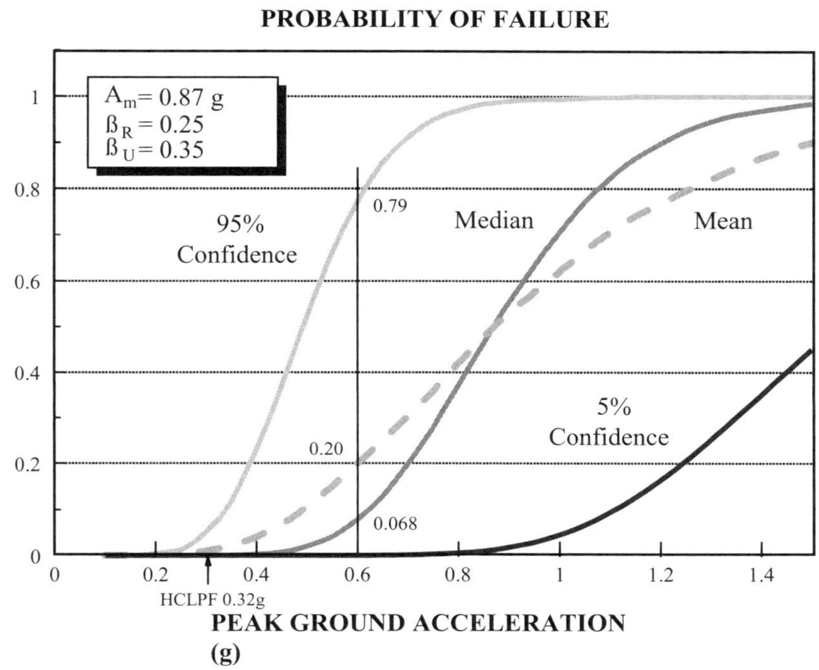

FIG. 8.12
TYPICAL FAMILY OF FRAGILITY CURVES FOR A COMPONENT

there have been some important advances in the methods for collecting and reviewing design documents, and in walkdown and documentation requirements. Performing probabilistic seismic response analysis is now relatively easy and cost-effective with personal computers. The variations in results among skilled analysts have decreased significantly over the years, especially since fragility calculation methods and data have been standardized with the publication of various EPRI documents [37], [38], and [39].

In addition, there is a strong data base, both test data and experiential data, to rely upon for certain categories of equipment. Considerable test data were developed years ago and remain relevant for some types of equipment, especially for identifying failure modes. Some very significant large-scale tests have been conducted and are being conducted in Japan. Earthquake experience data have been systematically collected for some equipment categories, and those data also aid in the identification of failure modes and in judging the seismic ruggedness for screening purposes. For structures, analysis methods are typically used to calculate fragilities and these methods are robust and widely used. These methods have been verified using large and small scale test data. The following discussion gives more details on containment capacity and fragility analysis including insights from some large-scale tests.

8.4.2.4 Systems Analysis and Quantification. In Section 8.4.2.1 above, systems analysis was described as the method "used to model the various combinations of structural and equipment failures that could initiate and propagate a seismic core damage sequence." A good description is found in Appendix 5A of the ASME-ANS PRA Combined Standard [35]:

The plant systems and sequence analyses used in seismic PRAs … can generally be summarized as follows:

(1) The analyst constructs fault trees or other systems models reflecting (a) failures of key system components or structures that could initiate an accident sequence and (b) failures of key system components or structures that would be called on to stop the accident sequence;

(2) The fragility of each such component (initiators and mitigators) is estimated;
(3) Fault trees or other systems models are used to develop Boolean expressions for severe core damage that lead to each distinct plant damage state sequence;
(4) Considering possible severe core damage sequences and containment mitigation systems (e.g., fan coolers and containment sprays), and Boolean expressions are developed for each release category.

Finally, a convolution is performed of the plant hazard curve(s) and, the fragility curve(s) for each accident sequences, and then the quantification is performed accounting for the likelihoods of both seismic-induced failures and non-seismic failures and human errors.

This methodology is similar to that used in internal-events PRA, although it involves some specialized work that most PRA systems analysts can easily be trained to perform [24].

8.4.3 Containment Capacity and Fragility Analysis

The geometry of containment structures, tall cylindrical shells, provides high resistance to seismic loading. Design loading conditions that include combined internal pressure load and seismic load introduce significant reserve capacity for containment structures beyond the demands of SSE loading. Large-scale seismic tests of containment structures were carried out by the Nuclear power Engineering Corporation (NUPEC) in 1980s and1990s [40], [41], [42], and [43]. Scaled tests of containment structures were performed under the sponsorship of the NRC at the Sandia National laboratories to establish the ultimate internal pressure capacity. A more detailed discussion of testing of containment structures and testing of large components attached to a containment structures is provided in (Chapter 6), Section 6.3 of this book. The following sections summarize major insights from these test programs along with some generic containment fragility information.

8.4.3.1 Seismic Capacity and Fragility Analysis. The Nuclear Power Engineering Corporation (NUPEC) conducted large-scale seismic proving tests of a reinforced concrete containment vessel (RCCV) and a pre-stressed concrete containment vessel (PCCV) from 1992 through 1999 at its high-performance shake table facility at the Tadotsu Engineering Laboratory. Summary of the test programs and results are given in References [40] and [41]. NRC cooperated with the NUPEC on analysis and analytical evaluation of test results. NRC evaluations are summarized in References [44] and [45].

Paraphrasing from References [40] and [41], the objective of proving tests was as follows:

> As the reactor containment vessel is supposed to be responsible for preventing radioactive substances from diffusing into an environment by forming a pressure barrier in cases of accidents, it must ensure its safety even if a severe earthquake occurs. Since some nuclear power plants may have a concrete containment vessel that is different in type from a steel one seismic reliability of which has already been proved, it is necessary to prove that a concrete containment vessel is seismically safe and reliable. As for concrete reactor containment vessels, this proving test is aimed at proving that a containment can retain its over-all safety and reliability in cases of earthquakes in view of structural integrity and leak tight function and further evaluating seismic margin by determining the ultimate strength of a containment by means of excitation until a test model fracture. The RCCV model was that of the containment structure used in the Advanced Boiling Water Reactors in Japan. This RCCV structure is shown schematically in Fig. 8.13 below. The ABWR RCCV is a cylindrical vessel composed of reinforced concrete foundation plate, shell section and top slab. The wall of the shell section is 2 meter thick and the vessel is lined with steel liner.

The following Table 8.3 compares the scaled model vs. actual RCCV dimensions.

In addition, the similarity rule was applied and added weight was placed on the top of test model so that stress and strain conditions of the shell leg may be equal to those of actual RCCV.

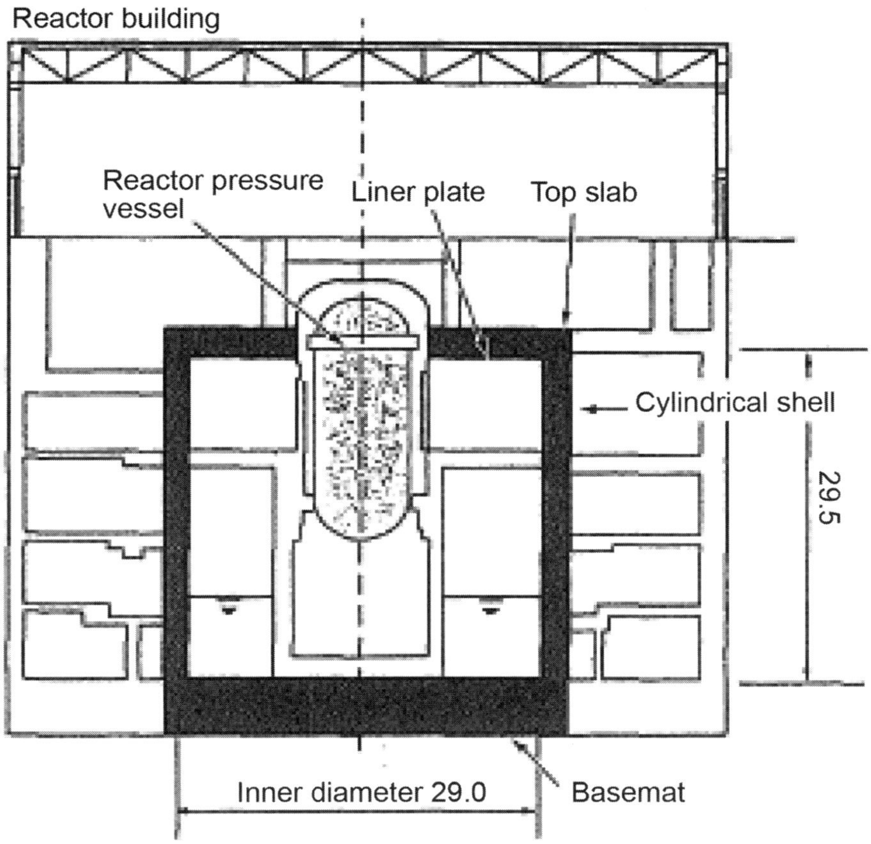

FIG. 8.13
SCHEMATIC DRAWING OF AN ABWR RCCV.

Very detailed description of test results and findings related to analytical methods are given in references [40] and [44]. The overall results indicated that the model withstood earthquake levels up to 5S2(H) and indicated that a very comfortable seismic margin existed for the scaled model (The S2 is a design level earthquake motion in the Japanese practice, somewhat equivalent to the safe shutdown earthquake in the United States. S2(H) is the horizontal component of the seismic motion. 5S2(H) is 5 times the earthquake motions associated with the S2 levels). The important point is that the vessel remained structurally intact and was able to resist additional earthquake loads during the 2S2(H), 3S2(H), 4S2(H), and 5S2(H) tests. Assuming accumulation of significant damage up to 5S2(H), test input was jumped to 9S2(H) and the model catastrophically failed.

TABLE 8.3
COMPARISON OF SCALED MODEL VS. RCCV PARAMETERS

	Actual RCCV Dimension (mm)	Test Model	
		A: Dimension (mm)	B: Reduced Scale
Internal diameter	29,000	3,625	1/8
Shell wall	2,000	200	1/10
Diameter of opening	4,300	538	1/8

The analytical evaluations carried out by Sandia for NRC indicated that the pretest analyses predict only general trends of the damage and failure regimes of the structure. The RCCV structural characteristics and test conditions introduced behavior regimes that required additional concrete material-model improvements. These included the dependence of shear stiffness, compressive stiffness, and viscous damping on the number of crack-open-close cycles. These modeling improvements had their greatest effect on the failure-level predictions and showed the analysis results to be in reasonably good agreement with test data. Both the NUPEC and NRC reports provide recommendations to improve predictability of the analytical methods.

The PCCV model was based on a containment structure for PWR plant designs in Japan. The outline of reactor building structure of a PWR plant is shown in Fig. 8.14. PCCV is a cylindrical building with a semi-spherical dome on the top, standing independently on a basemat. Figure 8.15 shows a photograph of the actual PCCV model on the shake table. The function of PCCV is broadly divided into structural integrity as containment vessel barrier (i.e., function of earthquake and pressure resistance) and leak tightness. Of such functions, function of earthquake and pressure resistance is secured by the concrete portion, while leak-tight function is ensured by the steel liner plate.

The following Table 8.4 shows the relationship between the scaled model and actual PCCV.

Very detailed description of test results and findings related to analytical methods are given in references [41] and [45]. Similar to the RCCV tests, the margin results indicated that there was a significant margin

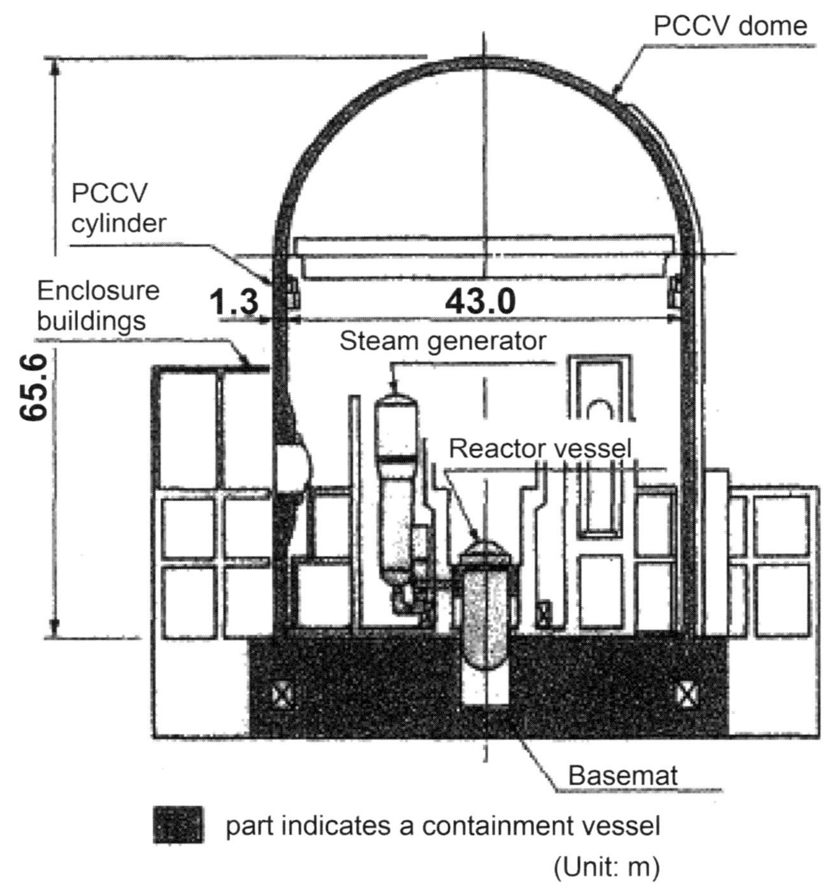

FIG. 8.14
SCHEMATIC DRAWING OF A PWR PCCV

FIG. 8.15
PCCV MODEL ON THE SHAKE TABLE

beyond the design basis. According to dynamic analysis and energy-based seismic marginal evaluation, the result of seismic margin was about 5S2. The test model suffered shear failure; however, leak preventive function was considered sound until the test model fractured. NRC's analytical evaluations [45] generally agreed with the test results and NUPEC's analysis results. The following Fig. 8.16, shows comparisons of analytical and test results.

Prior to the above concrete containment tests, NUPEC had also conducted seismic reliability proving tests of large-scale models of primary steel containments vessels of a PWR and a BWR in 1980s at its Tadotsu Engineering Laboratory. Results of these tests are presented in Refs. [42] and [43]. The BWR model was a Mark-II type with using a scale of 1/3.2. The PWR model used the scale of 1/3.7. Both the tests demonstrated soundness of structural strength and the leak tightness without deterioration of the test vessels under the design loading of S1 and S2. Furthermore, the test vessels also demonstrated sufficient strength margin when tested to increased seismic level in range of 1.3S2 to 1.5S2.

Based on the results of NUPEC tests and lessons learned from the analytical evaluations, structural seismic fragility evaluations were carried out in a NRC sponsored research programs for an actual RCCV (Surry Nuclear Power Plant) in Reference [46] and an actual PCCV (Zion Nuclear Power Station) in Reference [47].

The average failure levels for both types of containments were compared in the study, and are shown in Table 8.5. These failure levels are shown for three different characterizations of ground motion input. The following excerpts from [46] further describe results from the two studies.

TABLE 8.4
COMPARISON OF SCALED MODEL VS. PCCV PARAMETERS

Concrete Cylinder		Liner		
Size of Entire Configuration	Wall Thickness	Plate & Anchor Plate Thickness	Anchor Pitch	Anchor Webheight
1/10	1/8	1/4	1/4	1/8

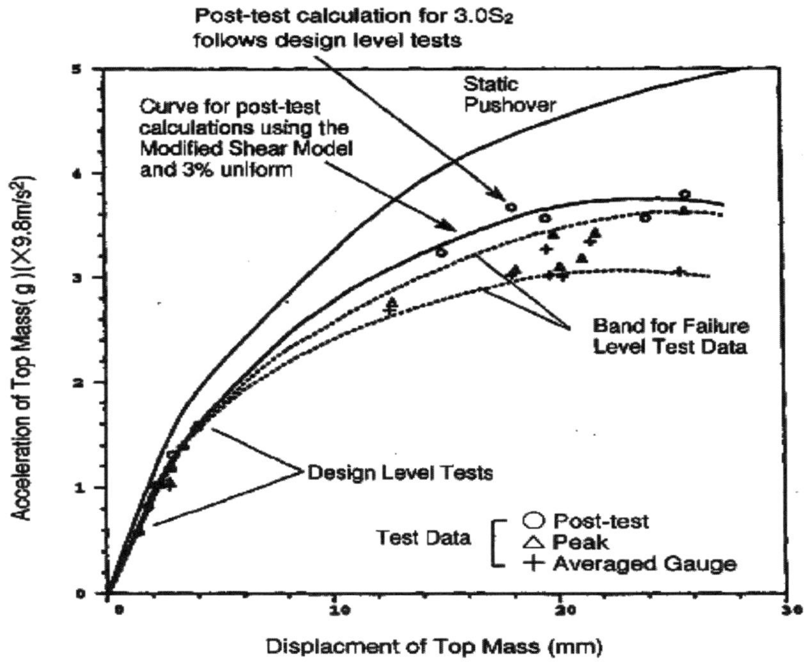

FIG. 8.16
ANALYSIS RESULTS FOR PCCV TESTS

The maximum average concrete shear strain in both the prestressed and reinforced concrete containments governed failure. Analysis of the reinforced concrete containment (Surry) resulted in a higher mean failure level for all cases, both with and without LOCA pressure and for all three categories of normalized earthquake inputs. Without an internal LOCA pressure applied, the mean failure levels were 30% to 40% higher for the reinforced containment (Surry) compared to the prestressed containment (Zion). With an internal LOCA pressure applied, the reinforced containment was only 6% to 12% higher than the prestressed containment. An example of this comparison between reinforced and prestressed concrete containment fragility curves is shown for one of the cases (no LOCA, earthquake normalized to PGA) in Fig. 8.17. This comparison between the analysis results of a prestressed concrete containment (Zion) and reinforced concrete containment (Surry) shows that, for these two specific containments, the reinforced containment is more robust, based on the average failure levels of both containments. The seismic design requirements were very similar at both sites; Zion was slightly higher, with a specified horizontal SSE of 0.17 g, whereas Surry's horizontal SSE was 0.15 g.

TABLE 8.5
COMPARISON OF AVERAGE FAILURE LEVEL AND UNCERTAINTIES

	No Internal Pressure				With Internal Pressure			
	Zion		Surry		Zion		Surry	
	Mean	COV	Mean	COV	Mean	COV	Mean	COV
Multiplication factor (MF) using RG 1.60 spectra	2.77	0.55	3.70	0.41	2.56	0.58	2.72	0.56
Multiples of g normalized to pga	2.34	0.48	3.22	0.36	2.17	0.50	2.42	0.55
Multiple of g normalized to spectral acceleration (2–5 Hz)	1.29	0.35	1.80	0.27	1.15	0.29	1.28	0.37

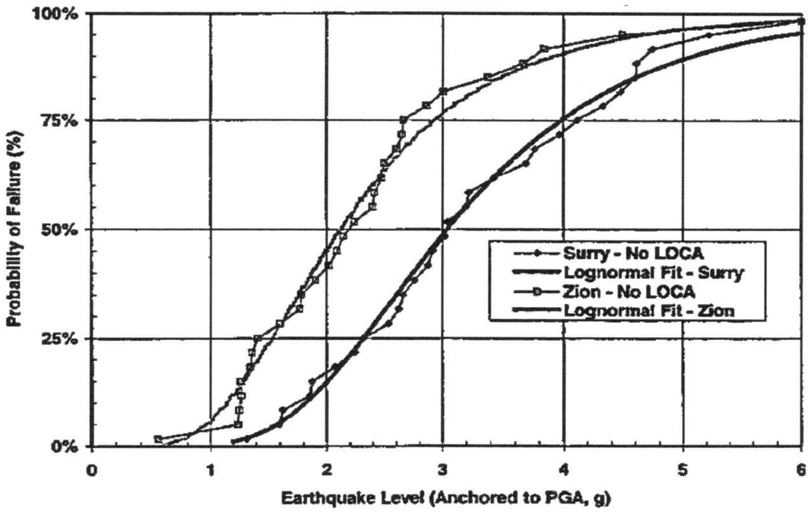

FIG. 8.17
FRAGILITY COMPARISON OF SURRY AND ZION PLANTS IN TERMS OF PGA

The robustness of this reinforced concrete containment is primarily because of the much thicker cylinder walls and because a reinforced containment without prestressing contains significantly more reinforcing steel than the prestressed containment. On the other hand, the reinforced containment will probably experience more concrete damage than the prestressed containment, particularly with an internal LOCA pressure applied. The results herein showed frequency reductions at increasing earthquake inputs were more significant in the reinforced concrete containment as a result of significant concrete cracking. This additional concrete cracking can be attributed to the lack of prestressing steel tendons, which yield at a much higher strain level, typically three to four times that of reinforcing bars.

As noted in the above studies, failure modes related to soil-structure interaction (SSI) effects and effects of basemat uplift were not included in the analyses. Consideration of these additional failure modes would be more representative of actual concrete containments in the U.S. Allowing basemat uplift and modeling SSI will significantly impact the frequencies, mode shapes, and even the input earthquakes. Based on the results of this report, failure will more likely be dominated by SSI effects and/or basemat uplift. In addition, analyzing containments with SSI and uplift would produce results more applicable for a probabilistic risk study because it would include most of the predominant failure modes of a concrete containment. However, some failure modes will always be plant-specific or require very detailed modeling.

8.4.3.2 Internal Pressure Capacity and Fragility Analysis. Experience with containment performance has been satisfactory during earthquakes, as well as during internal pressure from accidents. At TMI 2, the internal pressure due to hydrogen deflagration caused a pressure pulse of about 220 kPa (32 psig), but the containment integrity was never breached.

NRC sponsored a research program at the SNL to determine the ultimate capacity of various containment structures that were used in the existing nuclear power plants as protection against internal pressure from accidents. Discussions related to the Sandia tests are provided in (Chapter 6) Section 6.4 of this book. Although the following discussion overlaps somewhat with the discussions in Chapter 6, summaries of insights gathered from the tests are provided here for the convenience of the reader. NUREG/CR-6906 [48], “Containment Integrity Research at Sandia National Laboratories an Overview,” published in July 2006 is a very good source of information on the tests that were conducted and the conclusions drawn from them. A part of the executive summary is provided below.

A series of increasingly large and complex tests of scale models of containment structures and components were conducted at Sandia National Laboratories between 1983 and 2001. In conjunction with these experiments, an exhaustive analytical effort was conducted to predict the behavior of the models during the tests and to understand and simulate the actual behavior after the tests. The Sandia models, which were all subjected to static over-pressurization at ambient temperature. The following is a summary of the models tested:

- Four 1:32-scale models of steel containment structures typical of a large dry or ice condenser PWR containment
  - Two simple cylindrical shells with a hemispherical dome
  - One of similar geometry with hoop stiffeners
  - One of similar geometry with simulated penetrations
- A 1:8-scale model of a free-standing steel containment typical of ice-condenser PWR containments
- A mixed scale model of the steel containment of an improved BWR Mark II
  - The overall geometric scale was 1:10, the shell thickness scale was 1:4 to facilitate fabrication and to allow the use of prototypical materials
  - This model was constructed by the Nuclear Power Engineering Corporation (NUPEC) of Japan and tested by Sandia National Laboratories as part of a Cooperative Containment Research Program between NUPEC and the NRC.
- A nominal 1:6-scale model of a reinforced concrete PWR containment
- A 1:4-scale model of a large, dry PWR prestressed concrete containment
  - This model was also constructed by NUPEC and tested by SNL as part of the Cooperative Containment Research Program.

For most of these tests, various international nuclear research agencies, including regulatory bodies, designers, operators, universities and consultants, were invited to participate in test planning, pretest prediction and posttest analysis. Over 25 different agencies agreed to participate in the various tests. Their efforts complimented the analyses performed by Sandia and its subcontractors and supported the objectives of the program to demonstrate and improve the ability to predict containment performance up to and including failure. In addition to these tests of models of the containment structure, Sandia also tested full-scale and scale models of electrical penetration assemblies, personnel air locks, equipment hatches, drywell heads, penetration bellows and penetration seals and gaskets. Some general conclusions on predicting containment behavior can be made from this series of containment vessel model tests:

- Although establishing a generic margin of safety was not the purpose of the SNL program, the steel models have pressure capacities of four to six times, of the design pressure and the reinforced concrete models have pressure capacities of 2.5 to 3.5 times the design pressure.
- Global, free-field strains on the order of 2% to 3% for steel, 1.5% to 2% for reinforced concrete and 0.5 to 1.0% for prestressed concrete can be achieved before failure or rupture.
- Model (and presumably prototype) capacities are limited by high strains arising at local discontinuities which are present in both the model and the prototype.
- In the absence of a 'backup' structure, steel containment structure model capacities tend to be limited by gross structural failure or 'rupture'. Due to the inherent structural redundancy

of the liner and concrete system, steel lined concrete containments appear to be limited by functional failure (leakage). While the behavior that leads to tearing of the steel vessel or the steel liner is similar, i.e. local exceedence of the ductility limits of the steel at geometric discontinuities, the subsequent response of the vessels differs due to the presence of the surrounding structure.

- It seems reasonable to assume that with the added complexity of the actual containments, there is a higher probability that these local strain risers are present in, and possibly more severe than in any of the models tested. As a result, the capacities of the model, can, at best, be interpreted as an upper bound on capacity of prototypical containments.
- Analytical methods currently used are adequate to predict global response into the inelastic regime. One caveat on this statement is the discrepancy between predictions and observations of global yielding. Further investigation is required to understand the nature of this discrepancy (e.g. residual stresses, etc.) and its significance for calculation of prototypical containment capacities.
- Predictions of local failure mechanisms are highly dependent on the experience of the analyst, on the availability of accurate as-built information (geometry and material properties) at discontinuities, and on fabrication processes. Even if this information is available (not typical for actual containments), the prediction, a priori, of local failures is at best an uncertain proposition. The large scale model tests have, however, educated and sensitized the community to the types of details which may be critical in limiting containment capacities, and, hopefully, have improved the reliability of the predictions.
- These conclusions are predicated on failure of the containment structure. Any evaluation of the capacity of an actual containment must be based on the entire system, including mechanical and electrical penetrations and other potential leak paths.

An example of the containment fragility for beyond design basis internal accident pressure from the analysis of the Surry containment structure performed for the IPE program is shown in Fig. 8.18 [49]. At 0.5 MPa or 72.5 psi, the probability of failure due to internal pressure is near zero. Stated another way, the containment failure pressure is more than 2 times the design pressure.

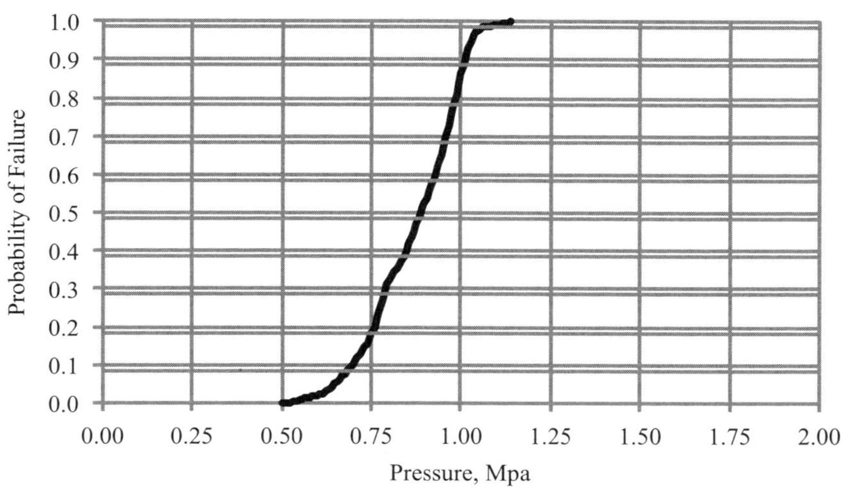

FIG. 8.18
SURRY IPE FRAGILITY

8.4.4 Examples and Insights Related to Containment Performance

As discussed earlier, what emerges from the large number of SPRA studies is that typically the seismic part of the overall reactor risk is a major contributor to the overall risk, sometimes dominant, almost always important, although sometimes negligible. The seismic-induced transients and small LOCAs contribute to core damage sequences with late failure of containments. The probability of seismic induced large LOCAs is small for several reasons: the primary piping loop has very high seismic capacity and extensive degradation has to take place to compromise this capacity. Other failure modes of primary systems, such as direct failure of containment structure or excessive movement of major primary loop components, such as steam generators because of support failures, also have high capacity. These aspects are examined in NUREG-1903, Seismic Considerations for the Transition Break Size [50] However, the sequences involving direct containment structural failures or movement of large primary loop components typically involve simultaneous initiation of core damage sequences with early failures of containment functions resulting in high consequence sequences. The following past examples involving scenarios related to failure of containments highlight the discussion above.

NUREG/CR-4334 [33] presents summary of fragilities of several containment structures. Most of these containments are large, dry prestressed or reinforced concrete containments with the exception of one steel containment. The results show that the median ground acceleration capacities of containments, involving shear failure mode, ranged from 2.5 to 9.2 g and HCLPF acceleration values ranged from 0.8 to 2.7 g. One point of caution, these examples involve analysis conducted in eighties, and may not reflect today's estimates of capacity or fragility. For example, characterization of ground motions in the central and eastern United States (CEUS) is significantly different then the understanding of early eighties (see Fig. 8.1, Comparison of Design Spectra). As discussed earlier, these high capacities are not surprising because the containments are designed conservatively for combination of SSE and design basis LOCA loads. Also, the containment wall thicknesses are controlled by the tornado generated missiles and thereby providing additional margins.

In one of the PRA studied in NUREG/CR-4334 [33], the plant damage state of "core damage followed by containment bypass" was shown to be an important seismic risk contributor. This can occur through failure of the containment crane wall which is taken to damage and produce a hole in the containment, or through seismic-induced rupture of a number of steam generator tubes causing containment bypass. This failure, however, occurs at peak ground accelerations much higher than 0.6 g, (HCLPF).

Another containment-failure mode generally considered in PRAs is the failure of containment isolation. Containment isolation is lost when the penetrations are damaged or if the main steam isolation valve (MSIV) or other containment isolation valves are stuck open in a seismic event. In fragility evaluation of containments, the analysts have judged that the major containment penetrations. (i.e., equipment hatch and personnel air locks) have seismic capacities larger than those of the rest of the containment shell. Small diameter piping and electrical penetrations have not been specifically analyzed in the seismic PRAs. It is judged that they have high seismic capacities.

Experience at Nevada Test Site supports this conclusion since penetrations which are typically short, stiff cylinders have not failed in extremely high ground shock tests. However, there were instances of leakage of small cable penetrations. Seismic capacities (HCLPF) of MSIVs have been estimated to be larger than 1g in the seismic PRAS.

The insights gained in the IPEEE with respect to the containment performance under seismic conditions have been summarized in NUREG-1742, Vol. 1 [34], as follows.

The primary purpose of a containment performance evaluation for a seismic event is to identify vulnerabilities that might cause early failure of containment functions. This includes an evaluation of containment integrity, containment isolation, prevention of containment bypass, and some specific systems depending on a containment design (e.g., igniters, suppression pools, ice baskets). The containment performance issues could be addressed either qualitatively by a systematic examination of the performance of the various containment functions under the review-level seismic condition, or, quantitatively, by a Level 2 PRA.

All SMA submittals contained qualitative assessments of seismic containment performance, as did most SPRA submittals. Of the 27 SPRA IPEEE submittals, 9 included a quantitative assessment of seismic containment performance. The relevant NUREG-1407 [51] guidance for SPRA quantitative containment performance analyses focuses on the assessment of containment failure modes that are significantly different from those encountered in the IPE. In some instances, the quantitative results provided in the IPEEE submittals are presented as frequencies of small and large releases (small and large are usually defined based on a release fraction of volatiles less than or greater than 10% of core inventory, respectively). In other cases, they are presented in the form of frequencies of small and large containment failures. (A definition of small and large failure was not provided in most cases. In one case, it was based on containment leakage less than or greater than that from a 3-inch diameter hole.) At least one SPRA IPEEE submittal (Kewaunee) also reported a containment HCLPF capacity for preventing large early releases.

The frequencies of early large release (or early release if the size of the release is not provided) obtained from the SPRA Level 2 analyses vary from less than 1E-7 to 1.6 E-5/ry. As a percentage of total seismic CDF, they vary from about 1% to more than 90%. Two of the nine plants have an early large release frequency greater than 1.0 E-6/ry. One plant, Pilgrim, is a BWR with a Mark I containment with a total frequency of early release determined as 1.6 E-5/ry, or about 27% of the seismic CDF. The dominant contributors to the early release frequency were identified in the submittal as: drywell liner melt-through, containment structural failure before core damage, and containment isolation failure. The other plant, Point Beach 1 and 2, is a PWR that has a large early release frequency of 1.3 E-5/ry, or more than 90% of total seismic CDF. This high release frequency resulted from the assumption that most core damage sequences would also result in the failure of the automatic containment isolation function, a potential seismic vulnerability. However, the licensee effectively invalidated this finding by assuming that the containment would be isolated manually by the operators at least 90% of the time in the event of core damage where automatic isolation failed. A qualitative assessment was also performed for this plant. This involved an examination of the containment safeguards systems significant to early large release, including containment integrity, mechanical penetrations, containment isolation, and containment cooling. The evaluation led the licensee to conclude that the containment and the system designed to ensure containment isolation are seismically sound, having no vulnerabilities.

Qualitative assessment of containment performance was essentially carried out the same way for analyses using either SPRA or SMA methods. These IPEEEs generally implemented a qualitative assessment of containment performance involving screening and/or walkdown examination of the following items:

(1) Containment structural integrity,

(2) Containment penetrations, hatches, and seals, and

(3) Containment cooling systems.

All licensees performed walkdowns of their containments. Licensees did not report any anomalous conditions with respect to containment structural integrity. However, in a few instances, they identified outliers pertaining to containment penetrations and containment cooling.

The NUREG-1742 includes tables listing findings associated with containment performance for each of the plants that performed IPEEE evaluations.

8.4.5 Consideration for New Reactors

For licensing generation III/III+ reactors (new reactors) in the United States, the NRC uses regulatory process of Title 10 of the *Code of Federal Regulations* (10 CFR) Part 52 [52], "Licenses, Certifications, and Approvals for Nuclear Power Plants," which provides requirements for early site permit, standard design certification (DC), and combined license (COL) applications. In this process, an application for a COL may incorporate by reference a design certification, an early site permit, both, or neither. This approach allows for the early resolution of safety and environmental issues. A COL review does not reconsider the safety issues that the design certification and early site permit processes have resolved. However, a COL application that incorporates a design certification by reference must demonstrate that pertinent site-specific parameters are confined within the envelopes established by the design certification. Although the NRC increasingly uses risk information and performance requirements to develop regulatory requirements and guidance, the agency recognizes that the design process of nuclear facilities is still inherently deterministic. To evaluate the robustness of new reactor designs, an assessment, such as SPRA/SMA, can enhance the safety performance of the design for new reactors.

In addressing severe accident prevention and mitigation for new reactors, NRC regulation 10 CFR 52.47(a)(27) [52] requires that the final safety analysis report for a design certification application describe the design-specific PRA and its results. The scope of this assessment should be a Level 1 and Level 2 PRA that includes internal and external hazards and that addresses all plant operating modes. However, the staff recognized that it is not practical for a design certification applicant to perform a site-specific seismic PRA because a design certification application would not contain site-specific seismic hazard information. As an alternative approach to a seismic PRA, the NRC staff proposed a PRA-based seismic margin analysis as described in NRC paper SECY-93-087, "Policy, Technical, and Licensing Issues Pertaining to Evolutionary and Advanced Light-Water Reactor (ALWR) Designs," dated April 2, 1993 [23]; the Commission approved this PRA-based seismic margin analysis in the corresponding staff requirements memorandum dated July 21, 1993 [54]. This analysis preserves key elements of a seismic PRA to the maximum extent applicable and estimates the design-specific plant seismic capacity in terms of high confidence of low probability of failure capacity. The goal is to demonstrate that this capacity is at least 1.67 times the ground motion acceleration of the design-basis safe-shutdown earthquake. Using this approach, the analysis can be used to evaluate the seismic risk for a design certification to show that it is acceptably low.

Accordingly, the NRC developed Interim Staff Guidance DC/COL ISG-020, "Interim Staff Guidance on Implementation of a Probabilistic Risk Assessment-Based Seismic Margin Analysis for New Reactors," dated March 22, 2010 [55], to provide a clear process for the implementation and the safety review of PRA-based SMA. The NRC based ISG-020 on the PRA standard [35] jointly developed by the American Society of Mechanical Engineers (ASME), the American Nuclear Society (ANS), and the applicable provisions of the NRC's Regulatory Guide 1.200 [53], "An Approach for Determining the Technical Adequacy of Probabilistic Risk Assessment Results for Risk-Informed Activities." The ISG-020 includes detailed elements with respect to the PRA-based SMA implementation in design certification applications and post-design certification updating activities, including COL updates to incorporate site- and plant-specific features and post-COL verifications.

8.4.6 Methods for Other Natural Hazards

So far the discussion has been centered on three natural hazards, namely seismic, flooding and wind, those affect most, if not all, of the sites and have to be considered in nuclear power plant designs. There are many other natural hazards that are very site-specific, and only have to be considered if they are applicable at a site. For example, sand storms and ice storms are only applicable in certain geographical areas. Also, concept of the design basis is not applicable to all of the natural hazards that may have to be considered. A list of natural hazards that are typically considered in a risk-evaluation is contained in [35] and [36]. Generally, a progressive screening approach is used to deal with these hazards. This screening approach begins with a review of the plant and site specific hazard data and licensing basis to identify potential hazards. Hazards with the acceptably low frequencies are not considered further. If the hazard frequency is not low, next step is to perform a bounding analysis. If bounding analysis is deemed too conservative, next steps involve successive refinements and eventually could lead to performing a PRA for the hazard. Part 6 of the ASME/ANS Standard [35] describes such an approach. One note of caution in using the screening step of low frequency; our understanding and knowledge of natural hazards are continually evolving and perceptions of the hazards may change. What was deemed a low frequency hazard at one time may not be deemed so in the future. The evolution of understanding of design basis flood is a good example of such a situation.

The ASME/ANS Standard includes requirements for conducting high wind, external floods, and other external hazards in Parts 7, 8, and 9, respectively.

8.5 CURRENT INITIATIVES FOLLOWING THE FUKUSHIMA EVENT

In aftermath of the Fukushima events, all of the countries that have nuclear power plants have undertaken activities and programs to address concerns, issues, and lessons that have emerged. There are common elements in the efforts directed at assurance of protection of nuclear installations against the effects from extreme external natural hazards.

Among other things, these elements include:

- Dealing with a long-term station blackout and loss of heat sink;
- Developing mitigation strategies for external events beyond design basis;
- Controlling of hydrogen deflagration and detonation;
- Implementing reliable spent fuel pool instrumentation;
- Re-evaluating external hazards specific to the nuclear installation sites, particularly seismic and flooding (which includes tsunami related flood effects); and
- Evaluating the plant responses to the re-evaluated hazards as well as to assess capability and the available safety margins to withstand potential higher level of hazards, including the communication of the associated risks

As stated earlier, the NRC formed a Near-Term Task Force (NTTF) to develop recommendations to address lessons learned. The Near-Term Task Force was established in response to Commission direction to conduct a systematic and methodical review of U.S. Nuclear Regulatory Commission processes and regulations to determine whether the agency should make additional improvements to its regulatory system and to make recommendations to the Commission for its policy direction, in light of the accident at the Fukushima Dai-ichi Nuclear Power Plant. The task force, in its report [11] made the following overarching recommendations in five categories:

Clarifying the Regulatory Framework

1. The Task Force recommends establishing a logical, systematic, and coherent regulatory framework for adequate protection that appropriately balances defense-in-depth and risk considerations.

Ensuring Protection

2. The Task Force recommends that the NRC require licensees to reevaluate and upgrade as necessary the design-basis seismic and flooding protection of SSCs for each operating reactor.
3. The Task Force recommends, as part of the longer term review, that the NRC evaluate potential enhancements to the capability to prevent or mitigate seismically induced fires and floods.

Enhancing Mitigation

4. The Task Force recommends that the NRC strengthen station blackout mitigation capability at all operating and new reactors for design-basis and beyond-design-basis external events.
5. The Task Force recommends requiring reliable hardened vent designs in BWR facilities with Mark I and Mark II containments.
6. The Task Force recommends, as part of the longer term review, that the NRC identify insights about hydrogen control and mitigation inside containment or in other buildings as additional information is revealed through further study of the Fukushima Dai-ichi accident.

7. The Task Force recommends enhancing spent fuel pool makeup capability and instrumentation for the spent fuel pool.
8. The Task Force recommends strengthening and integrating onsite emergency response capabilities such as emergency operating procedures, severe accident management guidelines, and extensive damage mitigation guidelines.

Strengthening Emergency Preparedness

9. The Task Force recommends that the NRC require that facility emergency plans address prolonged station blackout and multiunit events.
10. The Task Force recommends, as part of the longer term review, that the NRC pursue additional emergency preparedness topics related to multiunit events and prolonged station blackout.
11. The Task Force recommends, as part of the longer term review, that the NRC should pursue emergency preparedness topics related to decision-making, radiation monitoring, and public education.

Improving the Efficiency of NRC Programs

12. The Task Force recommends that the NRC strengthen regulatory oversight of licensee safety performance (i.e., the Reactor Oversight Process) by focusing more attention on defense-in-depth requirements consistent with the recommended defense-in-depth framework.

Most pertinent recommendation out of the twelve from the perspective of this chapter is Recommendation 2. NRC initiatives relative to this recommendation are further discussed in this chapter. There are three specific recommendations that are part of Recommendation 2 as follows:

> "The Task Force recommends that the Commission direct the following actions to ensure adequate protection from natural phenomena, consistent with the current state of knowledge and analytical methods. These should be undertaken to prevent fuel damage and to ensure containment and spent fuel pool integrity:
>
> 2.1 Order licensees to reevaluate the seismic and flooding hazards at their sites against current NRC requirements and guidance, and if necessary, update the design basis and SSCs important to safety to protect against the updated hazards.
>
> 2.2 Initiate rulemaking to require licensees to confirm seismic hazards and flooding hazards every 10 years and address any new and significant information. If necessary, update the design basis for SSCs important to safety to protect against the updated hazards.
>
> 2.3 Order licensees to perform seismic and flood protection walkdowns to identify and address plant-specific vulnerabilities and verify the adequacy of monitoring and maintenance for protection features such as watertight barriers and seals in the interim period until longer term actions are completed to update the design basis for external events."

In order to implement the Recommendations 2.1 and 2.3, the NRC issued a request for information to the licensees of all operating reactors on March 12, 2012 [56].

In its letter, the NRC indicated that a hazard evaluation consistent with Recommendation 2.1 will be implemented in two phases as follows:

Phase 1: Issue letters to all licensees to request that they reevaluate the seismic and flooding hazards at their sites using updated seismic and flooding hazard information and present-day regulatory guidance and methodologies and, if necessary, to request they perform a risk evaluation.

Phase 2: Based upon the results of Phase 1, the NRC staff will determine whether additional regulatory actions are necessary (e.g., update the design basis and SSCs important to safety) to provide additional protection against the updated hazards.

The following Fig. 8.19 shows key activities of Phases 1 and 2.

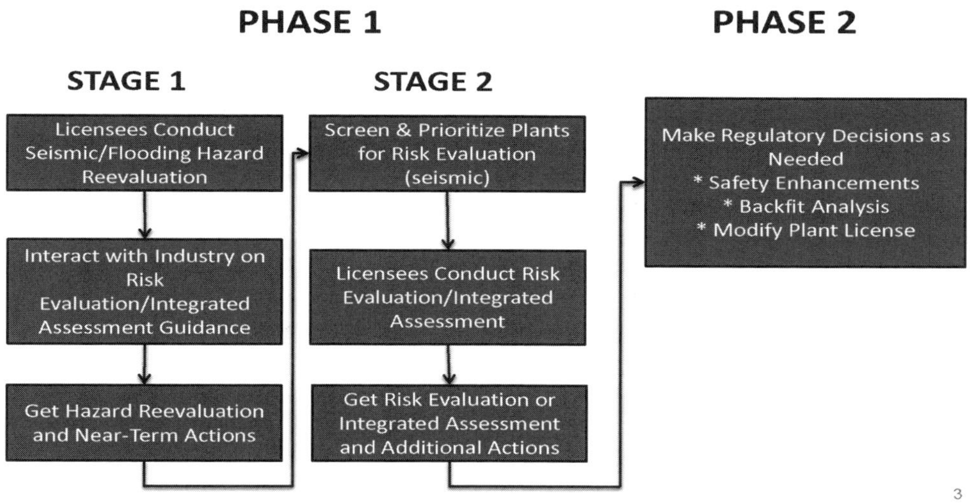

FIG. 8.19
NRC'S POST-FUKUSHIMA-RELATED ACTIVITIES

For the seismic hazard, the letter indicates that the addressees are requested to perform a reevaluation of the seismic hazards at their sites using present-day NRC requirements and guidance to develop a GMRS. Addressees are requested to submit, along with the hazard evaluation, an interim evaluation and actions planned or taken to address the reevaluated hazard where it exceeds the current design basis. For plants where the reevaluated hazard exceeds the current design basis, addressees may opt to perform an SPRA. In addition, an SPRA, rather than a SMA, may be necessary for cases where the SMA approach is not usable due to a higher reevaluated hazard (i.e., GMRS). For all other plants where the reevaluated hazard exceeds the current design basis, the NRC will provide guidance on when an SMA option can be used. According to the information request, factors that the staff will consider to determine whether a SPRA or a SMA is appropriate are (1) the extent to which the reevaluated hazard (GMRS) exceeds the current design basis (SSE), (2) the absolute seismic hazard based on an examination of the probabilistic seismic hazard curves for the site, and (3) previous estimates of plant capacity (e.g., IPEEE insights). The priority for the subsequent completion of the risk assessments by the addressees will also be based on the above factors.

The letter further requests that an evaluation of the spent fuel pool (SFP) integrity be also submitted along with an assessment of reactor integrity. The addressee's evaluation should consider all seismically induced failures that can lead to draining of the SFP. The evaluation should consider SFP walls, liner, penetrations (cooling water supplies or returns, drains), transfer gates and seals, seals and bellows between the SFP, transfer canal, and reactor cavity, sloshing effects (including loss of SFP inventory, wave-induced failures of gates, and subsequent flooding), siphon effects caused by cooling water pipe breaks, and other relevant effects that could lead to a significant loss of inventory of the SFP.

The letter also stated that while the seismic hazard reevaluation is being performed, NRC staff and stakeholders will continue interacting to develop strategies for screening, prioritization, and potential interim actions as well as implementation guidance for the risk evaluation.

For the flood hazard, the letter requested that addressees perform a reevaluation of all appropriate external flooding sources, including the effects from local intense precipitation on the site, probable maximum flood (PMF) on stream and rivers, storm surges, seiches, tsunami, and dam failures. It is requested that the reevaluation apply present-day regulatory guidance and methodologies being used for ESP and COL reviews including current techniques, software, and methods used in present-day standard engineering practice to develop the flood hazard. The requested information will be gathered in Phase 1 of the NRC staffs two phase process to implement Recommendation 2.1, and will be used to identify potential vulnerabilities.

For the sites where the reevaluated flood exceeds the design basis, addressees are requested to submit an interim action plan that documents actions planned or taken to address the reevaluated hazard with the hazard evaluation. Subsequently, addressees should perform an integrated assessment of the plant to identify vulnerabilities and actions to address them. The scope of the integrated assessment report will include full power operations and other plant configurations that could be susceptible due to the status of the flood protection features. The scope also includes those features of the ultimate heat sinks (UHS) that could be adversely affected by the flood conditions and lead to degradation of the flood protection (the loss of UHS from non-flood associated causes are not included). It is also requested that the integrated assessment address the entire duration of the flood conditions.

Similar to the seismic request, the NRC staff is to also work with the industry and other stakeholders to develop implementation guidance on the flood hazard and integration assessments.

There are some important points to be observed from the NRC's request for information. In both seismic and flood areas, the concept of risk assessments and integrated assessments are focusing on total integrated plant response to an external event. As discussed in this chapter, both the risk assessment and integrated assessment take into account design, operational, human, and available preventive and mitigative measures; thus identifying potential vulnerabilities, balance between prevention and mitigation, and insights into defense-in-depth. This type of information is essential to come up with most effective safety enhancement measures and appropriate regulatory actions.

There are some key differences in how the plants deal with seismic hazards and flood hazards. Each of the safety related SSCs are designed to seismic induced vibratory motion in accordance with the accepted standards and codes. As discussed earlier, the design process itself provides a significant margin beyond the design basis motion. The flooding protection is afforded in different ways. There are vast differences from plant to plant with regard to the flood protection features used [9]. Site flood protection may include incorporated, exterior, and temporary features with passive and active functions credited to protect against the effects of external floods. In addition to physical barriers, flood protection at nuclear power plants may involve a variety of operator manual actions. These operator manual actions may be associated with installation of features (e.g., floodgates, portable panels, placement of portable pumps in service), construction of barriers (e.g., sandbag barriers), and other actions.

Also, a margin beyond the design basis flood levels is not necessarily assured. As evidenced at the Fukushima-Daiichi, the flood hazard beyond certain level could lead to severe consequences, due to common cause effects and failing several defense-in-depth features at the same time. This particular effect is termed "cliff-edge" effects.

The use of terminology of "risk assessment" in connection with the seismic and "integrated assessment" in connection with the flooding analysis is by design. As described in detail in this chapter, the SPRA and SMA methodology are mature, having been in existence for more than thirty years, and applied to more than fifty power plants worldwide. The probabilistic flood hazard assessment methods for all flood causing mechanisms are not widely used. There are very few examples of external flood probabilistic risk assessments. The NRC approaches recognize these differences while maintaining the essential concept of an integrated plant response for both hazards.

The NRC staff has been working with the industry on developing implementation guidance for seismic risk assessment. The current version of the industry guidance is contained in the document titled: Screening, Prioritization and Implementation Details (SPID) for the Resolution of Fukushima Near-Term Task Force Recommendation 2.1: Seismic [57]. The NRC staff has also issued Interim Staff Guidance (ISG), JLD-ISG-2012-04 on seismic margin analysis [58]. The guidance on the integrated assessment for flooding is given in ISG, JLD-ISG-2012-05 [59]. The guidance on how to perform tsunami, surge, and seiche hazard assessments is provided in ISG, JLD-ISG-2012-06 [60].

All of the above guidance documents are important because they provide guidance that will be useful beyond the purpose of developing information to satisfy the NRC's request for information to implement NTTF recommendation related to the hazard reevaluation. The SPID [57] deals with seven technical areas: seismic hazard development, GMRS comparisons and screening of plants, seismic hazard and screening report, prioritization, seismic risk evaluation, and spent fuel pool integrity evaluation. It provides guidance on how some of the requirements of the ASME/ANS PRA Standard [35] can be implemented. The ISG on seismic margin analysis [58] consolidates all of the requirements in a single document and includes enhancements based on lessons learned from the past studies. The integrated assessment ISG [59] provides guidance on how to evaluate the total plant response to external flooding hazards. The guidance considers both the flood protection and mitigation capabilities of a plant. The guidance uses concepts of risk evaluations in-terms of looking at the integrated plant response, and includes consideration of physical flood protection features, flood-specific and other pertinent operating procedures, temporary protective measures, effectiveness of manual actions, and all available plant systems and components that can be used for prevention and mitigation. Some of the quantitative aspects of a traditional PRA are somewhat challenging to implement at present for several reasons. There are no commonly accepted practices for performing probabilistic flood hazard analysis for several flood causing mechanisms, particularly for low probability events that are of interest in a nuclear power plant PRA. Moreover, it may be challenging to quantitatively evaluate the reliability of the manual actions involved in the flood protection and mitigation. The ISG provides guidance on dealing with such issues, and as such, is a comprehensive and first-of-a-kind guidance to deal with evaluation of a plant's capability to provide protection against the flood hazard. The ISG on tsunami, surge, and seiche hazard assessments [60] reflects the experience gained and lessons learned during recent early site permit and combined license reviews conducted by the NRC. It reflects current practice and advancement in knowledge, tools, and data. Over the next few years, the industry will conduct hazard reevaluations and risk and integrated evaluations using the above guidance documents, in part.

In addition to responding the NRC initiatives, the US industry in response to the Fukushima has undertaken an initiative that deploys diverse and flexible coping strategies called FLEX [61]. The FLEX deals with the consequences of postulated beyond-design-basis external events (BDBEE) that are most impactful to reactor safety. These are loss of power and loss of the ultimate heat sink. As discussed in [61], the approach of adding diverse and flexible mitigation strategies — or FLEX — will increase defense-in-depth for beyond-design-basis scenarios to address a loss of ac power and loss of normal access to the ultimate heat sink occurring simultaneously at all units on a site.

According to ref. [61], FLEX consists of the following elements:

- Portable equipment that provides means of obtaining power and water to maintain or restore key safety functions for all reactors at a site. This could include equipment, such as portable pumps, generators, batteries and battery chargers, compressors, hoses, couplings, tools, debris clearing equipment, temporary flood protection equipment and other supporting equipment or tools.
- Reasonable staging and protection of portable equipment from BDBEEs applicable to a site. The equipment used for FLEX would be staged and reasonably protected from applicable site-specific severe external events to provide reasonable assurance that N sets of FLEX equipment will remain deployable following such an event.
- Procedures and guidance to implement FLEX strategies. FLEX Support Guidelines (FSG), to the extent possible, will provide pre-planned FLEX strategies for accomplishing specific tasks in support of Emergency Operating Procedures (EOP) and Abnormal Operating Procedures (AOP) functions to improve the capability to cope with beyond-design-basis external events.
- Programmatic controls that assure the continued viability and reliability of the FLEX strategies. These controls would establish standards for quality, maintenance, testing of FLEX equipment, configuration management and periodic training of personnel. The FLEX strategies will consist of both an on-site component using equipment stored at the plant site and an off-site component for the provision of additional materials and equipment for longer-term response.

As discussed in this chapter, there are several near-term and long-term activities are underway to implement lessons learned and to assure that the Fukushima types of consequences are avoided in the future.

8.6 SUMMARY AND GENERAL CONCLUSIONS

Our knowledge of natural hazard phenomena is highly uncertain and continually evolving. It is important that we have technical approaches and regulatory framework that allows efficient evaluations of impact on a plant safety and identify practical and necessary enhancements.

The structural engineering community and its standard writing committees have developed design, construction, and inspection and maintenance procedures for containment structures with the goal of ensuring that containment structures are effective against release of radioactive materials that can endanger public health and safety.

PRA methods provide a tool to evaluate an integrated response of a plant to natural hazards taking into consideration designs aspects, as-built and as-operated conditions, human actions, and all available protection and mitigation features. PRA methods have been demonstrated to be effective in identifying vulnerabilities and areas of potential improvement. Performance-based approaches allow incorporating risk-insights into deterministic design approaches.

Operating experience with the performance of containment structures in the U.S. has been satisfactory; however, it is essential for both the industry and the regulatory body to be vigilant and pro-active.

ACKNOWLEDGMENTS

The authors are greatly indebted to their past and present colleagues at the NRC for a very collaborative, stimulating, and learning experience. The authors have greatly benefitted from interactions with NRC colleagues and colleagues at the Department of Energy laboratories who have served as NRC contractors on many important activities. The authors' participation in collaborative activities with the domestic and international organizations, particularly with the Japan Nuclear Energy Safety Organization (JNES) formerly, Nuclear Engineering Power Corporation (NUPEC), International atomic Energy Agency, and Nuclear Energy Agency of Organization for Economic Co-operation and Development (OECD), has greatly enhanced authors understanding and knowledge. The large-scale seismic tests conducted by the JNES and NUPEC have provided very valuable data and insights. Finally, the authors want to acknowledge the pioneering contributions made to the nuclear safety by Dr. Robert Kennedy, Dr. Robert Budnitz, late Prof. C. Allin Cornell, Dr. M. K. Ravindra and Prof. Heki Shibata. The authors have been extremely fortunate to have opportunities to work with these leaders and learn from them in many ways.

REFERENCES

[1] American Society of Mechanical Engineers. Boiler and Pressure Vessel Code, Section III, "Code for Concrete Reactor Vessels and Containments," New York, NY.

[2] Regulatory Guide 1.221, 2001, "Design-Basis Hurricane and Hurricane Missiles for Nuclear Power Plants," U.S. Nuclear Regulatory Commission, Washington, DC.

[3] U.S. Code of Federal Regulations, Title 10 — Energy, Part 50, "Domestic Licensing of Production and Utilization Facilities," U.S. Nuclear Regulatory Commission, Washington, DC.

[4] Regulatory Guide 1.60, 1976, "Design Response Spectra for Seismic Design of Nuclear Power Plants," U.S. Nuclear Regulatory Commission, Washington, DC.

[5] U.S. Code of Federal Regulations, Title 10 — Energy, Part 100, "Reactor Site Criteria," U.S. Nuclear Regulatory Commission, Washington, DC.

[6] Regulatory Guide 1.208, 2007, "A Performance-Based Approach to Define the Site-Specific Earthquake Ground Motion," U.S. Nuclear Regulatory Commission, Washington, DC.

[7] NUREG 0800, 2007, "Standard Review Plan for Review of Safety Analysis Reports for Nuclear Power Plants: LWR Edition," U.S. Nuclear Regulatory Commission, Washington, DC, Revision 3.

[8] Regulatory Guide 1.59, 1977, "Design Basis Floods for Nuclear Power Plants," U.S. Nuclear Regulatory Commission, Revision 2.
[9] Regulatory Guide 1.102, 976, "Flood Protection for Nuclear Power Plants," U.S. Nuclear Regulatory Commission, Revision 1.
[10] Regulatory Guide 1.76, 1974, "Design-Basis Tornado and Tornado Missiles for Nuclear Power Plants," U.S. Nuclear Regulatory Commission, Revision 0.
[11] U.S. Nuclear Regulatory Commission, 2011, "Recommendations for Enhancing Reactor Safety in the 21st Century, the Near-Term Task Force Review of Insights from the Fukushima Dai-ichi Accident," Commission Paper SECY-11-0093, July 12.
[12] ASCE/SEI 43-05, 205, "Seismic Design Criteria for Structures, Systems, and Components in Nuclear Facilities," American Society of Civil Engineers/Structural Engineering Institute.
[13] Cornell, C. A., 1968, "Engineering Seismic Risk Analysis." Bull. Seism. Soc. Am., 58:1583–1606.
[14] Budnitz, R. J., Apostolakis, G., Boore, D. M., Cluff, L. S. Coppersmith, K. J. Cornell, C. A. Morris. P. A., 1997, "Recommendations for Probabilistic Seismic Hazard Analysis: Guidance on Uncertainty and Use of Experts," NUREG/CR-6372, Volumes 1 and 2, U.S. Nuclear Regulatory Commission, Washington, DC, April.
[15] Kammerer, A. M. and Ake, J. P., 2012, "Practical Guidelines foe SSHAC Level 3 and 4 Hazard Studies (NUREG-2117, Revision1)," NUREG-2117, U.S. Nuclear Regulatory Commission, Washington, DC, April.
[16] Reiter, L., 1990, Earthquake Hazard Analysis. New York: Columbia University Press.
[17] Kennedy, R. P., 2007, "Performance-Goal Based (Risk Informed) Approach for Establishing the SSE Site Specific Response Spectrum for Future Nuclear Power Plants," Invited Paper, Proceedings of Structural Mechanics in Reactor Technology Conference 19, Canada.
[18] Prasad, R., Hibler, L. F., Coleman, A. M., and Ward, D. L., 2011, NUREG/CR-7046, "Design-Basis Flood Estimation for Site Characterization at Nuclear Power Plants in the United States of America," U.S. Nuclear Regulatory Commission, Washington, DC.
[19] Regulatory Guide 1.76, 2007, "Design-Basis Tornado and Tornado Missiles for Nuclear Power Plants," U.S. Nuclear Regulatory Commission, Revision 1.
[20] Institute of Nuclear Power Operations, 2011, "Special Report on the Nuclear Accident at the Fukushima Daiichi Nuclear Power Station," Report INPO 11-005.
[21] U.S. Nuclear Regulatory Commission, 1975, "Reactor Safety Study: An Assessment of Accident Risks in U.S. Commercial Nuclear Power Plants [NUREG-75/014 (WASH-1400)]."
[22] U.S. Nuclear Regulatory Commission, 1995 "Use of Probabilistic Risk Assessment Methods in Nuclear Regulatory Activities; Final Policy Statement," Federal Register, Vol. 60, No. 158, August 16.
[23] U.S. Nuclear Regulatory Commission, 1993, "Policy, Technical, and Licensing Issues Pertaining to Evolutionary and Advanced Light-Water Reactor (ALWR) Designs," Commission Paper SECY-93-087, April 2.
[24] Budnitz, R. J., Chokshi, N. C., and Ravindra, M. K., 2011,"Use of Seismic PRA for Risk-Informed Decision Making By Utilities And Regulatory Agencies," International Topical Meeting on Probabilistic Safety Assessment and Analysis, PSA-11, Wilmington, NC, March 13–17.
[25] Smith, P. D., et al., "Seismic Safety Margins Research Program, Phase I Final Report," Report NUREG/CR-2015, in ten volumes, U.S. Nuclear Regulatory Commission, 1981.
[26] Kennedy, R. P., Cornell, C. A., Campbell, R. D., Kaplan, S., and Perla, H. F., 1980, "Probabilistic seismic safety study of an existing nuclear power plant," Nuclear Engineering and Design.
[27] Commonwealth Edison Company, 1981, "Zion Probabilistic Safety Study," Chicago, IL.
[28] Power Authority of the State of New York and Consolidated Edison Company of New York, Inc., 1982, "Indian Point Probabilistic Safety Study."
[29] Philadelphia Electric Company, 1983, "Severe Accident Risk Assessment, Limerick Generating Station," Philadelphia, PA.

[30] Northeast Utilities, 1983, "Millstone Unit 3 Probabilistic Safety Study," Hartford, Connecticut.

[31] "Seabrook Probabilistic Safety Study," Pickard, Lowe and Garrick, Inc., Newport Beach, CA, 1983.

[32] Electric Power Research Institute, 1991, "A Methodology for Assessment of Nuclear Power Plant Seismic Margin," EPRI Report NP-6041-SL, Revision 1, Palo Alto, California.

[33] Budnitz, R. J., Amico, P. J., Cornell, C. A., Hall, W. J., Kennedy, R. P., Reed, J. W., and Shinozuka, M., 1985, "An Approach to the Quantification of Seismic Margins in Nuclear Power Plants," NUREG/CR-4334, U.S. Nuclear Regulatory Commission.

[34] U.S. Nuclear Regulatory Commission, 2002, "Perspectives Gained From the Individual Plant Examination of External Events (IPEEE) Program," NUREG-1742.

[35] American Society of Mechanical Engineers/American Nuclear Society, 2009, "Standard for Level 1/Large Early Release Frequency Probabilistic Risk Assessment for Nuclear Power Plant Applications," Standard ASME/ANS RA-Sa-2009.

[36] U.S. Nuclear Regulatory Commission, 1982, "PRA Procedures Guide: A Guide to the Performance of Probabilistic Risk Assessments for Nuclear Power Plants," NUREG/CR-2300, December.

[37] Reed, J. W. and Kennedy, R. P., 1994, "Methodology for Developing Seismic Fragilities," Report TR-103959, Electric Power Research Institute.

[38] Electric Power Research Institute, 1994, "Methodology for Developing Seismic Fragilities," EPRI Report TR-103959, Palo Alto, California.

[39] Electric Power Research Institute, 2002, "Seismic Fragility Application Guide," EPRI Report 1002988, Final Report, Palo Alto, California, December.

[40] NUPEC Summary Report — (translation from excerpts), Proving Tests on Seismic Reliability for Nuclear Power Plant, Seismic Proving Test of Reinforced Concrete Containment Vessel (RCCV), Nuclear Power Engineering Corporation, 2001.

[41] NUPEC Summary Report — (translation from excerpts), Proving Tests on Seismic Reliability for Nuclear Power Plant, Seismic Proving Test of Prestressed Concrete Containment Vessel (PCCV), Nuclear Power Engineering Corporation, 2001.

[42] NUPEC Summary Report — (translation from excerpts), Proving Tests on Seismic Reliability for Nuclear Power Plant, Seismic Proving Test of BWR PCV, Nuclear Power Engineering Corporation, 1989.

[43] NUPEC Summary Report — (translation from excerpts), Proving Tests on Seismic Reliability for Nuclear Power Plant, PWR Reactor Containment Vessel, Nuclear Power Engineering Corporation, 1984.

[44] Cherry, J. L., James, R. J., Zhang, L., and Rashid, Y. R., 2001, "Seismic Analysis of a Reinforced Concrete Containment Vessel Model," NUREG/CR-6707, U.S. Nuclear Regulatory Commission.

[45] James, R. J., Zhang, L., Rashid, Y. R., and Cherry, J. R., 1999, "Seismic Analysis of a Prestressed Concrete Containment Vessel Model," NUREG/CR-6639, U.S. Nuclear Regulatory Commission.

[46] Klamerus, E. W., 2002, "Structural Seismic Analysis of the Surry Containment," NUREG/CR-6783, U.S. Nuclear Regulatory Commission.

[47] Klamerus, E. W. and Cherry, J. L., 2001, "Structural Seismic Fragility Analysis of the Zion Containment," NUREG/CR-6740, U.S. Nuclear Regulatory Commission.

[48] Hessheimer M. F. and Dameron, R. A., 2006, "Containment Integrity Research at Sandia National Laboratories an Overview," U.S. Nuclear Regulatory Commission, NUREG/CR-6906.

[49] Pilch, M. M., Allen, M. D., and Klamerus, E. W., 1995, "Resolution of the Direct Containment Heating Issues for All Westinghouse Plants with Large Dry Containments or Subatmospheric Containments," U.S. Nuclear Regulatory Commission, NUREG/CR-6338.

[50] Chokshi, N. C., Shaukat, S. K., Hiser, A. L., DeGrassi G., Wilkowski, G., Olson, R, and Johnson J. J., 2008, "Seismic Considerations for the Transition Break Size," U.S. Nuclear Regulatory Commission, NUREG-1903.

[51] U.S. Nuclear Regulatory Commission, 1991, "Procedural and Submittal Guidance for the Individual Plant Examination of External Events (IPEEE) for Severe Accident Vulnerabilities," NUREG-1407, June.

[52] U.S. Code of Federal Regulations, Title 10 — Energy, Part 52, "Licenses, Certifications, and Approvals for Nuclear Power Plants," U.S. Nuclear Regulatory Commission, Washington, DC.

[53] U.S. Nuclear Regulatory Commission, 2009, "An Approach for Determining the Technical Adequacy of Probabilistic Risk Assessment Results for Risk-Informed Activities," Regulatory Guide 1.200 Revision 2, March.

[54] U.S. Nuclear Regulatory Commission, 1993, "Staff Requirements Memorandum to SECY-93-087," July 21.

[55] U.S. Nuclear Regulatory Commission, 2010, "Interim Staff Guidance on Implementation of a Seismic Margin Analysis for New Reactors Based on Probabilistic Risk Assessment," Interim Staff Guidance DC/COL-ISG-020, March 15.

[56] U.S. Nuclear Regulatory Commission Letter to All Power Reactor Licensees et al., "Request for Information Pursuant to Title 10 of the Code of Federal Regulations 50.54(f) Regarding Recommendation 2.1, 2.3, and 9.3, of the Near-Term Task Force Review of Insights from the Fukushima Dai-ichi Accident", March 12, 2012.

[57] Electric Power Research Institute, "Seismic Evaluation Guidance: Screening, Prioritization, and Implementation Details (SPID) for the Resolution of Fukushima Near-Term Task Force Recommendation 2.1: Seismic," 1st Draft EPRI Report no. TR- 1025287 Palo Alto, California, 2012.

[58] U.S. Nuclear Regulatory Commission, 2012, "Guidance on Performing a Seismic Margin Assessment in Response to the March 2012 Request for Information Letter," Interim Staff Guidance, JLD-ISG-2012-04, November.

[59] U.S. Nuclear Regulatory Commission, 2012, "Guidance for Performing the Integrated Assessment for Flooding," Interim Staff Guidance JLD-ISG-2012-05, November.

[60] U.S. Nuclear Regulatory Commission, 2012, "Guidance for Performing a Tsunami, Surge, or Seiche Hazard Assessment," Draft Interim Staff Guidance JLD-ISG-2012-06, October.

[61] Nuclear Energy Institute, 2012, "Diverse and Flexible Coping Strategies (Flex) Implementation Guide," NEI 12-06, Washington, DC.

CHAPTER

9

EVOLUTION OF CONTAINMENT SYSTEMS FOR GEN III REACTORS

Jim Xu

The advanced reactors are often referred to as Generation III/III+ which are evolutionary or passive designs of pressurized water reactors (PWRs) and boiling water reactors (BWRs). In contrast to earlier generations of reactors, the advanced reactors rely on standardized designs which are simpler and more rugged and incorporate passive or inherent safety features, making them easier to operate and less vulnerable to operational upsets, therefore further reducing possibility of core melt accidents. The standardization process also facilitates expediting licensing, reducing capital cost, and shortening construction time.

This chapter provides an overview of the advanced reactors with particular emphasis on structural aspects of containment features. A discussion on the regulatory process for licensing advanced reactors in the United States is provided which is informative in understanding regulatory perspectives. Because Generation III/III+ are licensed under a different regulatory process compared to earlier reactors, particularly the provisions pertaining to standardized features, a number of challenges and technical issues pertain to the structural design and analysis, which are also discussed in this chapter. Finally, various salient structural features of containments associated with Generation III/III+ reactors in the United States are described.

Keywords: Generation III/III+ reactors, Design Certification (DC), Combined license (COL), Standard designs, Customized designs, Advanced boiling water reactor (ABWR), Economic simplified boiling water reactor (ESBWR), Advanced passive 1000 MWe pressurized water reactor (AP1000), US evolutionary pressurized water reactor (USEPR), US advanced pressurized water reactor (US-APWR)

9.1 INTRODUCTION

Since year 2000, the reactors have been loosely classified as Generation I to Generation IV reactors. Chapter 1 of this book describes the early reactors in Section 1.1, and these reactors can be characterized as Generation I reactors in the United States. The operating 104 reactors are characterized as Generation II reactors. Chapters 2 to 8 of the book are focused towards what has been developed relating to the containments of Generation II reactors. In this chapter, I am principally describing various aspects of Generation III reactors, also termed as "Advanced Reactors." and their containments, including their relation to the operating reactors.

9.2 REGULATORY PERSPECTIVES FOR GENERATION III/III+ REACTORS

10 CFR 52 (b) requires that unless otherwise specifically provided, the regulations in 10 CFR Chapter I, apply to a holder of or applicant for an approval, certification, permit, or license. A holder of or applicant for an approval, certification, permit, or license issued under this part shall comply with all requirements in 10 CFR Chapter I, i.e., the regulatory requirements discussed in Chapter 2 of this book, are applicable. A license, approval, certification, or permit issued under this part is subject to all requirements in 10 CFR Chapter I which, by their terms, are applicable to early site permits, design certifications, combined licenses, design approvals, or manufacturing licenses.

9.2.1 Part 52 Regulatory Process

As the U.S. Nuclear Regulatory Commission (NRC) is responsible for licensing and regulating the operation of commercial nuclear power plants in the United States, two regulations, namely 10 CFR Part 50 [1] and 10 CFR Part 52 [2], respectively, govern the licensing process [3]. The current operating nuclear power plants have been licensed under a two-step process described in Part 50. This process requires both a construction permit (CP) and an operating license (OL). The advanced reactors, often referred to as the Generation III and III+ reactors, are being licensed under a combined process as described in Part 52. Irrespective of the licensing process chosen, the NRC is responsible for the oversight of the construction and operation of a power reactor throughout its lifetime to assure compliance with NRC regulations for the protection of public health and safety, the common defense and security, and the environment.

In an effort to improve regulatory efficiency and to add greater predictability to the licensing process, the NRC in 1989 established an alternative licensing process which combined the construction permit and operating license into a single combined license (COL) with conditions in Subpart C of Part 52. A mandatory hearing is required for a COL. In addition, an applicant can apply an early site permit (ESP) under Subpart A of Part 52 and a design certification (DC) under Subpart B of Part 52. An ESP addresses and resolves the safety and environmental characteristics of the site and evaluates potential physical impediments to developing an acceptable emergency plan independent of a specific nuclear power plant design. The ESP also allows for a limited work authorization (LWA) per Part 50.35 to perform limited non-safety site preparation activities, subject to redress, in advance of issuance of a COL. The issuance of ESP requires a mandatory hearing, and is valid for 10 to 20 years and can be renewed for another 10 to 20 years.

A DC provides a standard nuclear power plant design independent of a specific site, which is approved and certified by the NRC through a rulemaking. The DC is valid for 15 years. An application for a DC must contain proposed inspections, tests, analyses, and acceptance criteria (ITAAC) for the standard design. In addition, a DC applicant must demonstrate compliance with the Commission's relevant regulations.

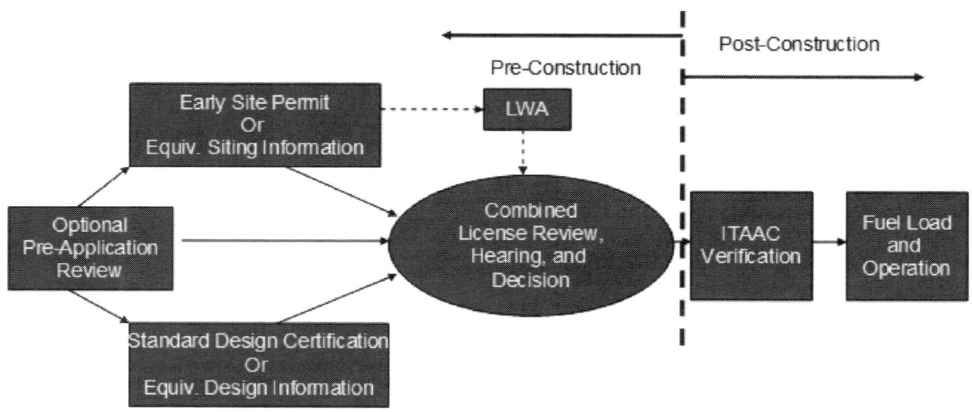

FIG. 9.1
KEY COMPONENTS OF PART 52 LICENSING PROCESS

Figure 9.1 illustrates the licensing process under Part 52. This process begins with an assumed generic site, and then completes a standard nuclear power plant design based on the assumed site parameters in DC applications. A COL can reference a DC and/or an ESP. The COL application reconciles the site characteristics with assumed site parameters to demonstrate that the referenced standard design is applicable to the site. The NRC will use the ITAAC verification process to ensure that the as-constructed plant conforms to the design. The loading of the fuel into the core and the commencement of the operation will begin after all ITAAC issues are resolved and the Commission makes a safety finding pursuant to Part 52.103(g). In contrast to Part 50, the advantage of the Part 52 process is that the issues resolved during the DC rulemaking and the ESP hearing processes are precluded from reconsideration later at the COL stage, therefore increasing the predictability of certain outcome for the licensing process.

9.2.2 Standardization of Reactor Designs

The nuclear power plants prior to Generation III were built (using Part 50 process) based on customized designs to meet the site-specific characteristics and plant-specific needs. Although the site investigation and design planning were conducted in the CP stage, detailed designs were often performed in parallel to the plant construction and safety issues identified during the design and construction should be resolved before the issuance of the OL. While the history of the safety operation of the current fleet of the U.S. commercial reactors serves as a testimony for the success of the Part 50 process, there exist uncertainties due to late identifications of safety issues during design and construction which could not be resolved among the stakeholders, thereby jeopardizing the future status of the plant. A textbook example to illustrate the detrimental impact of such uncertainty would be the Shoreham nuclear power plant [4]. The plant was located adjacent to the Long Island Sound in East Shoreham, New York. It was a General Electric (GE) BWR, which was begun construction in 1973 and was granted OL in 1989, but was decommissioned in 1994. The plant was never placed in service for generating electricity because of an impasse on issues raised during the construction associated with the plan for evacuation in events of accidents. The plant was eventually sold to the New York State for \$1 for decommissioning and the total cost incurred to tax payers was \$6 billion to end the two-decade saga.

A salient feature of Generation III and III+ reactors licensed under Part 52 process is the standardization of the reactor designs. An important lesson learned from Shoreham would be to streamline the licensing process to enable early resolution of safety issues, thereby ensuring predictable outcomes for design, construction and operation of a plant. The standardization based on Part 52 accomplishes this by requiring "essentially complete" design [5, 6] during a DC application. In essence, the essentially complete design need be sufficiently detailed to enable the NRC to judge the applicant's proposed means of assuring that construction conforms to the design and to reach a final conclusion on all safety questions before the certification is granted.

For structures, the essentially complete design is often interpreted as the complete design of critical structural sections, i.e., these structural sections which provide primary load-carrying functions; these are structural elements found in direct load paths as essential blocks for ensuring the safety of the structure. The design of critical sections has evolved gradually over years. Experiences gained from past constructions of nuclear facilities have been used in some DC applications to identify critical sections [7]. Recently, in support of the U.S. EPR DC application, AREVA NP developed a selection criteria based on a three-tier, systematic approach [8].

The selection methodology described in the reference [8] consists of three-tiered criteria — qualitative, quantitative, and supplemental — applied in a sequential manner. The qualitative criterion is used to select critical sections of so-called Nuclear Island (NI) structures that perform safety-critical functions. The quantitative criterion is used, together with a numerical algorithm and the finite element (FE) analysis of the NI structures, to identify critical sections that are highly stressed but are not chosen under the qualitative criterion. The quantitative criterion ensures that the structural elements essential to the structural load paths are adequately identified and considered. The supplemental criterion is based on engineering judgment and is intended to capture critical sections of seismic Category I structures that are not screened by the other two criteria but are necessary to obtain an adequate representation of all types of structural elements. Preliminary estimates discussed in the reference [8] indicate that the critical sections selected using this methodology are representative of approximately 77% of NI structures and 84% of all seismic Category I structures.

Following the selection methodology described in reference [8], or similar systematic methodologies that combine numerical evaluations with engineering judgment, would greatly simplify the selection process and ensure a consistent set of critical sections adequate for an essentially complete design.

Another important feature of the standard design is that the design is typically based on a well established generic site which generally encompasses a wide range of site characteristics so that the standard design upon certification can be built at multiple sites. A COL applicant can incorporate a standard design by reference upon demonstration that the site characteristics are appropriately bounded by the site parameters postulated for the generic site. The advantage of the standardization is that the safety issues which have been resolved during the certification process are not subject to future litigations, thereby ensuring more predicable outcomes in the licensing process.

Other significant features of standardized designs include but not limited to:

- Simpler and more rugged design, making them easier to operate and less vulnerable to operational upsets
- High reliability and longer operating life — 60 years
- Enhanced safety by further reducing possibility of core damage accidents
- Protection against an aircraft impact
- Better fuel efficiency
- Expedited licensing, reduced capital cost and shortened construction time

With continued appreciation in the reactor safety, emphasizing on better understanding the systemic and phenomenological characteristics as well as severe accident managements, advanced technologies based on evolutionary and passive designs with enhanced safety features have been deployed in Generation III/III+ reactors to provide better defense-in-depth and safety margins in accident preventions and mitigations than previous generations of reactors. Table 9.1 shows some of severe accident mitigating features seen in

TABLE 9.1
SEVERE ACCIDENT MITIGATING FEATURES AND RISK METRIC COMPARISON*

Feature	ABWR	ESBWR	AP1000	U.S. EPR	U.S. APWR
CDF/yr	$2E-7$	$2E-8$	$2E-7$	$3E-7$	$1E-6$
LRF/yr	$<1E-8$	$1E-9$	$2E-8$	$2E-8$	$1E-7$
CCFP (internal events at power)	<0.1	0.08	0.08	0.08	0.1
Core melt stabilization (CMSS) BiMAC or core catcher	core catcher passive core debris flooding	BiMAC, passive cooling of molten debris	IRWST flooding for cooling core debris in-vessel	core catcher passive cooling of molten debris	core debris trap
Reactor cavity flooding capability	AC-independent with alternate thermally actuated flooder system	GDCS with firewater as back-up	manual actuation	SAHRS and CMSS	dedicated cavity flooding pathway
Containment overpressure protection system (vent)	filtered vent	filtered vent		containment opening reserved for filter vent	
Containment combustible gas control	inerted containment	inerted containment	manual igniters, 2 PARs for LOCAs	PARs	glow type igniters
In-containment refueling water storage tanks			included	included	included
RCS severe accident depressurization	normal ADS	normal ADS	ADS valves	2 trains, manually actuated	1 train manually actuated
Reactor cavity concrete floor	sacrificial layer	sacrificial layer	large floor area with thick concrete layer	sacrificial layer	large floor area

*Information was taken from respective application FSAR or DCD.

various design certification applications in the United States. Also included in the table are comparisons of various risk metrics associated with different designs. The calculated core damage (CDF) frequencies for Generation III/III+ are typically in a range of 10^{-6} to 10^{-8} per year (internal events at full power) and corresponding large release frequencies (LRF) are between 10^{-7} and 10^{-8} per year, which are much lower than these seen in earlier reactors. For severe accident mitigations, new containment features include those such as: in-containment refueling water storage tanks (IRWST), core melt stabilization systems, containment overpressure protection systems, and containment combustible gas controls, etc. These features seen in standard designs provide much better measured safety performance than previous generations of reactors. In addition, all standard designs include evaluations to ensure that containments can withstand external impacts from commercial aircrafts in compliance with NRC regulation 10 CFR 50.150.

9.3 DESIGN AND ANALYSIS CONSIDERATIONS FOR STANDARD DESIGNS

9.3.1 Structural Aspects of Standardized Designs

The structural designs for Generation III/III+ reactors typically utilize the concept of a power block or nuclear island which is a large reinforced concrete common basemat shared by the containment and other safety-related structures to provide vital safety functions for the operation of the facility. A salient feature of the NI construction is to reduce the failures induced by the foundation differential settlement, especially for appurtenances and commodities that are connected between buildings. The nuclear island (NI) structures are designed to meet 10 CFR Part 50, Appendix A, General Design Criteria (GDC), including:

- GDC 1, "Quality Standards and Records," as it relates to the structures being designed, fabricated, erected, and tested to quality standards commensurate with the importance of the safety function to be performed.
- GDC 2, "Design Bases for Protection against Natural Phenomena," as it relates to the design of the safety related structures being able to withstand the most severe natural phenomena such winds, tornadoes, floods, and earthquakes and the appropriate combination of all loads.
- GDC 4, "Environmental and Dynamic Effects Design Bases," as it relates to the structures being appropriately protected against dynamic effects including the effects of missiles, pipe whipping, and discharging fluids due to blowdown loads associated with the loss-of-coolant accident (LOCA).
- GDC 5, "Sharing of Structures, Systems, and Components," as it relates to safety related structures not being shared among nuclear power units, unless it can be shown that such sharing will not significantly impair their ability to perform their safety functions.
- GDC 16, "Containment Design," as it relates to the containment being design with the capability to act as a leak-tight membrane to prevent the uncontrolled release of radioactive effluents to the environment.
- GDC 50, "Containment Design Bases," as it relates to the containment being designed with sufficient margin of safety to accommodate appropriate design loads.
- GDC 53, "Provisions for Containment Testing and Inspection," as it relates to the containment being designed to permit appropriate in-service inspections (ISI).

In addition, 10 CFR 50.44, "Combustible Gas Control for Nuclear Power Reactors," requires that the design demonstrates structural integrity of the containment against loads associated with combustible gas generation. Finally, the design of NI structures must satisfy the quality assurance criteria of 10 CFR Appendix B, "Quality Assurance Criteria for Nuclear Power Plants and Fuel Processing Plants."

These regulatory requirements for the design of safety related nuclear structures are typically met by following the NRC Standard Review Plan (SRP) [9], Section 3.8. The SRP provides an acceptable method for the structural design to comply with NRC regulations. Although the SRP is not a requirement, but according to

10 CFR 52.47 (9), applications based on methods that deviate from the SRP acceptance criteria need to demonstrate that the proposed alternative provides an acceptable method of complying with the NRC's regulations that underlie the corresponding SRP acceptance criteria. Therefore, the SRP criteria provide the framework for an acceptable standard to meet the regulations in the designs for nuclear power plant facilities.

The acceptance criteria for design of containments and other Category I structures and foundations are provided in SRP Section 3.8 to address the following technical aspects:

- Level of details in structural description
- Applicable codes, standards, and specifications
- Loads and load combinations
- Design and analysis procedures
- Structural acceptance criteria
- Materials, quality control, and special construction techniques
- Testing and in-service surveillance requirements
- ITAAC, technical specifications, and site interfaces

Containment can be of concrete or steel constructions. Concrete containments are further divided into reinforced concrete and prestressed concrete constructions. For prestressed containments, the current operating fleet of reactors in the United States typically utilized greased tendon technology except for H.B. Robinson steam electric plant which used grouted tendons for the containment in the vertical direction. For Generation III/III+ designs, an increasing number of PWR containments around the world including the U.S. EPR design have relied on grouted tendon deployments to withstand design basis LOCA loads, as the grouted tendons tend to provide stronger resistance to corrosions and better post tensioned performance than greased tendons.

Because of the dual function of the containment: (1) to house and to protect from design-basis hazards, safety-related structures, mechanical and electrical components, and distribution systems associated with a reactor coolant system; and (2) to serve a primary function as an engineered safeguard to contain the postulated radiological consequences of a LOCA accident, the containment structure is typically designed for a design pressure as high as 4 atmospheres (420 kPa, 60 psi) and a design temperature of 170°C (350°F) for a short duration design basis LOCA, and 377°C (650°F) for local hot spots with the concurrent occurrence of a 10^{-4} year or 10,000 year return period earthquake with design-basis mean peak ground surface accelerations that range from 0.1 to 0.75 g (1.0 g equals 980 gals). Concrete containments are also required to resist the effects of tornadoes with maximum wind speeds ranging from 386 km/h (kph) [240 miles/hours (mph)] to 579 kph (360 mph) concurrent with differential pressure drops and design basis tornado missiles. Hurricane wind effects are typically bounded by the design basis tornadoes; however, a recent study [10] indicated that for a number of coastal areas of the United States, the hurricane wind could exceed the tornado wind. Therefore, the standard designs are also required to evaluate the design basis hurricane wind effect in the design of concrete containments as well as other Category I structures [11]. Additionally, generation III/III+ reactors have also incorporated additional protections such as a shield structure to ensure that the containment structures can withstand the impact of a commercial jetliner crash as required by 10 CFR 50.150.

Since the design requirements to contain pressure and temperature effects are more stringent (except for earthquake-induced membrane shear and possibly membrane tension in a concrete wall segment) than those required to protect the safety-related components, concrete containment structures tend to follow pressure vessel design code (ASME) more closely than they do with building design practice (i.e., ACI 318). Concrete containments are designed to the requirements of the ASME B&PV Code, Section III, Division 2, Subsection CC, "Code for Concrete Reactor Vessels and Containments," also known as ACI Standard 359-01, American Concrete Institute [12]; while steel containments including containment penetration steel assemblies are designed to the ASME Code, Section III, Division 1, Subsection NE, "Class MC Components" [13]. Internal structures of the containment, which are typically comprised of concrete walls and slabs providing support to the reactor systems, are generally subject to differential pressure loads during the blowdown of the reactor coolant system. Since they are not required to remain leak-tight, the ACI-349 Code [14] is normally used for their designs. Internal structures could be supported on 2- or 3-ft-thick deformed-bar, reinforced-concrete slabs locate above the containment base mat liner or anchored via penetrations through the containment base liner directly to the containment base mat.

The current fleet of operating reactors were designed and constructed by customizing the design to the specific site conditions pursuant to provisions of 10 CFR Part 50. In this approach, the design for structures, systems, components (SSCs) considers site-specific natural phenomenon hazards such as: earthquake, tornado, hurricane and flooding, based on the physical characteristics of the site including geology, seismology, meteorology and hydrology of the site. For sites located in low seismicity zones, the customized design may not be controlled by seismic loads; rather, the design may be governed by other hazards such as tornadoes or hurricanes if the plant is located near a coast. And vise versa.

Standard designs, on the other hand, are generic in nature and are generally intended for multiple site conditions. Therefore, standard designs generally do not aim at addressing any particular site; rather, they focus the design to satisfy a broad range of site characteristics, so that any site can incorporate the standard design by reference without major modifications to the design. The site characteristics for a standard design are typically established by defining a generic site. In 10 CFR Part 52, the generic site is established by identifying associated site parameters. According to Part 52.47 (a) (1), a DC must include the site parameters postulated for the standard design, and an analysis and evaluation of the design in terms of those site parameters. Since the generic site dictates the extent of sites to which the standard design can apply, the scope of the site parameters should adequately cover the range of site conditions intended for future constructions of the standard design. SRP Section 2.0 provides guidance for developing site parameters for a standard design.

As discussed previously, a customized design needs to consider natural phenomena hazards such as: earthquakes, tornadoes, hurricanes, and flooding, which are most severe based on historic data for the particular site, and to ensure that the structures can withstand these hazards without loss of the capability to perform their safety functions. In fact, one can apply the performance-based approach such as ASCE 43 [15] for the seismic design by establishing a performance goal or objective in terms of annual probability of exceeding a design limit state. For nuclear structures, the design limit state typically corresponds to the onset of significant inelastic deformation or essentially elastic limit state, which means that the structure's stress or strain state are within elastic limits globally while allowing limited inelastic excursions in certain confined regions of geometric discontinuities where localized high stress concentrations are expected. The same design principles can also apply to the standard designs. However, instead of considering the most severe of site-specific natural phenomenon hazards as is the case for a customized design, a standard design typically considers the most severe of natural phenomenon hazards for a wide range of sites or regions. For instance, because of the consideration for potential future sites, a standard design may need to include in the design the most severe snow loads, tornado loads, hurricanes and seismic loads across a region such as central and eastern United States. The design is performed to the envelope of all these loads. Therefore, the final design tends to be much more conservative than any individual sites targeted for the design. This would ensure the standardization of the plant with substantial benefits in enhanced safety as well as the use of modular construction techniques, streamlined process for constructions, procurements of systems and components, etc.

Lastly, a discussion is worthwhile on aspects of the constructions to ensure that as-constructed plant conforms to the design. The structural design is completed in a DC application with respect to the so called "critical sections." These critical section designs are typically presented in terms of design drawings with sufficient details. To ensure that the critical sections and other structural elements which the critical sections represent are constructed in a manner that is consistent with the design drawings in the FSAR. 10 CFR Part 52 institutes the ITAAC verification process which will allow the NRC to verify through ITAACs established for the critical sections that the as-built structures are consistent with the DC design before loading the fuels into the reactor. Any deviations or departures from the standard design are controlled through rigorous regulatory processes in 10 CFR Part 52 to ensure the proposed changes to the design will enhance safety and result in an efficient and effective process for the constructions.

9.3.2 Technical Considerations and Challenges in Structural Designs and Analyses

Since the generation III/III+ reactors utilize Part 52 process based on standard designs, the containment and civil structural designs and analyses would require somewhat different approaches from Part 50 reactors. As described in previous section, in contrast to a customized plant for which the design is done to address hazards and conditions for a specific site, the standard plants are generic in nature and are typically designed to the envelope of natural phenomenon hazard effects encompassing a broad range of site conditions envisioned for potential sites for the design. Designing power plant structures for generic plants would encounter different technical issues and challenges for the designers and reviewers alike. Generally, nuclear power plant (NPP) structures need to be designed to withstand both internally and externally initiated hazards. For internal events such as internal floods, jet impingement, blowdown loads from loss-of-coolant accidents (LOCA), etc., the design treatments for containment and other Category I structures are essentially the same between Part 50 and Part 52 facilities. However, for externally initiated events such as winds, snows, earthquakes, etc., significantly different approaches to structural analyses and designs are required for Part 52 reactors as opposed to Part 50 facilities, due to the distinct differences in the licensing processes. Moreover, externally initiated events such as earthquakes, etc., are inherently site and frequency dependent and the structural designs are better suited where the site is well defined as for Part 50 facilities. However, recognizing that the external hazards are strongly site dependent, the design for civil structures is challenging for the standard plants licensed under Part 52 process.

Recognizing that a standard plant based on Part 52 process is designed to a generic site, a well-defined set of site parameters for the generic site is essential to the design as well as the NRC review of the standard plant. For containment and Category I structures and foundations, site parameters include those which affect establishing natural phenomenon loads due to earthquakes, winds, tornadoes, hurricanes, and floods. These parameters should be postulated which bound potential sites for the plant. For instance, if a standard plant is intended for the coastal region of the eastern United States, hurricane effect based on historical data should be adequately reflected in the generic site [11]. Similar scenarios can be postulated for other natural phenomenal hazards. It is recognized that in general, the standard design based on this generic site being defined on the bounding concept would be more conservative than any particular site because it is less likely that any particular site would be exposed to the most severity of all hazards; however, the conservative aspects of the standard design are considered to be a small cost compared to the benefits of achieving the standardization for the plant.

Because the seismic response has strong dependency on the characteristics of ground motion inputs and the geotechnical properties both of which are site-specific in nature, the seismic design for structures based on the generic site is particularly challenging for the standard plant [16, 17]. The following paragraphs provide a discussion of seismic issues which were identified as challenging during the application reviews. The materials included in the discussion are mostly extracted from a co-authored paper [17].

Some background information of how the seismic design for nuclear power plants (NPPs) evolved may be beneficial in providing insights and put the challenges in the seismic design for the standard plants in the proper perspectives. Until the 1970s, the seismic analysis and design of NPPs was based on the concept of "safe-shutdown earthquake" (SSE), which is defined in Appendix S, "Earthquake Engineering Criteria for Nuclear Power Plants," to 10 CFR Part 50 and Appendix A, "Seismic and Geologic Siting Criteria for Nuclear Power Plants," to 10 CFR Part 100, "Reactor Site Criteria" [18] as the vibratory ground motion for which safety-related SSCs must be designed to remain functional and within applicable stress, strain, and deformation limits. The NRC subsequently published a series of regulatory guides (RGs) in support of 10 CFR Part 50, Appendix S, and 10 CFR Part 100, Appendix A, including RG 1.60, "Design Response Spectra for Seismic Design of Nuclear Power Plants" [19]. The seismic analysis process for NPPs thus required site-specific investigations involving seismology, geology, and geotechnical engineering that were used to estimate site-specific, hazard-based, peak ground accelerations (PGA). The site-specific SSE was then specified in terms of broadband smooth response spectra, analogous to the RG 1.60 spectra, anchored to the PGA and defined at the ground surface in the free field. The corresponding seismic analysis and design of safety-related SSCs typically followed the acceptance criteria of the SRP, which provides guidance to the NRC staff for performing its safety reviews.

To address the seismic issues associated with the 10 CFR Part 52 licensing process, pertaining to the seismic analysis and design of new NPPs, the nuclear industry initiated the New Reactor Seismic Issues Resolution Program to coordinate its effort with the NRC. The program addressed two critical issues: (1) the implementation of a performance-based approach for developing site-specific ground motions and (2) an evaluation methodology for considering the effects of high-frequency ground motions.

The NRC implemented the performance-based approach for developing site-specific ground motions through the publication of RG 1.208, "A Performance-Based Approach to Define the Site-Specific Earthquake Ground Motion" [20], which provides a method for developing risk-consistent site-specific ground motions based on a probabilistic seismic hazard analysis (PSHA). The performance-based approach with a stipulated target performance goal is consistent with the risk goals described in the NRC's policy statement for the operation of NPPs.

To address the high-frequency seismic issues in DC and COL applications, the NRC developed Interim Staff Guidance (ISG)-01, "Interim Staff Guidance on Seismic Issues Associated with High Frequency Ground Motion," issued 2009 [21], which describes an acceptable process for assessing the effect of high-frequency ground motion on the seismic design of new NPPs. ISG-01 supplements the guidance provided in SRP Section 3.7.1, "Seismic Design Parameters," Revision 3.

To illustrate the challenges for seismic design for the standard plants, several topics important to standard designs are discussed below.

9.3.2.1 Certified Seismic Design Response Spectra (CSDRS) and Associated Generic Site Conditions for Design Certification. The CSDRS represent the design basis ground motion inputs as part of the site parameters postulated by the DC applicant for seismic analysis and design. In contrast to a customized design which focuses on a well characterized site typically with a narrow range of frequencies of interest to the design in both ground motion inputs and soil profiles (geotechnical properties), a standard design often relies on a generic site with sufficiently broad range frequencies of interest in selecting ground motion inputs and soil profiles to ensure the applicability of the certified design to the broadest range of potential sites. The DC vendors typically select a CSDRS of sufficient magnitude and broad frequency range to ensure a robust seismic design. DC applicants typically specify spectral shape similar to RG 1.60 anchored at 0.3 g peak ground acceleration (PGA) with high frequency enhancement. While this definition for the CSDRS imposes a very significant seismic demand, the amplified frequency region of the RG 1.60 spectral shape does not envelope the hard rock high-frequency (HRHF) spectra for certain central and eastern United States (CEUS) sites. To ensure the seismic design adequacy of certified designs for HRHF spectra, DC applications considered several approaches, as described below.

Approach 1

The DC applicant defined its CSDRS as an envelope of RG 1.60 spectra anchored at 0.3 g PGA and representative HRHF spectra that enveloped all potential candidate CEUS sites at the time of submittal of the DC application. The HRHF spectra define the CSDRS envelope above 10 Hertz (Hz) while the RG 1.60 spectra define the CSDRS envelope below 10 Hz. A COL applicant would demonstrate that site-specific GMRS are enveloped by the CSDRS and that the site soil profile is bounded by the generic soil profiles analyzed for the CSDRS demand.

Approach 2

The DC applicant defined the CSDRS using a set of spectra. Response spectra similar to RG 1.60 spectra anchored at 0.3 g PGA but modified to account for soft, medium, and stiff soil conditions were selected, in conjunction with a set of HRHF spectra for CEUS sites, to establish the CSDRS for seismic analysis and design. The seismic analyses were then performed for each design spectra and associated generic soil profiles (ranging from soft and medium to stiff soil conditions, as well as rock conditions related to HRHF) to determined the seismic demands for the design. A COL applicant would have to demonstrate that site-specific GMRS are enveloped completely by one of the CSDRS spectra and that the site soil profile is bounded by the generic site profile(s) associated with the selected CSDRS spectra.

Approach 3

The DC applicant defined its CSDRS as RG 1.60 spectra enhanced at high frequencies and anchored at 0.3 g PGA. An evaluation of the design based on the CSDRS and assumed generic soil profiles was subsequently performed to demonstrate the adequacy of the design for representative HRHF spectra in accordance with the guidance of ISG-01. A COL applicant would have to demonstrate that site-specific GMRS are enveloped either by the CSDRS or by the representative HRHF spectra, and that the site soil profile is bounded by the applicable generic assumptions.

Once the CSDRS are defined, the next important consideration is the selection of a set of soil profiles for the generic site. To maximize the applicability of the standard design to the broadest range of potential sites, the applicant typically specifies a range from "soft soil" with a minimum shear wave velocity of 305 m/s (1000 ft/s) to "hard rock" with a shear wave velocity of 2440 m/s (8000 ft/s) or higher. Between these bounds, there are theoretically an infinite number of site conditions that can be specified for soil-structure interaction (SSI) analysis. The objective is to define a reasonable number of generic site conditions (typically 10 or less) for SSI analysis, such that the envelope of responses will be directly applicable to a large number of potential sites.

In general, the response characteristics of the ground motion input and associated soil profiles are interdependent for a site. For instance, site-specific GMRS represent the response of the soil profiles to the uniform hazard spectra (UHS) and are developed by directly convolving the UHS through appropriate set of soil profiles to the surface of the free field. The SSI analyses are performed to propagate the ground motion input through the site profiles to develop the structural responses to be used in the design. These SSI analyses are typically based on a mean profile, and a lower and upper bound profiles to ensure that the frequency content of ground motion input is appropriately captured. However, the selection of soil profiles for a generic site is much more complex because the development of the CSDRS is not directly linked to any particular site, but considering a collection of sites with broad range of site characteristics. Nonetheless, the principles for determining the site profiles for the SSI analyses in a customized design also apply to the generic design. Therefore, the selection of the generic soil profiles should consider the following two aspects. First, the postulated site profiles to be used in the seismic SSI analysis are defined to capture the scope of the site conditions for a DC application. Second, the CSDRS should be shown to be appropriate for these postulated site profiles in frequency content by demonstrating that the frequencies for the amplified portion of the CSDRS are consistent with the site profile column frequencies. Otherwise, the postulated site profiles will not be able to propagate the CSDRS in the SSI analysis, and thereby, will not subject the SSCs to the amplified response over the frequency range of interest to the SSI. If the CSDRS are intended to be the envelope of the GMRS for a broad range of site conditions, de-convolving the CSDRS through the generic site profiles should produce an envelope to the UHS for the same set of sites. Because of the hysteretic nature of soils, the characteristics of site profiles (modulus, damping, frequencies) are dependent upon the specific motions imposed on them. It is not possible for generic site profiles associated with the CSDRS to capture exactly the site profiles for any specific site since the CSDRS is not typically selected to match any individual GMRS; rather, the generic profiles could be interpreted as bounding for multiple site conditions. This bounding concept is sometimes misconstrued as the numerical values of site-specific properties being confined within the numerical values of the generic profiles. Rather, because the CSDRS and associated generic profiles together represent the response characteristics of the generic site, a particular site is considered being bounded by the generic site only if the site response characteristics typified by the GMRS and site-specific profiles are consistent with the CSDRS and associated generic profiles. To this end, site-specific analyses may be unavoidable, but the extent of the analysis to be performed varies, depending on the degree to which the site response characteristics are closely represented by the generic site.

9.3.2.2 Structural Models for Seismic Analysis. The issue associated with structural modeling approach for seismic analysis for the standard design arose from the fact that the objective of the standard design is intended to encompass a broad range of site conditions involving a wide range of frequencies. Since a customized design addresses a specific site, the seismic modeling may focus on the specific site conditions typically with much narrower range of frequencies being considered than that for the generic design. Therefore, for a soil site with dominant frequencies in a range of less than 10 Hz, a lumped mass stick model representing containment structure may be suffice and is typically used for the seismic analysis. A standard design, on the other hand, may need to consider a generic site encompassing a much broader range of frequencies than for a customized design. Therefore, the mathematical models of seismic Category I structures t should have sufficient refinement to capture all vibration modes of interest. This is typically up to 33 Hz for spectra comparable to RG 1.60 spectra and up to 50 Hz for eastern United States hard rock high frequency (HRHF) spectra. Especially for high frequency responses, the local (out-of-plane) flexibility of floors, walls, and ceilings may result in amplified local responses that need to be considered in structural design and in developing the ISRS for the seismic analysis of subsystems and components. Such high frequency responses cannot be captured through a lumped mass stick model, albeit it may be sufficiently detailed. Therefore, a detailed finite element model may be required to adequately capture the high frequency responses. Depending on the response interest to the design, the local regions in need of refinement should be identified during the model development stage and considered by the analyst in the development of computationally efficient models for conducting SSI analyses. Recent DC applicants have recognized the need to consider local flexible regions which could potentially be excited by high frequency inputs and have included local model refinement in their finite element models for the seismic analysis [22].

In addition, cracking can affect the seismic response of concrete structures in a significant manner. In general, concrete cracking reduces the stiffness of the structural elements and determines a corresponding reduction in the fundamental SSI frequencies. Depending on the frequency content of the ground motion, this frequency shift can result in an increase in seismic demands, if the SSI frequencies are located on the descending slope of the input spectra.

The severity and extent of concrete cracking depends on the state of stress of the structural elements for the various design load combinations. Therefore, cracking is not uniformly distributed throughout the structure but, rather, is most significant in the regions with the largest stress levels. The complexity of concrete cracking makes it impractical to realistically include these effects in the seismic analysis model for a standard design. Industry standards such as American Society of Civil Engineers (ASCE) 4-98, "Seismic Analysis of Safety-Related Nuclear Structures and Commentary" [23] (see Section 3.1.3 and commentary), and ASCE 43-05, "Seismic Design Criteria for Structures, Systems, and Components in Nuclear Facilities" [15] (see Section 3.4.1 and commentary), provide pertinent guidance on how to modify the structural stiffness to account for the effects of concrete cracking, in a simplified manner, for different types of structural elements and different stress levels.

DC applications typically account for concrete cracking in a simplified manner by globally reducing the concrete modulus of elasticity of all structural elements by a certain amount. This is done as the envelope of demands calculated using a seismic analysis model with (1) the full value of the concrete modulus of elasticity and (2) a 50% reduced value of the concrete modulus of elasticity. Another approach which has been used by DC applications is to justify for using a reduced value for the concrete modulus of elasticity (for example, 80%) in a linear analysis model to account for concrete cracking effect. The linear analysis using the reduced concrete modulus is then justified based on a comparison to a detailed nonlinear analysis that incorporated concrete cracking in the finite element model.

9.3.2.3 Stability Evaluation for Seismic Design. An important aspect of the design for NPP structures is to demonstrate stability for seismic loads. SRP Section 3.8.5, Revision 3 provides acceptance criteria for evaluating the stability of seismic Category I structures in terms of margins of safety — expressed as factors of safety (FS) — against sliding and overturning that are greater than 1.1. The implicit assumption is that FS are calculated based on the ratio of minimum seismic resistance divided by maximum seismic demand, where both resistance and demand are unique numbers derived from site-specific design parameters and simple, equivalent-static, calculations.

For standard designs, however, it is often difficult to demonstrate the seismic stability using simple calculations based on the pseudo-static equilibrium. The reason for this difficulty is discussed below. In a standard design, the analysis and design of SSCs are performed based on postulated design parameters that are bounding for future potential sites. Therefore, design parameters (e.g., seismic ground motions, environmental loadings, soil conditions) are often conservatively specified, representing envelopes of the site-specific characteristics to be expected at potential sites. This poses certain challenges to the structural design, since design demands tend to be overestimated. In addition, the lack of specific design parameters may lead to a design approach based on postulated worst-case scenarios, which may appear to be overly conservative but are necessary to establish design bounds. Therefore, the seismic demands tend to be overestimated while the seismic resistance tends to be conservatively underestimated. To demonstrate seismic stability, DC applicants have relied on performing more realistic time-history seismic analyses in place of the pseudo-static approach that, in some instances, explicitly incorporated the nonlinearities caused by sliding, uplift, and lateral soil pressures at the perimeter of embedded structures. Real time FS are calculated from these dynamic analyses and are compared against the SRP criteria to establish the stability for the structural designs. The results of these analyses often indicate that small amounts of sliding or uplift may occur during a seismic ground motion. To meet the intent of the SRP, the safety review needs to determine, on a case-by-case basis, whether the small amounts of sliding or uplift are acceptable.

Shear keys are often added under the foundations of structures to increase seismic resistance against sliding and to improve overall seismic performance. It is noted that, for typical NPP structures with relatively low aspect ratios, sliding stability is more difficult to demonstrate than overturning stability.

To evaluate seismic stability using a time-history analysis, time-dependent FS against sliding and overturning are computed and compared against a value of 1.1. In some situations, the computed FS of less than 1.1 may still be acceptable if these FS correspond to durations that are judged too short to cause physical movement of the structure. The time-history analysis should explicitly model the friction at the foundation-soil interface when evaluating the duration, magnitude, and extent of potential sliding; particularly whether sliding corresponds to localized relative displacements at the interface or whether there is a tendency for uniform, rigid-body sliding. The model should explicitly include uplift at the foundation-soil interface to establish whether there is a tendency for rocking. The analysis should also account for lateral soil pressures, including both active and passive pressures; however, it is cautioned that passive pressures can only be activated if soil displacements are greater than a geotechnical threshold that is a function of the soil type. Since it is particularly difficult to model lateral soil pressures in a realistic manner, it may be preferable to use conservative simplifications in the analysis. When the multiple time history approach is applied, the seismic stability should be evaluated for each seed input to ensure the seismic demands are established for the seismically stable structure.

9.3.2.4 Considerations of Settlement Effect in Standard Designs. Soil settlements either during the construction or during the design life of the structure, if not properly considered in the design and construction, could severely impact the performance of the structure. In accordance with SRP Section 3.8.5, Revision 3, seismic Category I structures (foundations and superstructures) should be designed to take into account the additional forces and moments that are induced by differential settlements of the soil under the foundation, as well as by the effects of the construction sequence. Past experience and current industry codes and standards (e.g., American Concrete Institute (ACI) 349-06, "Code Requirements for Nuclear Safety-Related Concrete Structures (ACI 349-06) and Commentary," issued 2006 [14]) indicate that these are important design considerations. It should be noted that, in recent NPP designs, the most important seismic Category I structures (e.g., the containment and shield buildings, fuel buildings) are grouped together on a nuclear island (NI) that is supported on a single large concrete foundation mat.

To investigate the effect of soil settlements on the structural performance of the NI foundation mat and connecting superstructure walls, it is necessary to distinguish between the settlements that are expected to occur during construction and post-construction settlements caused by soil consolidation and compaction — the soil stiffness for these two cases can be substantially different. For a generic design based on Part 52 process, two issues need to be considered regarding: (1) how the effects of these settlements are accounted for in the standard design, where assumptions need to be made regarding generic soil parameters, and (2) how the specifications are established such that the COL applicant can demonstrate that, for a particular construction sequence, forces and moments induced by predicted and measured settlements at a particular site are bounded by those considered in the standard design.

To consider settlement effects, the standard design should adequately address the effects of the construction sequence and associated soil settlements in the design. On the other hand, these issues are often site-specific per se. Therefore, it can be a significant challenge for a DC applicant to establish a practicable interface that allows for the standard design to account for appropriate construction sequences and settlement loads while permitting the COL applicant to verify that these loads are not exceeded during or after construction. One way to address the interface issue could be that the standard design considers (1) a postulated set of soil stiffness parameters for the construction phase, (2) a postulated set of soil stiffness parameters for the post-construction phase, and (3) a postulated construction sequence and corresponding set of construction loads. To account for construction sequence and settlement loads in the standard design, it may be necessary to perform a detailed, sequential, finite element (FE) analysis of the NI foundation and superstructure with realistic modeling of the supporting soil stiffness and construction sequence, including anticipated effects through the end of the operating life of the NPP. The sequential FE analysis should be based on the postulated soil conditions and validated with geotechnical soil settlement analyses. The envelope of forces and moments computed during the sequential FE analysis can then be compared with corresponding forces and moments obtained from a “reference” FE analysis that does not include construction sequence or settlement effects. From this comparison, any difference in forces and moments could be taken as a separate construction sequence/soil settlement load case, to be considered in the structural design of the foundation and superstructure in addition to all other load cases, in accordance with ACI 349-06, Section 9.2.2. The settlement profiles at all stages of the sequential FE analysis should also be computed; the COL applicant can then use these profiles for verification purposes, as described below.

The soil conditions postulated for settlement effects in the standard design would correspond to a relatively soft soil to bound potential sites. However, soft clayey soils behave differently from soft sandy soils, especially regarding long-term settlements. The sequential FE analysis should consider the soil conditions, clay or sand, that result in greater induced moments and forces on the foundation and superstructure. In addition, the possibility that a stiffer soil could result in greater induced moments or forces in certain areas of the foundation or superstructure should also be investigated.

In the above discussion, it is assumed that the total magnitude of soil settlements is not a design consideration during either construction or post-construction phases. It should be noted that it is not so much the magnitude of the settlements that affects the structural performance of the NI foundation and superstructure; rather, it is the relative shape of the settlement profile in terms of slope and curvature. This last statement is only valid when considering an individual structure; total settlement is clearly of interest when considering adjacent structures connected by appurtenances (nonflexible commodities, such as piping and conduit), which would need to be addressed in the design.

Based on the settlement profiles established in the DC, the COL applicant should perform a site-specific geotechnical investigation to determine predicted settlement profiles during construction and post-construction phases, based on the actual construction sequence to be used. If the predicted settlement profiles compare favorably to the DC settlement profiles — in terms of slope and curvature, not necessarily in absolute magnitude — then it is inferred that the forces and moments induced by the predicted settlements are bounded by the forces and moments considered in the standard design. This comparison may be made in terms of the “angular distortion” concept, as described in U.S. Army Corps of Engineers Manual No. 1110-1-1904, “Engineering and Design: Settlement Analysis,” issued 1990 [24]. In addition to the predictive calculations, a settlement monitoring program needs to be established to verify whether measured settlements are consistent with predicted settlements during the operating life of the NPP.

9.4 CONTAINMENT FEATURES OF GENERATION III/III+ REACTORS

This section highlights advances of generation III/III+ reactor designs in comparisons to previous generations of reactors. The descriptions of containment features for typical generation III/III+ designs are also provided which are anticipated for constructions in the United States. There are five generation III/III+ reactor designs which either have been approved or in the process of being reviewed by the NRC. They are: (1) Advanced Boiling Water Reactor (ABWR), (2) Economic Simplified Boiling Water Reactor (ESBWR), (3) Advanced Passive Pressurized Water Reactor 1000 (AP1000), (4) U.S. Economic Pressurized Water Reactor (US EPR), and (5) U.S. Advanced Pressurized Water Reactor (USAPWR). Other generation III/III+ reactor designs may also be considered in the future for the U.S. market, such as: South Korea's APR-1400 Advanced PWR design which evolved from the US System 80+ with enhanced safety and seismic robustness and was earlier known as the Korean Next-Generation Reactor and the Advanced CANDU Reactor (ACR-1000) designed by Atomic Energy of Canada Ltd (AECL) which is a light-water PWR that incorporates features of both CANDU pressurized heavy water reactors (PHWRs) and advanced light-water reactors.

9.4.1 Advances of Generation III/III+ Reactor Designs

Containments for LWRs are constructed as part of engineered safeguards in assuming an important role in the safety of the NPP operation. Traditional containments are designed to be shell type of steel or reinforced concrete structures focusing on the dual functions of: (1) protection of NSSS and safety related systems and equipment and (2) leak-tight membrane to contain the postulated radiological release to the environment. The generation III/III+ reactors have incorporated design features such as adding a second reinforced concrete shield structure over the primary containment for protection against the effects of the potential use of large commercial aircraft as a weapon against a nuclear installation. In addition, the new reactor containment designs have increasingly utilized more reliable evolutionary and passive safety systems which can be actuated by gravity, convective natural circulations that use fewer or no active safety systems, therefore are less prone to human errors.

Compared to previous generations of reactor design, the advances in technologies utilized in generation III/III+ may be highlighted as follows:

1. Employ highly reliable, less complex safe shutdown systems, particularly ones which use inherent or passive safety features, and reduce the possibility of core melt accidents, resulting in an enhanced safety;
2. Incorporate simplified and better understood systems which allow more straightforward engineering analysis and design, operate with fewer operator actions to significantly reduce potential human errors, and increase operator comprehension of reactor conditions;
3. Use standardized designs to enable identification and resolution of safety issues related to design and constructions early and allow for concurrent resolution of safety and security requirements, resulting in an overall security systems that requires fewer human actions;
4. Include features which prevent a simultaneous breach of containment and loss of core cooling from an aircraft impact, or that inherently delay any radiological release;
5. Consist of features which maintain spent fuel pool integrity following an aircraft impact;
6. Design for higher availability and longer operating life — typically 60 years as opposed to 40 years for current operating reactors, and;
7. Incorporate features to allow the use of higher burn-up and burnable absorbers to extend fuel life resulting in increased fuel efficiency and reducing the amount of nuclear waste.

9.4.2 Generation III/III+ BWR Designs

9.4.2.1 ABWR. The ABWR was originally designed by General Electric in 1990s and the standard design was certified by the NRC in 1997. ABWRs are the evolutionary design of the generation III. Several NPPs had been constructed by Hitachi and Toshiba and currently are in commercial operation in Japan and two ABWRs are in construction is Taiwan. The South Texas Project Electric Generating Station (STP) plans to build COL a Toshiba ABWR design with rated power of 1350 MWe; the STP COL is currently being reviewed by the NRC. The ABWR design incorporates features of the BWR designs in Europe, Japan, and the United States, and uses improved electronics, computer, turbine, and fuel technology. The design is expected to show improvement in plant availability, operating capacity, safety, and reliability. Improvements include the use of internal recirculation pumps, control rod drives that can be controlled by a screw mechanism rather than a step process, microprocessor-based digital control and logic systems, and digital safety systems. The design also includes safety enhancements such as containment over-pressure protection, passive core debris flooding capability, an independent water makeup system, three emergency diesels, and a combustion turbine as an alternate power source.

The ABWR containment utilized the same pressure suppression concept as GE's previous generations of BWR designs, and was evolved from the Mark II type of containment (Fig. 9.2).

The ABWR containment is significantly smaller that the Mark III containment because the elimination of the recirculation loops translates into a significantly more compact containment and reactor building [25]. The containment structure is made of reinforced concrete with a steel liner from which it derives its name — RCCV, or reinforced concrete containment vessel (Fig. 9.3). The RCCV consists of an Upper Drywell, a Lower Drywell, a Suppression Pool Chamber (also called wetwell), and the vents connecting the drywells and wetwell. The height of the containment is 31.8 m measured from the top of the basemat to the top of the upper drywell with an inside diameter of 29 m. The ABWR containment is about 10 m shorter than the Mark II design.

As shown in Fig. 9.3, the upper drywell (UD) surrounds the reactor pressure vessel (RPV) and houses the steam and feedwater lines and other connection of the reactor primary coolant system, safety/relief valves (SRVs) and drywell HVAC coolers. The UD is a cylindrical, reinforced concrete structure with a removable

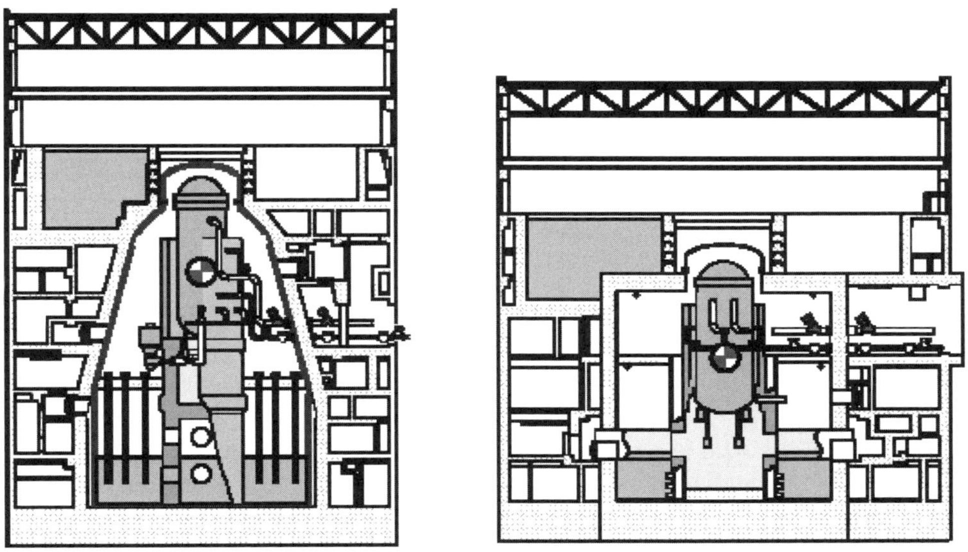

(a) Mark II containment (b) ABWR containment

FIG. 9.2
COMPARISON BETWEEN MARK II AND ABWR CONTAINMENTS [25]

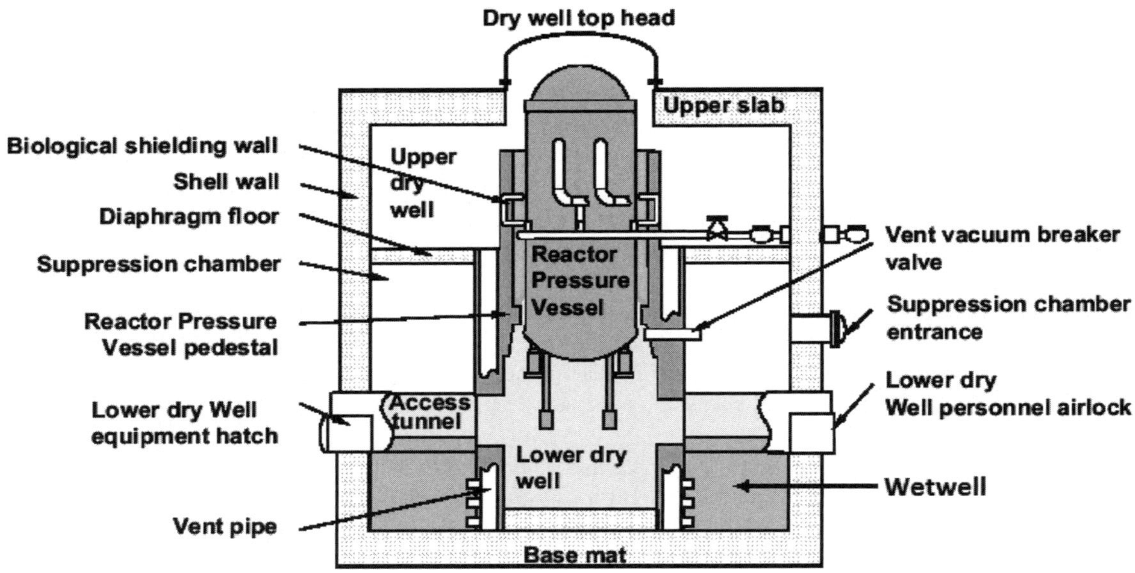

FIG. 9.3
RCCV, REACTOR INTERNALS AND MAJOR PENETRATIONS FOR ABWR CONTAINMENTS [25]

steel head and a reinforced concrete diaphragm floor. The lower drywell (LD) accommodates the reactor internal pumps, under vessel components and servicing equipment. The pedestal consisting of cylindrical prefabricated concrete-filled steel structure, which supports the RPV, is connected rigidly to the diaphragm floor and separates the LD from the wetwell. Ten drywell connecting vents (DCVs) are built into the pedestal and connect the UD and LD. The DCVs are extended downward via steel pipes, each having three horizontal vent outlets into suppression pool (Table 9.2). In addition, the LD floor is topped with a think basaltic concrete pad under the RPV. In an event of ex-vessel relocation of molten core melt through the lower head of

TABLE 9.2
KEY DESIGN CHARACTERISTICS OF CONTAINMENT FOR ABWR [20]

Containment	
Primary	
Type	Pressure suppression
Construction	Reinforced concrete with steel liner
Drywell	Concrete cylinder
Wetwell	Concrete cylinder
Design pressure, MPa	0.31
Containment ultimate pressure capacity, Mpa	0.97
Design leak rate, % free volume/day	0.5 — Excluding MSIS leakage
Drywell free volume, m^3	7350
Wetwell free volume, m^3	5960
Suppresion pool water volume, m^3	3580
Number of vertical vents	10
Vertical vent diameter, m	1.2
Number of horizontal vents/vertical vent	3
Horizontal vent diameter, m	0.7
Secondary	
Type	Controlled leakage
Construction	Reinforced concrete with steels

the RPV, the heated core melt will be both caught and held on the basaltic pad, which is subsequently flooded and cooled by wetwell's water supplied through fusible links within the wall separating the wetwell from the lower drywell. The pressure retaining concrete walls of the RCCV are lined with leak-tight steel plates and for the pressure boundary of the primary containment system.

In addition, the RCCV is structurally integrated with the reactor building and forms the a major structural part of this building as shown in Fig. 9.2b. The reactor building is separated into four compartments, three house the three independent divisions of the emergency core cooling system (ECCS), and one is reserved for non safety systems. Because the RCCV and reactor building are monolithically connected, the overall structural stiffness is increased, therefore is capable of resisting higher dynamic and shear loads. To illustrate this, the reactor building is designed for a safe shutdow earthquake (SSE) of 0.3 g peak ground acceleration (PGA); however, a seismic margin assessment indicated that the ABWR plant level seimsic capacity in terms of high confidence and low probability of failure (HCLPF) is estiamted at two times the SSE at the sequence level. This ensure very little possibility of a core damage event as a result of earthquake.

In addition, the ABWR containment incorporates the containment overpressure protection system (COPS). The COPS protects the containment if a less likely severe accident occurs which increases the containment pressure to a point where containment structural integrity is threatened. The COPS will release this containment pressure through a line connecting the wetwell atmosphare to the plant stack, thus providing a filtered release path from the wetwell airspace to preclude an uncontrolled containment failure due to over pressurization.

The COPS consists of a relief line connecting the wetwell airspace to the plant, containing two rupture disks in series. The COPS is normally closed even at the design pressure. When the containment is over-pressurized above the design pressure but below a setpoint pressure of approximately 1.0 MPa, corresponding to the ASME Service Level C capability of the containment structures, the rupture disks would open, allowing the pressure to be relieved in a manner that forces escaping fission materials to pass through the suppression pool water, thereby preventing the release of radioactive materials to the environment. After the containment pressure has been reduced and normal containment heat removal capability has been regained, the operator can close two normall open air-operated valves in the relief path to reestablish containment integrity. Note that the COPS is completely passive, and no power is required for initiation or operation of the COPS.

9.4.2.2 ESBWR. The ESBWR, which is designed by GE Hitachi, is a generation III+ reactor that utilizes complete passive features and natural circulation principles for accident preventions and mitigations. The ESBWR is rated at 1560 MWe. The reliance on natural circulation and passive safety systems enhances the plant performance and simplifies the design for the ESBWR. The passive safety features eliminate the need for safety-grade pumps and AC powers. Design simplification also results on a reduction in building volume compared to the ABWR.

The ESBWR containment also used the pressure suppression concept as GE's BWR designs but allowed different configurations and arrangements from the ABWR containment to accommodate natural circulations and passive safety systems inside the containment (Fig. 9.4).

Most noticeably are the elevated pools to allow for natural circulations and the ECCS system within the containment which consists of gravity driven cooling system (GDCS) and automatic depressurization system (ADS). Similar to the ABWR, the ESBWR containment structure consists of a reinforced concrete cylindrical structure with a steel liner from which it derives its name — RCCV, or reinforced concrete containment vessel (Fig. 9.4). The RCCV consists of an Upper Drywell, a Lower Drywell, three GDCS pools, a Suppression Pool Chamber (also called wetwell), and the vents connecting the drywells and wetwell. The RCCV is capped off with a steel head and is submerged in the PCCS/IC pools which are part of the reactor building. As shown in Fig. 9.5, the RCCV and the reactor building are structurally integrated through reinforced concrete walls and slabs. Therefore, the reactor building can serve as a secondary barrier to accidental release of radioactive materials to the environment.

The upper drywell of the RCCV consists of reinforced concrete cylindrical structure with an inside diameter of 36 m and is 19.95 m high. The wall and roof slab have a thickness of 2 m and 2.4 m, respectively. It

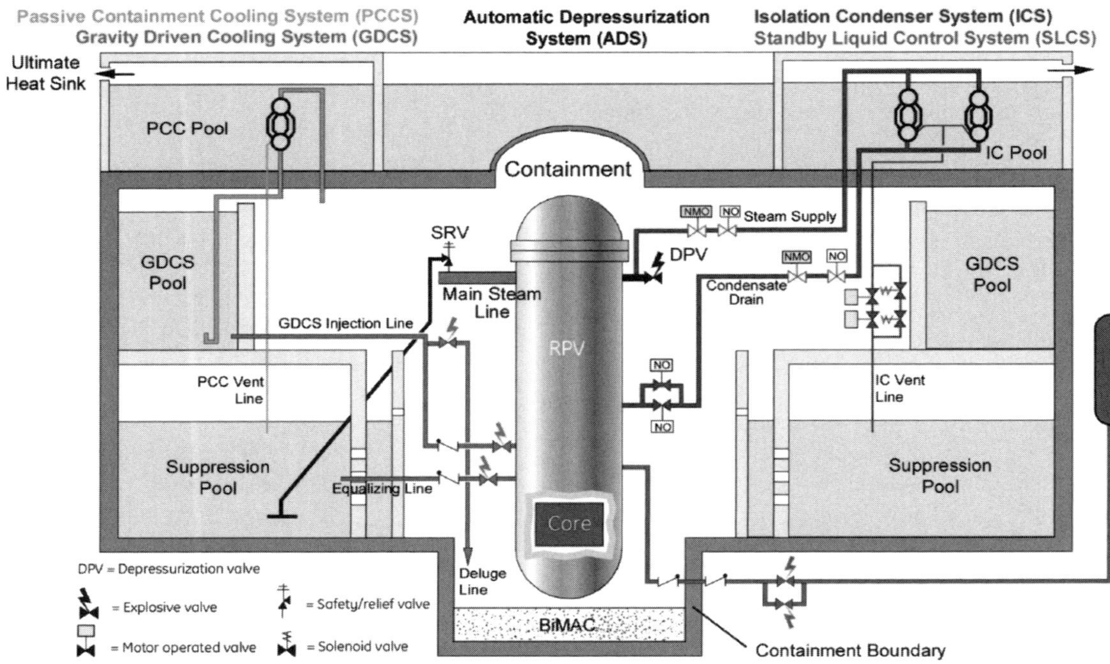

FIG. 9.4
SCHEMATIC VIEW OF CONTAINMENT SYSTEM FOR ESBWR [26]

surrounds the RPV and houses three GDCS pools and a large suppression pool. It also has a large airspace to allow the accumulation of large amount of steams released as a result of a postulated reactor accident. The lower drywell is a cylindrical airspace directly below the RPV and is surrounded by the reactor pedestal on the side and basemat at the bottom. The lower drywell has an inside diameter of 11.2 m and is 15.05 m high. The foundation mat is 5.1 m thick. The reactor pedestal is a 2.5-thick reinforced concrete cylindrical structure which is part the containment boundary and also provides support to the RPV. In addition, at the bottom of the lower drywell cavity and above the basemat is a BiMAC (basemat internal and melt arrest and coolability) device (Fig. 9.6) which consists of a sacrificial refractory concrete slab with embedded inclined pipes which are connected to GDCS pools. In a severe core melt event, the molten core melt would breach the bottom of the RPV and relocate in the lower drywell cavity. BiMAC is passive system actuated by thermocouples which detect the relocated core melt, which triggers the opening of the squib valves that allow the GDCS pool water flow through the BiMAC pipes to cover and cool the molten core melt. It should be noted that the design of BiMAC is not available for other designs and therefore is unique for the ESBWR.

The ESBWR design consists of four main passive systems [26, 27]:

1. Automatic depressurization system (ADS) — The ADS consists of 10 safety relief valves (SRVs) mounted on top of the main steamlines which discharge steam to the suppression pool, and 8 depressurization valves (DPVs) that discharge steam to the drywell airspace;
2. Gravity driven cooling system (GDCS) — The GDCS provides the makeup water through gravity flow into the vessel after the ADS depressurizes the RPV. The GDCS and ADS together become the plant's ECCS;
3. Isolation condenser system (ICS) — The ICS removes delay heat from the reactor following transient events involving reactor scram including station blackout. The ICS consists of four independent high pressure loops, each containing a heat exchanger which condenses steam on the inside of the tube wall.

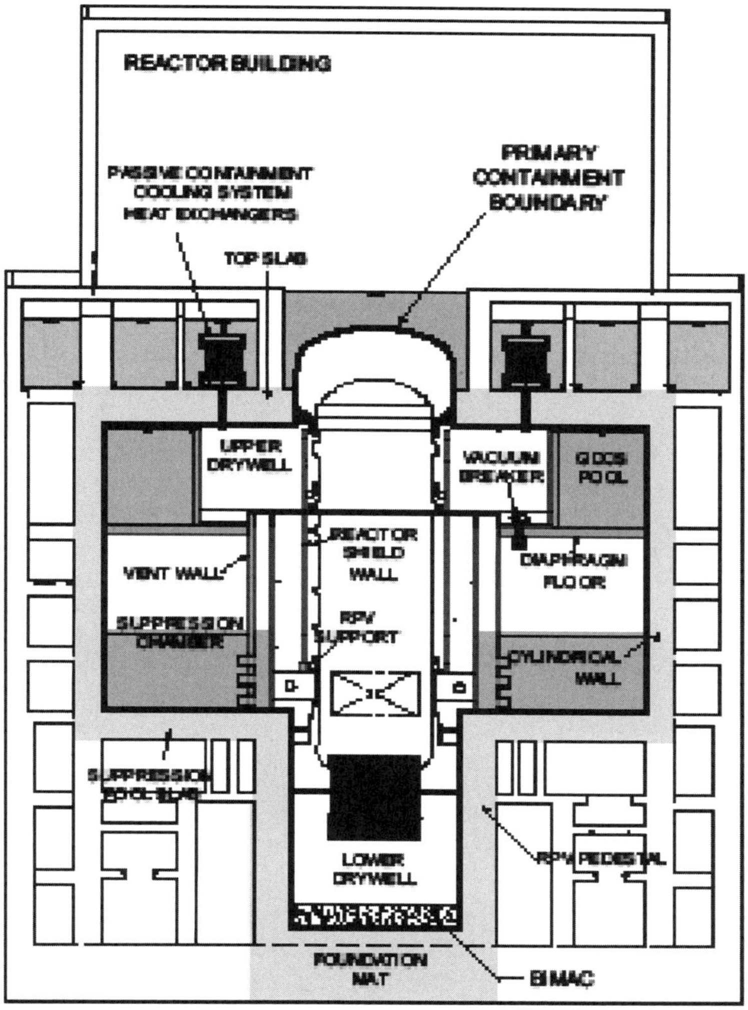

FIG. 9.5
SCHEMATIC VIEW OF REACTOR BUILDING AND CONTAINMENT FOR ESBWR [26]

The tubes are located in a large pool above the RCCV. The system uses natural circulation to remove decay heat; and

4. Passive containment cooling system (PCCS) — The PCCS removes heat from the RCCV following a LOCA. The system consists of four safety-related low pressure loops. Each loop has a heat exchanger open to the containment, a condensate drain line and a vent discharge line submerged in the suppression pool. Similar to the isolation condensers, the PCCS heat exchangers are located in cool pools external to the containment.

The drywell structure is designed for the pressure and temperature transients associated with the rupture of any of the primary system piping inside the drywell. It is also designed for the negative differential pressures associated with containment depressurization events, when the steam in the drywell is condensed by the PCCS, the GDCS, the fuel and auxiliary pool cooling system (FAPCS), and cold water cascading from the break following post-LOCA flooding of the RPV.

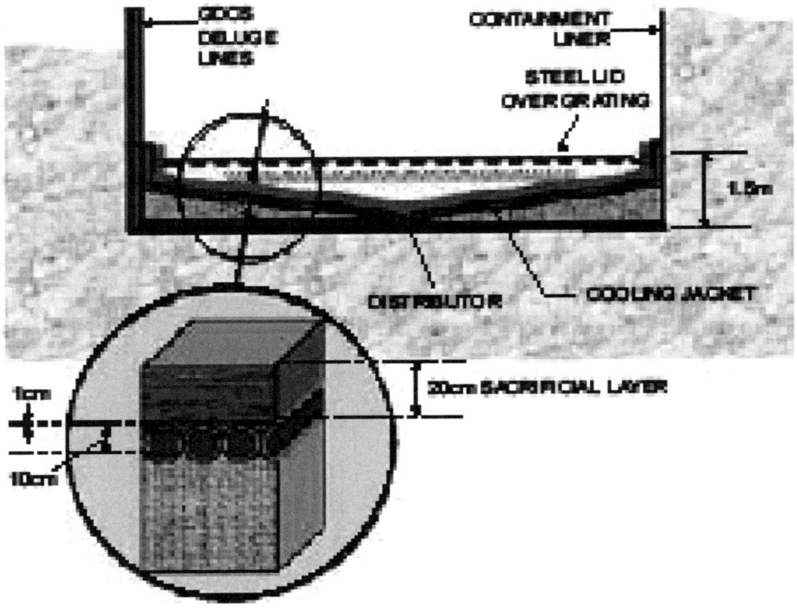

FIG. 9.6
CONCEPTUAL VIEW OF BIMAC DEVICE FOR ESBWR [26]

In a LOCA event, the increased pressure inside the drywell forces a mixture of non-condensable gases, steam and water vapor though the PCCS vertical and horizontal vent pipes into the suppression pool where the steam is rapidly condensed. The non-condensable gases are contained in the airspace of the wetwell. The containment design pressure is 0.31 MPa (45 psig) (Table 9.3).

As described above, the RCCV is an integral part of reactor building structure which also houses structures, systems, and components above the RCCV, the buffer pool which is used to store the dryer,

TABLE 9.3
KEY DESIGN CHARACTERISTICS OF CONTAINMENT FOR ESBWR [26]

Containment	
Primary	
Type	Pressure suppression
Construction	Reinforced concrete with steel liner
Drywell	Concrete cylinder
Wetwell	Concrete cylinder
Design pressure, MPa	0.31
Containment ultimate pressure capacity, Mpa	1.0
Design leak rate, % free volume/day	0.5 — excluding MSIS leakage
Drywell free volume, m^3	7206
Wetwell free volume, m^3	5432
Suppresion pool water volume, m^3	4424
Number of vertical vents	12
Vertical vent diameter, m	1.2
Number of horizontal vents/vertical vent	3
Horizontal vent diameter, m	0.7
Secondary	
Type	Controlled leakage
Construction	Reinforced concrete with steels

the equipment storage pool for storing the chimney partitions and the separator, and steam tunnels, refuel area and other supporting systems. The reactor building is a rigid box type shear wall structure, constructed of reinforce concrete. Similar to the ABWR, the ESBWR RCCV is also monolithically connected to the reactor building. The reactor buidling not only increases the structural performance of the RCCV but also provide a secondary barrier to accidental release of radioactive fission products into the environment.

The reactor building is designed for a safe shutdown earthquake (SSE) of 0.3 g peak ground acceleration (PGA); however, a seismic margin assessment indicated that the ESBWR plant level seismic capacity in terms of high confidence and low probability of failure (HCLPF) is estimated at 1.67 times the SSE at the sequence level. This ensure very little possibility of a core damage event as a result of earthquake.

In addition, the ESBWR containment incorporates the containment overpressure protection system (COPS). The COPS protects the containment if a less likely severe accident occurs which increases the containment pressure to a point where containment structural integrity is threatened. The COPS will release this containment pressure through a line connecting the wetwell atmosphare to the plant stack, thus providing a filtered release path from the wetwell airspace to preclude an uncontrolled containment failure due to over pressurization.

The COPS consists of a relief line connecting the wetwell airspace to the plant via a remote manual valve. The COPS is normally closed even at the design pressure. The containment pressure builtup would take some time to reach above the design pressure but below a setpoint pressure of approximately 1.0 MPa, corresponding to the ASME Service Level C capability of the containment structures, which would allow time for the operator to manually open the valve allowing the pressure to be relieved in a manner that forces escaping fission materials to pass through the suppression pool water, thereby preventing the release of radioactive materials to the environment. This function of the containment inerting system is called the manual containment overpressure protection system (MCOPS).

9.4.3 Generation III/III+ PWR Designs

9.4.3.1 AP1000. This section provides a description of the AP1000 reactor features, particularly the features relating to the containment design, based on the information from publicly available documents [28, 29]. The AP1000 is a two-loop PWR built and improved upon the established of technology of major components (steam generators, fuels, pressurizer, reactor vessels, etc.) used in current Westinghouse-designed plants with proven, reliable operating experience over the past 35 years. The AP1000 design with rated power at 1117 MWe includes advanced passive safety features and extensive plant simplifications to enhance the safety, construction, operation, and maintenance of the plant. The safety systems use natural circulations, gravity flow, pressurized gas, and convective flow, but do not rely on active components such as pumps, fans, and diesel generators as were seen in most current operating reactors. The AP1000 standard design was certified by the NRC in 2005. Subsequently, the Westinghouse applied an amendment to the AP1000 to revise but not limited to the shield building for utilizing advanced modular structural designs for better performance against the effect of commercial aircraft impact and for taking advantage of advanced construction technologies. The AP1000 amendment was approved by the NRC in 2011.

Some of the high-level design characteristics of the AP1000 are highlighted below:

- Net electrical power of at least 1117 MWe; and a thermal power of 3415 MWt
- Rated performance is achieved with up to 10% of the steam generator tubes plugged and with a maximum hot leg temperature of 321°C (610°F)
- Core design is robust with at least a 15% operating margin on core power parameters
- Short lead time (5 years from owner's commitment to commercial operation) and construction schedule (3 years)
- No plant prototype is needed since proven power generating system components are used
- Major safety systems are passive; they require no operator action for 72 hours after an accident and maintain core and containment cooling for a protracted time without ac power

- Predicted core damage frequency of 2.4E – 07/yr is well below the 1E – 05/yr requirement and frequency of significant release of 1.95E – 08/yr is well below the 1E – 06/yr requirement
- Standard design is applicable to anticipated US and international sites
- Occupational radiation exposure expected to be below 0.7 man-Sv/yr (70 man-rem/yr)
- Core is designed for a 18-month fuel cycle
- Refueling outages can be conducted in 17 days or less
- Plant design life of 60 years without replacement of the reactor vessel
- Overall plant availability greater than 93%, including forced and planned outages; the goal for unplanned reactor trips is less than one per year
- Leak-before-break on primary lines > 6 in. and on main steamlines
- Seismic based on 0.3 g ground acceleration
- Security enhanced with all safe shutdown equipment located in safety reinforced concrete Nuclear Island buildings
- In-vessel retention of core debris following core melt which significantly reduces the uncertainty in the assessment of containment failure and radioactive release to the environment due to ex-vessel severe accident phenomena
- No reactor pressure vessel penetrations below the top of the core. This eliminates the possibility of a loss of coolant accident by leakage from the reactor vessel, which could lead to core uncover
- Containment design pressure is 0.407 MPa and containment pressure capacity at severe accident conditions is estimated about 0.9 MPa

The AP1000 uses the nuclear island to support Seismic Category I structures including: the containment vessel, the shield building and the auxiliary building. The nuclear island structures are designed to withstand the effects of postulated internal events, as well as the effects of natural phenomenon hazards such as winds, tornadoes, hurricanes, and earthquakes. The containment vessel is a freestanding cylindrical steel shell structure with a wall thickness of 4 mm (1.75 in.), and a diameter of 40 m (130 ft.), and is surrounded by the shield building (Fig. 9.7).

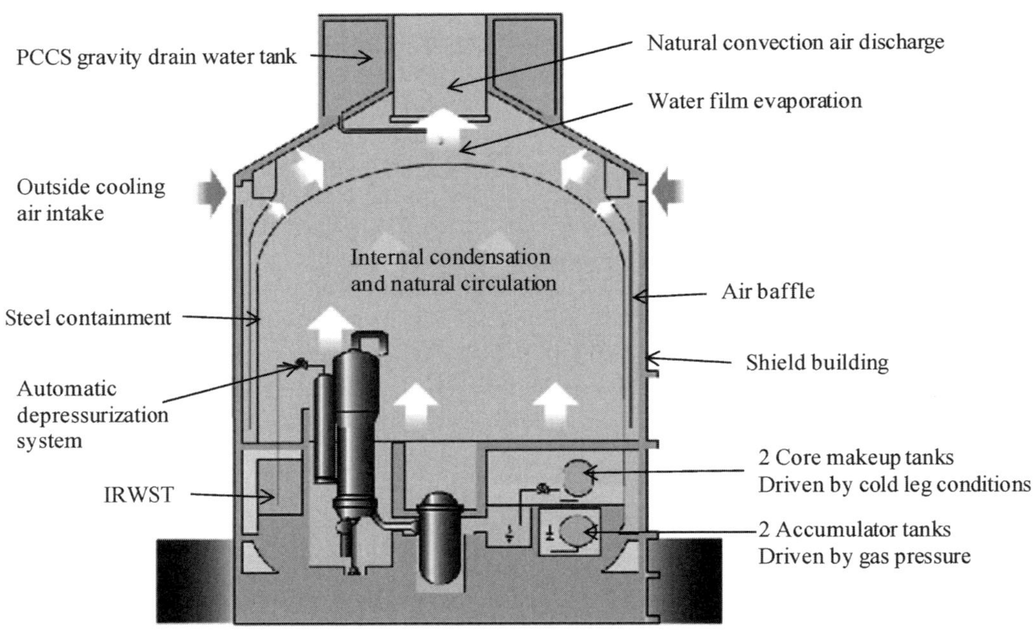

FIG. 9.7
SCHEMATIC VIEW OF CONTAINMENT AND CONTAINMENT SYSTEMS FOR AP1000 [28]

The principle systems located inside the containment include the reactor coolant system, the passive core cooling system (PCCS), and the reactor coolant purification portion of the chemical and volume control system. Both the containment vessel and the shield building are an integral part of the PCCS.

The AP1000 containment contains a 4.9-m (16 ft) diameter main equipment hatch and a personnel airlock at the operating deck level, and a 4.9-m (16 ft) diameter maintenance hatch and a personnel airlock at grade level. Because of their locations, these large hatches significantly enhance accessibility to the containment during outages and, consequently, reduce the potential for congestion at the containment entrances. These containment hatches, located at the two different levels, allow activities occurring above the operating deck to be unaffected by activities occurring below the operating deck.

The containment vessel provides protection against the uncontrolled release of radioactive fission materials to the environment. It is designed with a leakage rate of 0.1 wt.% per day of the containment air mass which is present at the start of a design basis accident and the resulting containment isolation.

As shown in Fig. 9.7, the shield building surrounds the primary steel containment vessel and provides the natural convective annulus for containment cooling. During normal operations, the shield building, in conjunction with the internal structures of the containment vessel, provides the required shielding for the reactor coolant system and all the other radioactive systems and components housed in the containment. During accident conditions, the shield building provides an additional barrier for radioactive airborne materials that may be dispersed in the containment as well as radioactive particles in the water distributed throughout the containment.

The PCCS air baffle is located in the upper annulus area of the shield building. The function of the PCCS air baffle is to provide a pathway for natural circulation of cooling air in the event that a design basis accident results in a large release of energy into the containment. In this event the outer surface of the containment vessel transfers heat to the air between the baffle and the containment shell. This heated and thus, lower density air flows up through the air baffle to the air diffuser and cooler and higher density air is drawn into the shield building through the air inlet in the upper part of the shield building.

Another function of the shield building is to protect the containment building from external events. The shield building protects the containment vessel and the reactor coolant system from the effects of tornadoes and tornado produced missiles.

The shield building design in the original design certification is a cylindrical, reinforced concrete structure. It has a conical roof which supports the PCCS water storage tank and air diffuser. The PCCS is also referred to as PCS (passive cooling system). During the amendment application, the Westinghouse revised the shield building to substantially improve the structural performance of the shield building against the effect of potential commercial aircraft impact. The new design incorporates a steel-plate composite (SC) walls for the upper portion of the shielding building, which are connected to the traditional reinforced concrete (RC) walls in both the roof and the lower portion of the shielding building (Fig. 9.8).

Although the applications of SC constructions have seen in other countries, it is the first time that such a non-conventional design feature is proposed for nuclear structures in the United States. The design for traditional RC structures typically meets the provisions of ACI codes (such as ACI-318 for building designs, ACI-349 for nuclear structures) to ensure the adequate quality for the design. However, the design codes had not addressed the conventional designs such as SC constructions. During the amendment application review, the NRC staff raised a number of technical concerns pertaining to the adequate design and testing with respect to the SC constructions. Subsequently, the Westinghouse and its consultants made substantial design improvements to the design SC and its connections to RC sections, and also performed a number of testing to confirm adequate structural performance of the SC design [30–32]. To address the apparent lacking of code requirements, the American Institute of Steel Construction (AISC) formed a sub-committee to develop an appendix to AISC N690 focusing on SC walls.

A significant technological advance of the AP1000 over the previous PWRs is that the defense-in-depth of the plant relies on completely passive safety features without the need for either pumps or AC powers. These safety features include: (1) the passive core cooling system (PXS), (2) the passive containment cooling system (PCS), (3) the main control room emergency habitability system (VES), (4) in-vessel retention of molten core melt.

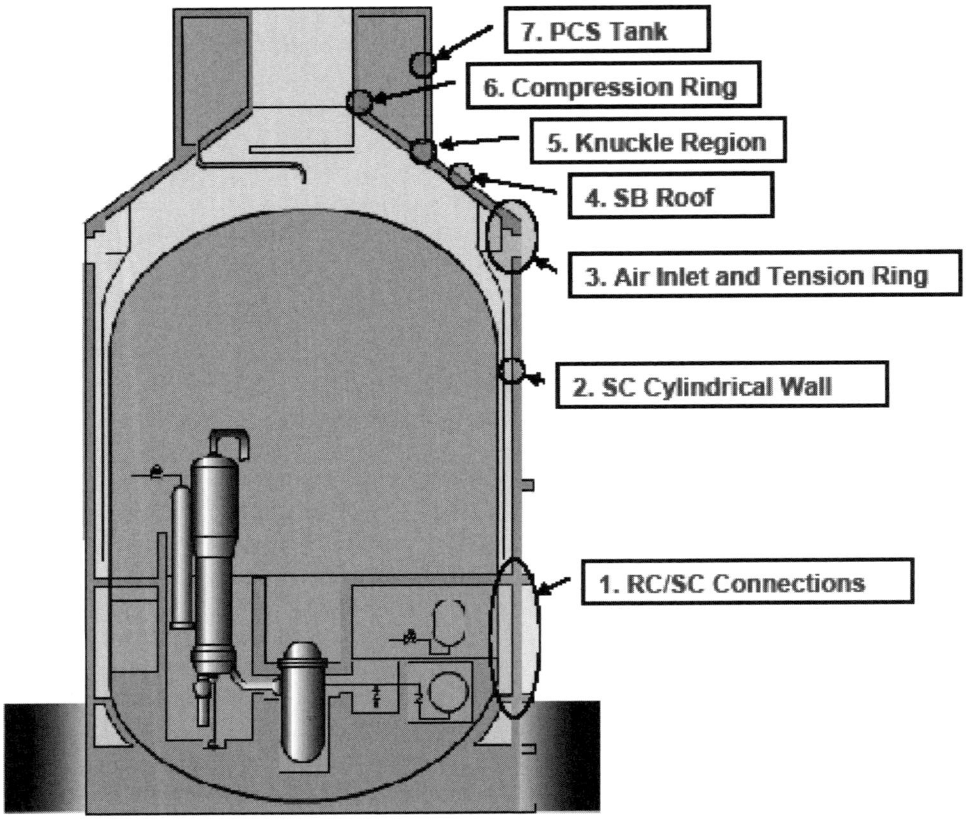

FIG. 9.8
HIGHLIGHTS OF MODIFICATION FOR SHIELD BUILDING DESIGN FOR AP1000 [28]

Safety injection and depressurization — The PXS (Fig. 9.9) is the ECCS to protect the plant against RCS leaks and ruptures of various sizes and locations, and is designed to perform safety injection, reactor depressurization, and passive residual heat removal (PRHR). The PXS performs safety injections to cool the core relying on water from three passive sources: the core makeup tanks (CMT), the accumulators, and the in-containment refueling water storage tank (IRWST), which all located within the primary containment, as compared to conventional PWRs which locate these tanks outside of the containments.

The PXS injection sources are directly connected to two nozzles on the reactor vessel through dedicated direct vessel injection (DVI) lines so that no injection flow can be spilled for the main reactor coolant pip breaks. The high pressure safety injection is accomplished by the two CMTs which effectively replace the high pressure injection systems in conventional PWRs. The CMTs consist of large volume stainless steel tanks with inlet lines that connect the RCS cold legs to the top of the CMTs and outlet lines connecting to the DVI line. The DVIs are connected to the reactor vessel downcomer. Each CMT is filled with borated water. The CMT inlet valve is normally open and therefore the CMT is normally at primary system pressure. The CMT outlet line is normally closed, preventing natural circulation during normal operation. When the reactor is tripped, the CMT outlet valve is open, establishing a natural circulation path. The cold borated water is injected in the reactor by gravity flow and hot primary fluid flows upward into the top of the CMT.

The intermediate pressure injection is done in a manner similar to current PWRs through accumulators. The PXS uses accumulators which are large spherical tanks approximately three-quarters filled with cold

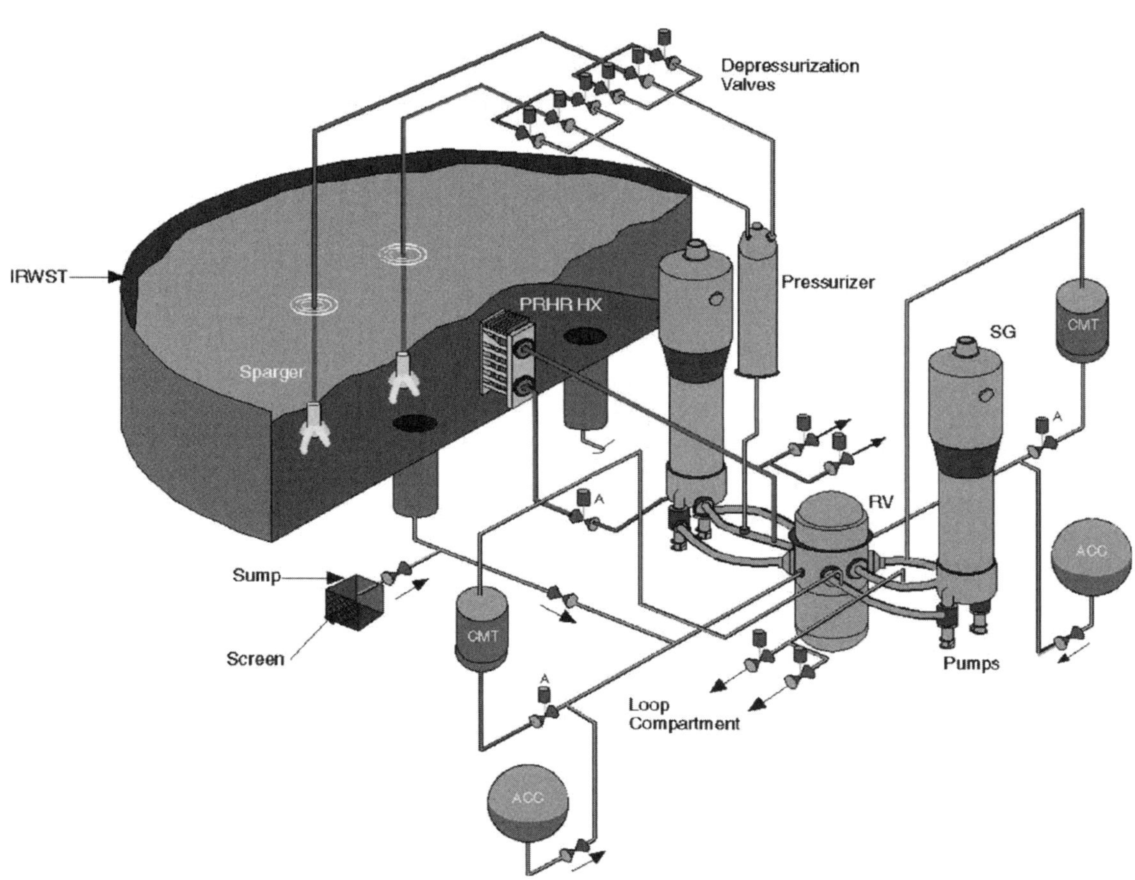

FIG. 9.9
AP1000 RCS AND PASSIVE CORE COOLING SYSTEM (PXS) [28]

borated water and pressurized to 4.83 MPa (700 psig) with nitrogen. A pair of check valves prevents injection flow during normal operation conditions. When system pressure drops below the accumulator pressure, the check valves open allowing cold borated water flow into the reactor downcomer via DVI line.

The low pressure injection is supplied by gravity flow of coolant makeup from the IRWST which is located at a height above the RCS loops. The IRWST is a very large concrete pool (Fig. 9.9) filled with borated water at atmospheric pressure and the injection occurs only when the RCS is depressurized to near atmospheric pressure. The AP1000 automatic depressurization system (ADS) has four stages of valves sufficient for the controlled reduction of primary system pressure. The three stages consist of two trains of valves connected to the top of the pressurizer. The ADS 1-3 valves discharge primary system steam into a sparger line than vents into the IRWST. The steam is condensed by direct connect with the highly subcooled borated water in the IRWST. The fourth stage of the ADS consists of two large valves attached to ADS lines on each hot leg. The ADS-4 valves open on low CMT liquid level and bring the primary system pressure to the containment conditions. The ADS-4 valves vent directly into the containment airspace.

The IRWST also serves as the heat sink for the PRHR heat exchanger (PRHR HX) which is submerged in the IRWST. The PRHR HX removes the decay heat and protects the plant against transients that upset the normal steam generator feedwater and steam systems. The PRHR HX satisfies the safety criteria for loss of feedwater, feedwater line breaks and steam line breaks. The PRHR HX and the passive containment cooling system provide indefinite decay heat removal capability without operator actions.

Passive containment cooling — The PCS provides the safety-related ultimate heat sink for decay heat and protect the containment from overheating and exceeding the design pressure. As shown in Fig. 9.7, the PCS remove decay heat through convective natural circulation. The steel containment vessel provides the heat transfer surface that removes heat from inside the containment into the annulus space between the containment and the shield building. Outside air is pulled in through orifices near the top of the shield building cylindrical walls and circulates around the baffle and through the annulus to remove the heat from steel containment surface, and then discharges the heated air upward out of the shield building through the central opening in the shield building roof. In addition, the PCS gravity drain water tank located on the top of the shielding building drains by gravity to cover the outer surface of the primary steel containment vessel.

Main control room emergency habitability system — The VES provides fresh air, cooling, and pressurization to the main control room (MCR) following a plant accident. Operation of the VES is automatically initiated upon receipt of a high MCR radiation signal, which isolates the normal control room ventilation path and initiates pressurization. Following system actuation, all functions are completely passive. The VES air supply is contained in a set of compressed air storage tanks. The VES also maintains the MCR at a slight positive pressure, to minimize the infiltration of airborne contaminants from the surrounding areas.

In-vessel retention (IVR) of molten core melt — The AP1000 manages the severe accident molten core debris cooling utilizing the IVR strategy. During postulated severe accidents, the accident management strategy is to flood the reactor cavity by IRWST water to submerge the reactor vessel. The water cools the external surface of the vessel and prevents molten debris in the lower head from failing the vessel wall and relocating into the containment. Retaining the debris in the reactor vessel protects the containment integrity by preventing ex-vessel severe accident phenomena, such as ex-vessel steam explosion and core-concrete interaction, which have large uncertainties with respect to containment integrity.

The passive AP1000 design is uniquely suited to in-vessel retention because it contains features that promote external cooling of the reactor vessel. Figure 9.10 provides a schematic of the AP1000 reactor vessel, vessel cavity, vessel insulation and vents configuration that promotes IVR of molten core debris.

9.4.3.2 US EPR. The US evolutionary pressurized water reactor (USEPR) is developed based on the European pressurized water reactor (EPR) design derived from the German Konvoi types with features from the French N4 reactors. The USEPR design certification application was submitted to the U.S. NRC in 2007 and is currently under NRC review. This section provides a brief description of the USEPR reactor features as they relate to containment designs, based on the information from publicly available information documents [33–37].

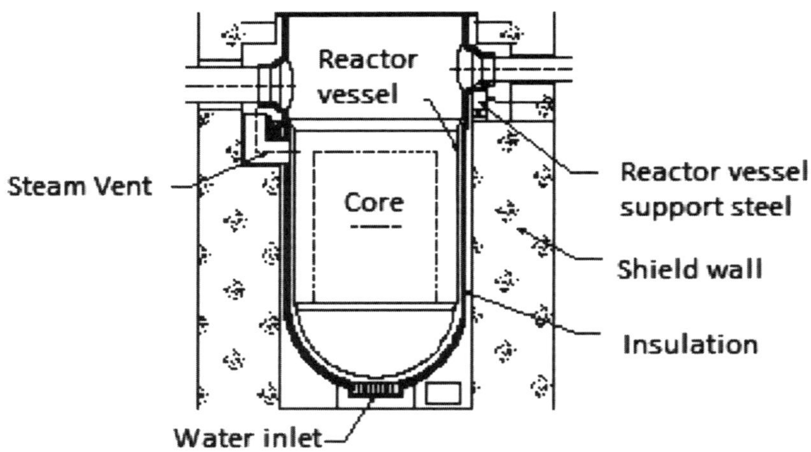

FIG. 9.10
AP1000 IN-VESSEL RETENTION OF MOLTEN CORE DEBRIS [28]

The USEPR is a four-loop evolutionary generation III+ PWR with rated power at about 1600 Mwe. The USEPR design is built and improved upon the established technologies from the well-proven N4 and Konvoi reactors in France and Germany, and incorporates a combination of active and passive safety features to ensure safety and reduces the probability of a core accident at power by a factor of 10 compared to previous generations of reactors. The main design features for the USEPR include:

- Proven four-loop RCS design with increased NSSS component volumes allowing extended time to response to transients and accidents
- Four completely independent trains of safety systems housed in four divisions of safeguard buildings
- Double shell containment building to protect against commercial aircraft impact and other external hazards
- Increased redundancy and physical separation of safety trains
- Top-mounted in-core instrumentation system eliminates penetrations on RPV lower head
- Four emergency diesel generators located in two physically separated buildings to provide AC power to four trains of safety systems
- Prestressed post-tensioned concrete containment with grouted tendons to withstand internal design basis pressure of 0.43 MPa (62 psi) with ultimate pressure capability of 0.82 MPa (119 psi) at Service Level D limits
- In-containment refueling water storage tank (IRWST) as water source for both safety injections and severe accident mitigations
- Dedicated primary depressurization system to prevent high-pressure core melt in event of severe accidents
- Passive ex-vessel melt stabilization, conditioning, and cooling using core catcher and IRWST
- Passive containment spray system
- Control of hydrogen concentration using passive autocatalytic recombiners
- Predicted core damage frequency less that 1 E-06/yr and large and early release frequency less than 1E – 07/yr
- Core is designed for 12- to 24-month fuel cycle
- Refueling outages can be conducted in 17 days or less
- Advanced cockpit control room design
- Plant design life of 60 years without replacement of the reactor vessel

The USEPR containment and other primary Category I structures are built on a common basemat called the nuclear island (NI), including: the reactor containment building (RCB), the containment shield building (RSB), the fuel building, and four safeguard buildings (Fig. 9.11). The NI basemat is a cruciform shape with dimensions of 108 m (360 ft) by 108 m (360 ft) by 3.048 m (10 ft) thick, and embedded below grade by approximately 12.2 m (40 ft). Also indicated in Fig. 9.16 are two Category I emergency power generating buildings (EPGB) housing four emergency diesel generators. The EPGBs are separated from the NI basemat structures.

Not shown in Fig. 9.11 are four Category I emergency service water buildings (ESWB) housing four trains of the essential service water system which cools the component cooling water system (CCWS) heat exchangers with water from the ESWS tanks as the ultimate heat sink during normal operating conditions and accident conditions.

The RCB is a large, post-tensioned reinforced concrete shell structure consisting of an upright cylinder capped with a spherical dome and a free volume of 79,287.2 m^3 (2.8 million cubic feet), and is located within the RSB (Fig. 9.12). The design leak-rate is less than 0.25% per day at design pressure of 0.43 MPa (62 psi). The containment is 66.45 m (218 ft) high and has an inner diameter of 47 m (153.5 ft) with the wall thickness of 1.3 m (4.3 ft) and a steel liner plate of a thickness of 6.35 mm (¼ inches) covers the entire inner surface of the containment to provide a leak-tight membrane to airborne radioactivity following postulated design

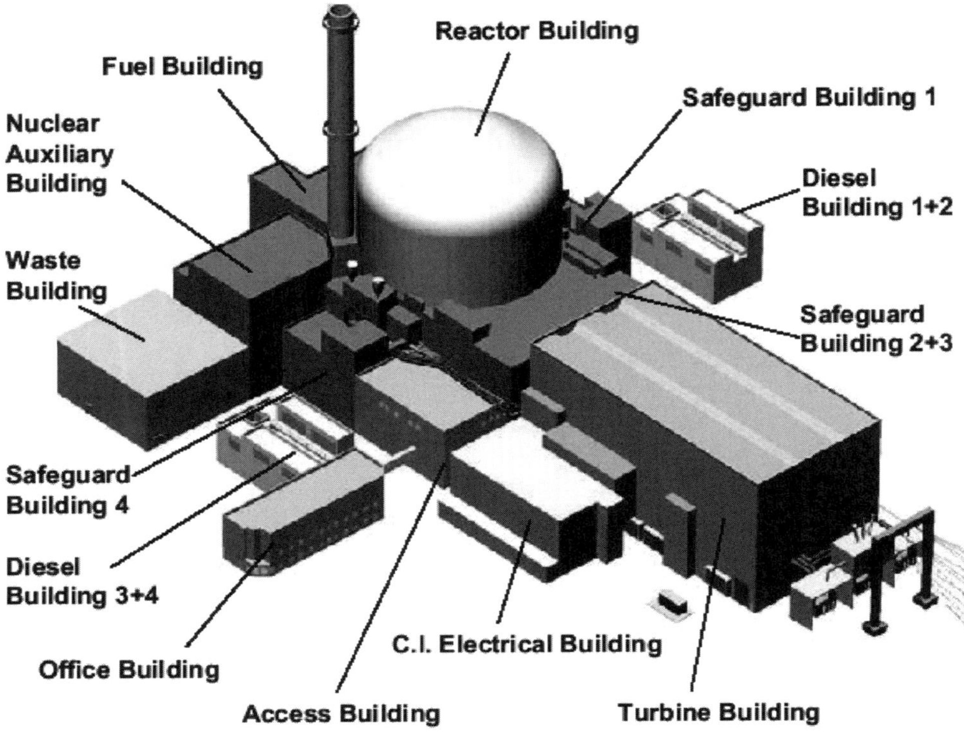

FIG. 9.11
US EPR PLANT BUILDING LAYOUT [33]

basis accidents (DBA). The containment houses the NSSS components separate from the rest of containment using a "two room" concept. The major NSSS components inside the equipment space are located in separate compartments; concrete wall are erected between the individual coolant loops and between the hot leg and cold leg piping of each loop. This design feature reduces possibility of hydrogen concentrations. In addition, the passive autocatalytic hydrogen recombiners (PARs) are installed in different compartments to keep the average concentration below 10% at all times to avoid the risk of deflagration or detonation.

The wall and dome of the RCB are post-tensioned with hoop, vertical, and gamma tendons to withstand the design basis internal pressure. Reinforced steel bars are also placed in RCB concrete wall and dome for crack control and to provide strength to accommodate seismic and other loads. The USEPR is the only standard design in the United States to date which utilizes post-tensioned prestressing concrete containment with fully grouted tendons. The RCB uses the Freyssinet C-range 55C15 tendon system, each tendon consisting of 55 seven-wire strands. Three groups of tendons are provided for the RCB: horizontal hoop, vertical, and gamma tendons. The horizontal hoop tendons are provided around the cylindrical shell of the RCB. The tendons terminate at the three vertical buttresses around the outside of the containment wall at azimuths of 0-, 112-, and 230-degree apart. Each hoop tendon extends the full circumference of the wall to allow the termination of tendons to alternate among the three vertical buttresses. The vertical tendons extend up along the cylindrical shell and terminate at the top of the ring girder that is provided at the transition of the wall to the spherical dome. Two groups of gamma tendons placed at 90-degree apart also extend vertically up through the wall where they then wrap around over the dome and terminate at the ring girder on the opposite side of the wall. The bottom of both the vertical and the gamma tendons terminate at the tendon gallery located directly under the circumference of the cylindrical containment wall below the NI foundation basemat. This gallery provides access for installing and maintaining the lower terminations of the tendons.

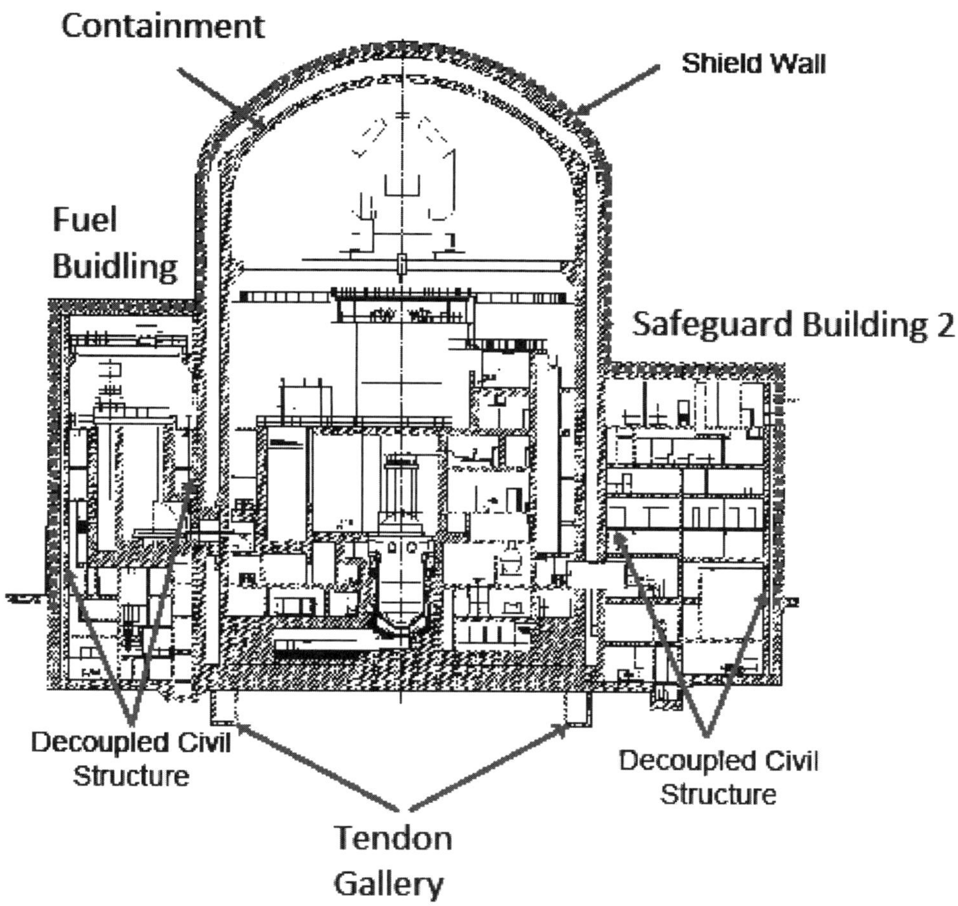

FIG. 9.12
US EPR REACTOR CONTAINMENT BUILDING AND CONTAINMENT SHIELD BUILDING [33]

After tensioning of the strands, the tendon sheath is injected with cement which acts as corrosion inhibitor. Both the quality of the cement material and the tendon installation process follow the procedure provided in Regulatory Guide 1.107 [38]. In addition, the USEPR employs Alternative B of Regulatory Guide 1.90 [39] for monitoring deformation under pressure during the periodic in-service inspections as required by General Design Criterion (GDC) 53, "Provisions for Containment Testing and Inspection." Membrane compression is maintained and the maximum stress in the tensile reinforcing bars will be limited to one-half the yield strength of the reinforcing steel (0.5 fy) under the peak expected pressure for in-service inspection tests.

Another salient feature of the containment is the design of the reactor pit area consisting of the transfer channel, the corium spreading area and the IRWST pool, located in the RCB lower part below the reactor pressure vessel (Fig. 9.13). The USEPR severe accident mitigation management relies upon the ex-vessel retention strategy to minimize the possibility of an energetic corium water interaction. In a severe accident scenario where core melt is assumed to have occurred, a passive melt plug at the bottom of the RPV opens to flush the full core of corium to the reactor pit and through the transfer channel into the corium spreading area (Fig. 9.14). The corium in the spreading area is passively cooled by gravity injection of water from the IRWST. The cooling water flows directly under the spreading area to protect the containment basemat.

Furthermore, the RCB is surrounded by the containment shield building (RSB) for protection against external hazards such as winds, tornadoes and hurricanes, as well as missile impact of a commercial aircraft crash. The

FIG. 9.13
CONCRETE STRUCTURAL COMPARTMENTS FOR IRWST AND CORIUM SPREADING AREA [33]

RSB is a heavily reinforced concrete structure consisting of a cylindrical shell and a dome. The RSB, which is approximately 56.7 m (186 ft) in diameter by 70 m (230 ft) high, completely encloses the RCB. The RSB also serves as an additional preventive barrier to the release of radiation from the RCB. In addition, the plant layouts as shown in Fig. 9.11, especially the spatial separations of the four safeguard buildings, ensure the safe shutdown of the plant in the event of any postulated external hazards including a commercial aircraft impact.

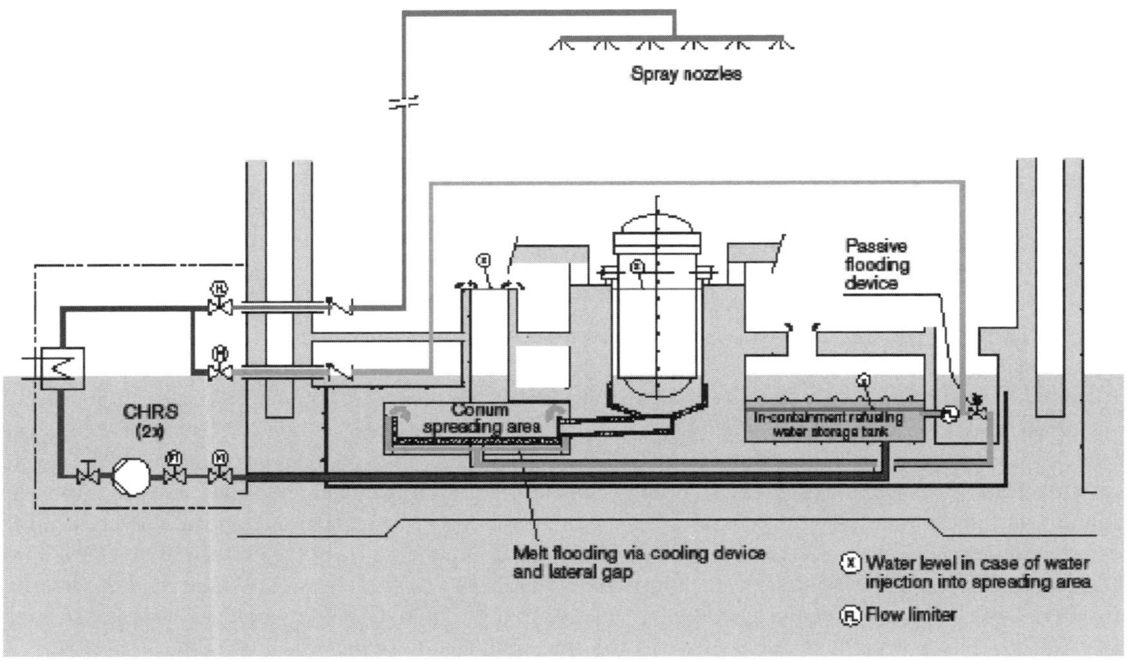

FIG. 9.14
US EPR PASSIVE CORIUM RETENTION EX-VESSEL [33]

9.4.3.3 US APWR. The US advanced pressurized water reactor (US-APWR) is designed by the Mitsubishi Heavy Industries, LTD (MHI) based and improved on the Japanese APWR. The US-APWR design certification application was docketed at the U.S. NRC in 2008 and is currently under NRC review. This section provides a brief description of the US-APWR reactor features as they relate to containment designs, based on the information from publicly available information documents [40–42].

The USAPWR is a four-loop evolutionary generation III+ PWR with rated power at about 1700 Mwe, the largest reactor among all standard designs to date. The US-APWR design is built and improved upon the established technologies and experiences in the Japanese PWR reactors, and incorporates a combination of active and passive safety features to ensure safety and reduces the probability of a core accident at power by a factor of 10 compared to previous generations of reactors. The main design features for the US-APWR include:

- Proven four-loop RCS design with increased NSSS component volumes allowing extended time to response to transients and accidents
- Four 50% safety train systems protected in Category I structures with train separation and segregation barriers
- Advanced accumulator system which eliminates need for low head safety injection pumps
- Containment is surrounded by reinforce concrete reactor building for protection against commercial aircraft impact and other external hazards
- Increased redundancy and physical separation of safety trains
- Top-mounted in-core instrumentation system eliminates penetrations on RPV lower head
- Emergency gas turbine generators in lieu of conventional diesel generators to provide AC power to four trains of safety systems
- Prestressed post-tensioned reinforced concrete containment with greased tendons to withstand internal design basis pressure of 0.47 MPa (68 psi) with ultimate pressure capability of 1.10 MPa (159 psi) at Service Level D limits
- In-containment refueling water storage pit (IRWSP) as water source for both safety injections and severe accident mitigations
- Dedicated primary depressurization system to prevent high-pressure core melt in event of severe accidents
- Passive ex-vessel melt stabilization, conditioning, and cooling using core catcher and IRWSP
- Passive containment spray system consisting of four independent trains
- Control of hydrogen concentration using passive autocatalytic recombiners
- Predicted core damage frequency less that 1E – 06/yr and large and early release frequency less than 1E – 07/yr
- Core is designed for up to 24-month fuel cycle
- Main control room emergency habitability system design
- Plant design life of 60 years without replacement of the reactor vessel

The US-APWR power block consists of reinforced concrete reactor building (R/B), prestressed concrete containment vessel (PCCV) and containment internal structure founded on a common basemat (Fig. 9.15). The layout of the US-APWR building structures is designed to facilitate plant operation and maintenance, minimize personnel radiation exposure, and protect safety-related systems and components from internal and external hazards including the effect of a commercial aircraft impact by optimizing separation and segregation barriers. Provisions including redundant train separation and segregation barriers have been made to ensure that the functions of the safety-related systems are maintained in the event of postulated incidents such as fires, floods, and high-energy pipe break events.

The PCCV and an annulus enclosing the containment penetration area provide an efficient leak-tight barrier and radiation protection under all postulated conditions, including loss of coolant accident (LOCA). The PCCV is designed to withstand the peak pressure under LOCA conditions. Access galleries are provided for a periodic inspection and testing of circumferential and axial pre-stressing tendons. The US-APWR

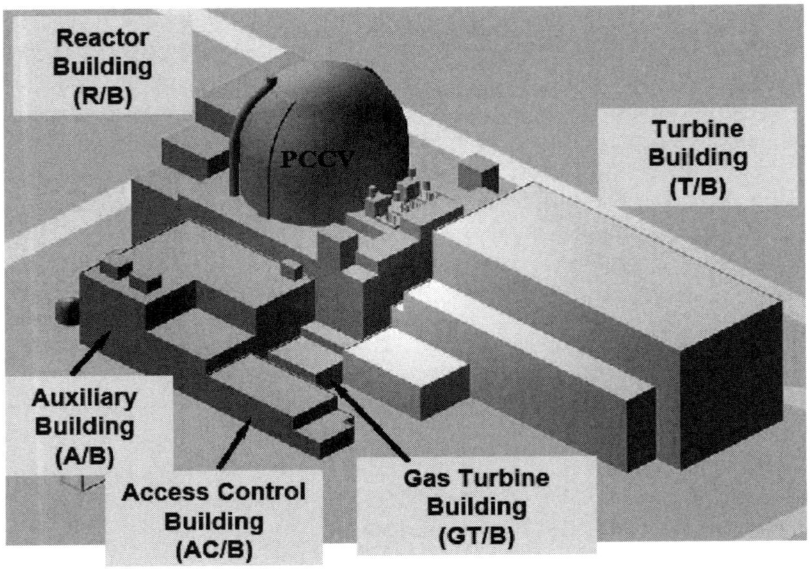

FIG. 9.15
US-APWR PLANT BUILDING LAYOUT [40]

safety-related structures, systems, and components are designed to withstand the effects of natural phenomena, including earthquakes, without jeopardizing plant safety.

The PCCV has an inside diameter of 45.12 m (149 ft, 2 in.) and an inside height of 69 m (226 ft, 5 in.) (Fig. 9.16). The thickness is 1.32 m (4 ft, 4 in.) for the cylinder and 1.11 m (3 ft, 8 in.) for the dome. Areas around the large openings are thickened to provide additional strength and provide space for the prestressing tendons that are deflected around the openings. The PCCV consists of a prestressed concrete shell containing unbonded (greased) tendons and reinforcement steel. Prestressing is obtained through post-tensioning — a method of prestressing in which tendons are tensioned after concrete has hardened. Two types of tendons are used for the US-APWR PCCV: horizontal hoop tendons and vertical inverted U tendons. The horizontal hoop tendons are used in the cylinder and the lower part of the dome. The tendons wrap around the entire circumference, and are anchored at two vertical buttresses 180° apart. The anchors for the horizontal tendons are staggered such that adjacent tendons are anchored on opposite buttresses. The horizontal tendons anchored at the two vertical buttresses are accessed for servicing through vertical chases provided in the R/B at each buttress.

The inverted U tendons run from the tendon gallery located within the containment basemat vertically up the cylinder, over the dome in a non-radial mesh pattern, and down to the tendon gallery on the opposite side. These inverted U tendons, approximately configured in the form of an inverted "U," are anchored at each end in a tendon gallery. The circular tendon gallery allows for servicing and installation of the inverted U tendons and is located entirely within the reinforced concrete basemat foundation. The tendon gallery is accessed through a hallway, which passes horizontally through the basemat to the exterior plant yard.

Reinforcing steel is also provided overall in the cylinder and dome. Additional reinforcement is provided at discontinuities, such as the cylinder-basemat interface, around penetrations and openings, at buttresses, and at other areas. The concrete shell inner surface is lined with a minimum 1/4-in. carbon steel plate that is anchored to the concrete shell and dome to provide the required pressure boundary leak tightness. Areas around penetrations, support brackets, inner walls, and heavy components bases have thickened steel liner plates.

The standard seismic design for the US-APWR standard plant is based on the CSDRS based on RG 1.60 spectra enhanced above 9 Hz to address the U.S. hard rock, high frequency (HRHF) characteristics. The peak ground acceleration (PGA) of the CSDRS is anchored to 0.3 g for the two horizontal directions and the vertical direction.

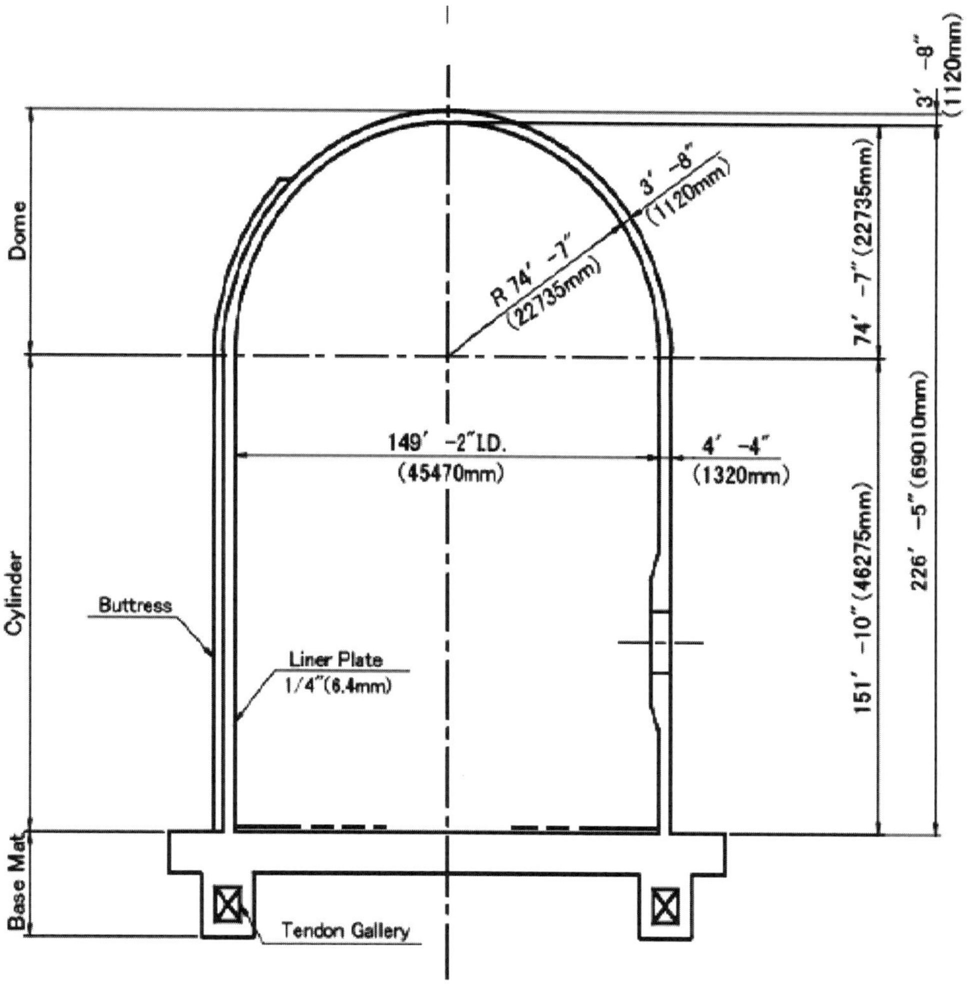

FIG. 9.16
US-APWR REACTOR CONTAINMENT VESSEL [40]

Similar to the USEPR, the US-APWR designs the refueling water storage pit (RWSP) located at the lowest part of containment, with four recirculation sumps installed directly below the pit (Fig. 9.17). This configuration provides a continuous suction source for the safety injection pumps (SIP) and the containment spray/residual heat removal pumps (CS/RHRP), thus eliminating the conventional connection from the refueling water storage pit (outside containment) to the containment recirculation sump. The RWSP also serves as the source of water to mitigate the consequences of hypothetical severe accidents such as a reactor vessel failure. The fundamental concept of the US-APWR for severe accident mitigation is to flood the reactor cavity with coolant water and to keep the molten fuel within the reactor cavity. In order to achieve this, the US-APWR is provided with a very reliable reactor cavity flooding system, which consists of two diverse independent coolant water supply systems.

Other salient safety features of the US-APWR include four-train safety injection with advanced accumulators, emergency gas turbine generators, passive containment spray system, and main control room emergency habitability system. The engineered safety features of the US-APWR use a four-train configuration (Fig. 9.18). Required safety functions are performed following an accident assuming a single failure in one train with a

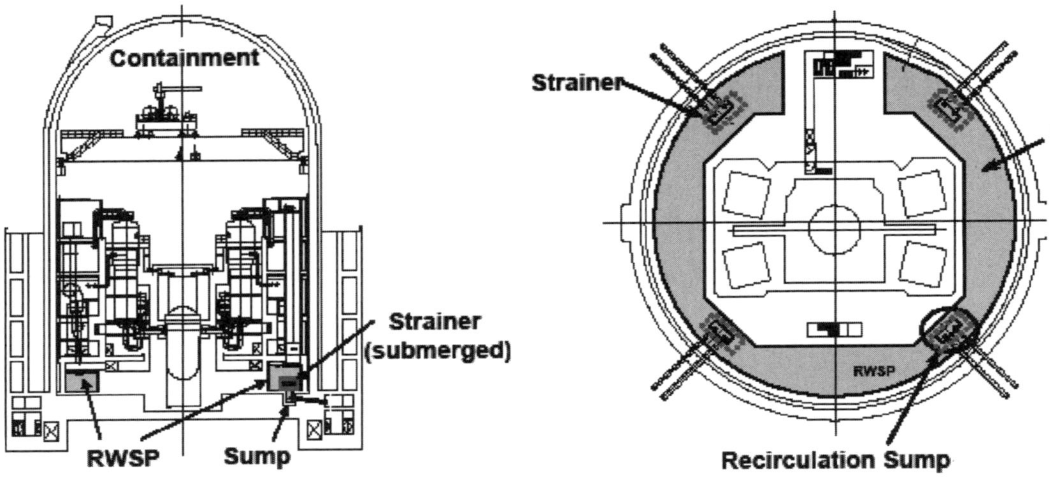

FIG. 9.17
US-APWR IRWSP AND CONTAINMENT SUMP [40]

second train out of service for maintenance. The four-train direct vessel injection (DVI) system is simple and compact and reduces the capacity of each train from 100% to 50%. The accumulator system consists of four advanced accumulators (ACCs) and the associated valves and piping, one for each reactor coolant system (RCS) loop. The system is connected to the cold legs of the reactor coolant piping and injects borated water when the RCS pressure falls below the accumulator operating pressure. It is a completely passive system.

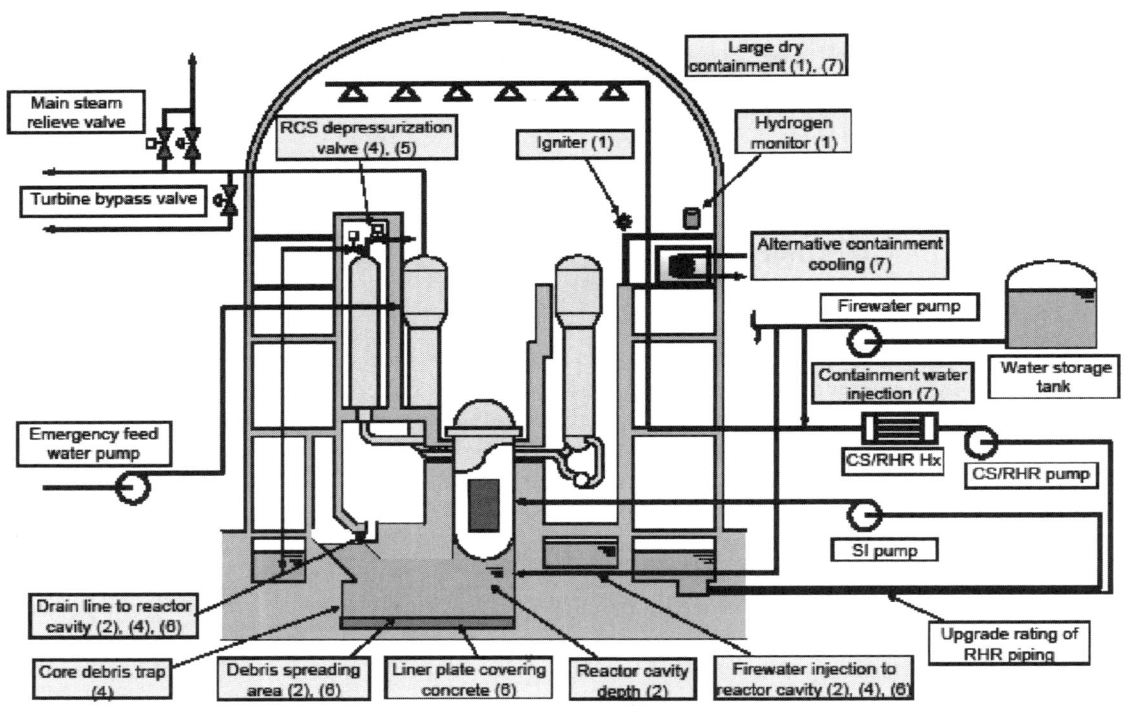

FIG. 9.18
SCHEMATICS OF US-APWR SAFETY SYSTEMS [40]

The advanced accumulator incorporates an internal passive flow damper. Its function is to inject a large flow of water to refill the reactor vessel, and then reduce the flow as the water level in the accumulator drops. When the water level is above the top of the standpipe, water enters the flow damper through both the top of the standpipe and the side of the flow damper, thus it is injected at a high flow rate. When the water level drops below the top of the standpipe, water enters the damper through the side inlet only, reducing the flow rate. Reduced flow injection is performed during core re-flooding, in association with the safety injection pumps. The combined performance of the accumulator system and the high head injection system eliminates the need for a conventional low head injection system.

Furthermore, the US-APWR designs the safety-related electrical systems consisting of four 50% systems, which means that two of four trains are required for safe shutdown of the plant, a configuration which allows maintenance to be done online. Gas turbine generators (GTGs) are used instead of diesel engines for the emergency power supply system, as they are easier to maintain, have a smaller footprint and have fewer auxiliary systems. The gas turbine generators take longer to start up than diesels, but use of the advanced accumulator technology (see above) allows a start up time of 100 seconds adequate for starting the GTGs on demand.

The passive containment spray system (CSS) consists of four independent trains, each containing a containment spray/residual heat removal (CS/RHR) heat exchanger, a CS/RHR pump, spray nozzle and valves. The CSS is automatically actuated by containment spray signal. The GTGs will supply the AC power in case of a loss of offsite power. The IRWSP provides a continuous suction source for the CS/RHR pumps thus eliminating the realignment from the RWST to the containment sump in conventional PWRs.

Finally, the main control room (MCR) HVAC system incorporates an emergency filtration system which maintains the pressure of the control room envelope (CRE) higher than that of adjacent areas during a radiological accident, thus allowing an operator to stay safely inside the CRE both in normal and accident conditions.

REFERENCES

[1] U.S. Code of Federal Regulations, Title 10 — Energy, Part 50, "Domestic Licensing of Production and Utilization Facilities," U.S. Nuclear Regulatory Commission, Washington, DC.

[2] U.S. Code of Federal Regulations, Title 10 — Energy, Part 52, "Licenses, Certifications, and Approvals for Nuclear Power Plants," U.S. Nuclear Regulatory Commission, Washington, DC.

[3] NUREG/BR-0298, rev. 2, 2004, "Nuclear Power Plant Licensing Process," U.S. Nuclear Regulatory Commission, Washington, DC, July.

[4] Long Island Power Authority, 2012, "Introduction & History of the Shoreham Nuclear Power Plant," Long Island, New York.

[5] NRC Office of Public Affairs, 2005, "Backgrounder on Nuclear Power Plant Licensing Process," U.S. Nuclear Regulatory Commission, Washington, DC, July.

[6] NRC Office of Public Affairs, 2008, "Backgrounder on New Nuclear Plant Designs," U.S. Nuclear Regulatory Commission, Washington, DC, June.

[7] Miranda, M., Braverman, J., Wei, X., Hofmayer, C., and Xu, J., 2011, "Structural Design Challenges in Design Certification Applications for New Reactors," Proceedings of the ASME 2011 Pressure Vessels & Piping Division Conference, Baltimore, Maryland, July.

[8] Jung, S.-K., Harold, J., Goodman, A., and Alchaar N., 2010, "Critical Section Selection Methodology for the U.S. EPR Standard Nuclear Power Plant," Proceedings of the ASME 2010 Pressure Vessels & Piping Division Conference, Bellevue, WA, July.

[9] NUREG 0800, 2007, "Standard Review Plan for Review of Safety Analysis Reports for Nuclear Power Plants: LWR Edition," U.S. Nuclear Regulatory Commission, Washington, DC, Revision 3.

[10] NUREG/CR-7005, 2011, "Technical Basis for Regulatory Guidance on Design-Basis Hurricane Wind speeds for Nuclear Power Plants," U.S. Nuclear Regulatory Commission, Washington, DC.

[11] Regulatory Guide 1.221, 2011, "Design-Basis Hurricane and Hurricane Missiles for Nuclear Power Plants," U.S. Nuclear Regulatory Commission, Washington, DC.
[12] ASME Boiler and Pressure Vessel Code, Section III, "Rules for Construction of Nuclear Power Plant Components," Division 2, "Code for Concrete Reactor Vessels and Containments," 2001 edition through 2003 addenda, American Society of Mechanical Engineers, New York, NY, also known as ACI Standard 359-01, American Concrete Institute, Farmington Hills, MI.
[13] ASME, Section III, Division 1, Subsection NF, "Class MC Components."
[14] ACI 349-01, 2001, "Code Requirements for Nuclear Safety Related Concrete Structures," American Concrete Institute Committee 349.
[15] ASCE/SEI 43-05, 2005, "Seismic Design Criteria for Structures, Systems, and Components in Nuclear Facilities," ASCE Standard.
[16] Gupta, S. et al., 2011, "Importance of Geotechnical Parameters in Design of Nuclear Structures," Transactions, SMiRT 21, New Delhi, India.
[17] Morante, R., Miranda, M., Braverman, J., and Xu, J., 2011, "Seismic Analysis Issues in Design Certification Applications for New Reactors," Proceedings of the ASME 2011 Pressure Vessels & Piping Division Conference, Baltimore, Maryland, July.
[18] 10 CFR Part 100, "Reactor Site Criteria," U.S. Nuclear Regulatory Commission, Washington, DC.
[19] Regulatory Guide 1.60, 1976, "Design Response Spectra for Seismic Design of Nuclear Power Plants," U.S. Nuclear Regulatory Commission, Washington, DC.
[20] Regulatory Guide 1.208, 2007, "A Performance-Based Approach to Define the Site-Specific Earthquake Ground Motion," U.S. Nuclear Regulatory Commission, Washington, DC.
[21] Interim Staff Guidance (ISG)-01, 2009, "Interim Staff Guidance on Seismic Issues Associated with High Frequency Ground Motion," U.S. Nuclear Regulatory Commission, Washington, DC.
[22] Tabatabaie, M. et al., "Detailed Finite Element Modeling of US EPR Nuclear Island for Seismic SSI Analysis," Transactions, SMiRT 21, New Delhi, India, 2011.
[23] ASCE 4-98, "Seismic Analysis of Safety-Related Nuclear Structures and Commentary," ASCE Standard.
[24] U.S. Army Corps of Engineers Manual No. 1110-1-1904, "Engineering and Design: Settlement Analysis," 1990.
[25] General Electric, Inc, The ABWR Plant General Description, GE Energy Nuclear Marketing, Wilmington, NC, USA, 2006.
[26] General Electric, Inc, The ESBWR Plant General Description, GE Energy Nuclear Marketing, Wilmington, NC, USA, 2007.
[27] IAEA-TECDOC-1624, "Passive Safety Systems and Natural Circulation in Water Cooled Nuclear Power Plant," ISSN 1011-4289, IAEA, 2009.
[28] Westinghouse Electric Co., LLC, "The Westinghouse AP1000 Advanced Nuclear Plant," AP1000 Brochure, 2003.
[29] Schulz, T. L., "Westinghouse AP1000 Advanced Passive Plant," International Journal of Nuclear Engineering and Design, Elsevier, 2006.
[30] Varma, A. H. et al., "Steel-Plate Composite (SC) Walls for Safety Related Nuclear Facilities: Design for In-Plane and Out-of-Plane Demands," Transactions, SMiRT 21, Division-VI, New Delhi, India, 2011.
[31] Varma, A. H. et al., 2011, "Steel-Plate Composite (SC) Walls: Analysis and Design Including Thermal Effects," Transactions, SMiRT 21, Division-VI, New Delhi, India.
[32] Varma, A. H. et al., 2011, "In-Plane Shear Behavior of SC Composite Walls: Theory vs. Experiment," Transactions, SMiRT 21, Division-VI, New Delhi, India.
[33] Stack, T., 2008, U.S. EPR: Evolutionary Design, Proven Technology, AREVA NP, Inc.
[34] Stack, T., 2006, Reactor Building and Associated Systems, AREVA NP, Inc.
[35] AREVA Brochure, 2005, "EPR," Framatome ANP.
[36] AREVA Brochure, 2009, "The 1600+ Mwe Reactor," AREVA SA.

[37] U.S. EPR Final Safety Analysis Report, Revision 3, AREVA NP, Inc., 2011.
[38] Regulatory Guide (RG) 1.107, 2011, "Qualification for Cement Grouting for Prestressing Tendons in Containment Structures," U.S. Nuclear Regulatory Commission, Washington DC, Revision 2.
[39] Regulatory Guide (RG) 1.90, 2012, "Inservice Inspection of Prestressed Concrete Containment Structures with Grouted Tendons," U.S. Nuclear Regulatory Commission, Washington DC, Revision 2.
[40] Mitsubishi Nuclear Energy System, Inc., "2007, US-APWR Overview," June.
[41] Kaneda, M. et al., 2006, "US-APWR for Deployment in the United States," Mitsubishi Heavy Industries, Ltd. Technical Review, Vol. 43 No. 4.
[42] US-APWR Design Certification Control Document (DCD).

APPENDIX A GLOSSARY OF NPP-RELATED TERMS

Most of the terms explained in the glossary are sorted from the glossary provided in NRC's Information Digest (NID) 2010 to 2011. An attempt has been made to make it relevant to the content of the book. Some terms may not be in NID and have been added to provide an explanation of the terms relevant to the book.

Applicant: A company, organization, institution, or other entity that has applied for a construction permit to build a reactor facility or applied for the combined license.

Atomic Energy: The energy that is released through a nuclear reaction or radioactive decay process. The process of particular interest is the process known as fission, which occurs in a nuclear reactor and produces energy, usually in the form of heat. In a nuclear power plant, this heat is used to boil water in order to produce steam that can be used to drive turbines. This, in turn, activates generators to produce electrical power. Atomic Energy is more correctly called *Nuclear Energy*.

Background radiation: The natural radiation that is always present in the environment. It includes cosmic radiation which comes from the sun and stars, terrestrial radiation, which comes from the Earth, and internal radiation which exists in all living things. The typical average individual exposure in the United States from natural background sources is about 3 mSv (300 millirem) per year.

Boiling-Water Reactor (BWR): A nuclear power reactor design in which water flows upward through the core of the reactor, where it is heated by fission and allowed to boil in the reactor vessel. The resulting steam then drives turbines, which activate generators to produce electrical power (see Fig. 1.2). BWRs operate similar to electrical plants using fossil fuel, except that the BWRs are powered by 370 to 800 nuclear fuel assemblies in the reactor core.

Capability: The maximum load that a generating unit, generating station, or other electrical apparatus can generate under specified conditions for a given period of time without exceeding the design limits of temperature and pressure.

Capacity: The amount of electric power that a generating unit can produce. The amount of electric power that a manufacturer rates its generator, turbine transformer, transmission circuit, or system, is able to produce.

Capacity factor: The ratio of the available capacity (defined as the amount of electrical power actually produced by a generating unit) to the theoretical capacity (defined as the amount of electrical power that the generating unit could theoretically produce, if it operates continuously at full power) during a given time period.

Cask: A heavily shielded container used for the dry storage or shipment (or both) of radioactive materials, such as spent nuclear fuel or other high-level radioactive waste. Casks are often made from lead, concrete, or steel. Casks must meet regulatory requirements. The cask is not intended for long-term disposal in a repository.

Containment Structure: A gas-tight shell or an enclosure that surrounds a nuclear reactor to confine fission products that otherwise might be released to the atmosphere in the event of an accident. Such enclosures are usually dome-shaped and made of steel or reinforced concrete.

Contamination: Undesirable radiological, chemical, or biological material with potentially harmful effects. The contamination could be either airborne, or deposited in, or on the surface of structures, objects, soil, water, or living organisms in a concentration that makes the medium unfit for its next intended use.

Criticality: The normal operating condition of a reactor, in which nuclear fuel sustains a fission chain reaction. A reactor achieves criticality (and is said to be critical) when each fission event releases a sufficient number of neutrons to sustain an ongoing series of fission reactions.

Defense-in-depth: An approach to designing and operating nuclear facilities that prevents and mitigates accidents that release radiation or hazardous materials. The key is creating multiple independent and redundant layers of defense to compensate for potential human and mechanical failures so that no single layer, no matter how robust, is exclusively relied upon. Defense in depth includes the use of access controls, physical barriers, redundant and diverse key safety functions, and emergency response measures.

Design-Basis Threat (DBT): A profile of the type, composition, and capabilities of an adversary. The NRC uses the DBT as a basis for designing safeguards systems to protect against acts of radiological sabotage and to prevent the theft of special nuclear material. Nuclear facility licensees are expected to demonstrate they can defend against the DBT.

Design Certification (DC): Certification and approval by the NRC of a standard nuclear power plant design independent of a specific site or an application to construct or operate a plant. A design certification is valid for 15 years from the date of issuance but can be renewed for an additional 10 to 15 years.

Dry Cask Storage: A method for storing spent nuclear fuel above ground in special containers known as casks. After fuel has been cooled in a spent fuel pool for at least 1 year, dry cask storage allows approximately one to six dozen spent fuel assemblies to be sealed in casks and surrounded by inert gas. The casks are large, rugged cylinders, made of steel or steel-reinforced concrete [approximately 50 cm (20 in.) thick]. They are welded or bolted closed, and each cask is surrounded by steel, concrete, lead, or other material to provide leak-tight barrier and radiation shielding. The casks may be placed horizontally in aboveground concrete bunkers, or vertically in concrete vaults or on concrete pads.

Early Site Permit (ESP): A permit through which the NRC resolves site safety, environmental protection, and emergency preparedness issues, in order to approve one or more proposed sites for a nuclear power facility, independent of a specific nuclear plant design or an application for a construction permit or combined license. An ESP is valid for 10 to 20 years, but can be renewed for an additional 10 to 20 years.

Economic Simplified Boiling-Water Reactor (ESBWR): A 4500-MWt nuclear reactor design, which has passive safety features and uses natural circulation (with no recirculation pumps or associated piping) for normal operation. GE-Hitachi Nuclear Energy (GEH) submitted an application for final design approval and standard design certification for the ESBWR on August 24, 2005.

Efficiency, NPP: The percentage of the total energy content of a power plant's fuel that is converted into electricity. The remaining energy is lost to the environment as heat.

Electric Power Grid: A system of synchronized power providers and consumers connected by transmission and distribution lines and operated by one or more control centers. In the continental United States, the electric power grid consists of three systems—the Eastern Interconnect, the Western Interconnect, and the Texas Interconnect. In Alaska and Hawaii, several systems encompass areas smaller than the State.

Emergency Classifications: Sets of plant conditions that indicate various levels of risk to the public and which might require response by an offsite emergency response organization to protect citizens near the site. Both, nuclear power plants and research and test reactors, use the emergency classifications (see 10 CFR Part 50, Section 50.47).

Emergency Preparedness (EP): The programs, plans, training, exercises, and resources necessary to prepare emergency personnel to rapidly identify, evaluate, and react to emergencies, including those arising from terrorism or natural events such as hurricanes. EP strives to ensure that nuclear power plant operators can implement measures to protect public health and safety in the event of a radiological emergency. As a condition of their licenses, plant operators must develop and maintain EP plans that meet NRC requirements.

Energy Information Administration (EIA): The agency, within the U.S. Department of Energy, that provides policy-neutral statistical data, forecasts, and analyses to promote sound policymaking, efficient markets, and public understanding regarding energy and its interaction with the economy and the environment.

Entomb: A method of decommissioning, in which radioactive contaminants are encased in a structurally long-lived material, such as concrete. The entombed structure is maintained and its surveillance is continued until the entombed radioactive waste decays to a level permitting termination of the license and unrestricted release of the property. During the entombment period, the licensee maintains the license previously issued by the NRC.

Event Notification (EN) System: An automated event tracking system used internally by the NRC's Headquarters Operations Center to track incoming notifications of significant nuclear events with an actual or potential effect on the health and safety of the public and the environment. Significant events are reported to the Operations Center by the NRC's licensees, Agreement States, other Federal agencies, the public and other stakeholders.

Exposure: Absorption of ionizing radiation or ingestion of a radioisotope. Acute exposure is a large exposure received over a short period of time. Chronic exposure is exposure received over a long period of time, such as during a lifetime. The National Council on Radiation Protection and Measurements (NCRP) estimates that an average person in the United States receives a total annual dose of about 6.2 mSv (620 millirem) from all radiations sources, a level that has not been shown to cause humans any harm.

Of this total, natural background sources of radiation—including radon and thoron gas, natural radiation from soil and rocks, radiation from space and radiation sources that are found naturally within the human body account for approximately 50%. Medical procedures such as computed tomography (CT scans) and nuclear medicine account approximately for another 48%. Other small contributors of exposure to the U.S. population include consumer products and activities, industrial and research uses, and occupational tasks. The maximum permissible yearly dose for a person working with or around nuclear material is .05 Sv (5 rem). For more information on this subject, the readers should refer to NCRP Report No. 160 (2009), available on the NCRP website at www.NCRPpublications.org.

Federal Emergency Management Agency (FEMA): A component of U.S. Department of Homeland Security, responsible for protecting the nation and reducing the loss of life and property from all hazards, such as natural disasters and acts of terrorism. FEMA leads and supports a risk-based, comprehensive emergency management system of preparedness, protection, response, recovery, and mitigation. FEMA also administers the National Flood Insurance Program.

Fissile Material: A nuclide that is capable of undergoing fission after capturing low-energy thermal (slow) neutrons. Although sometimes used as a synonym for fissionable material, this term has acquired its more-restrictive interpretation with the limitation that the nuclide must be fissionable by *thermal neutrons*. With that interpretation, the three primary fissile materials are uranium-233, uranium-235, and plutonium-239. This definition excludes natural uranium and depleted uranium that have not been irradiated or have only been irradiated in thermal reactors.

Fission (fissioning): The splitting of an atom, which releases a considerable amount of energy (usually in the form of heat) that can be used to produce electricity. Fission may be spontaneous, but is usually caused by the nucleus of an atom becoming unstable (or "heavy") after capturing or absorbing a neutron. During fission, the heavy nucleus splits into roughly equal parts, producing the nuclei of at least two lighter elements. In addition to energy, this reaction usually releases gamma radiation and two or more daughter neutrons.

Force-on-Force (FOF): Inspections designed to evaluate and improve the effectiveness of a licensee's security force and ability to defend a nuclear power plant and other nuclear facilities against a design-basis threat (DBT). An essential part of the security program, instituted by the NRC, is a full force-on-force inspection spanning 2 weeks that includes tabletop drills and multiple simulated combat exercises between a mock commando-type adversary force and the plant's security force.

Fragility-Containment: Fragility of the containment structure is the conditional probability of its failure at a given hazard input level. The input is the internal hazard (e.g., pressure–temperature) resulting from the initiation of a beyond design basis accident.

Fragility-General Definition: Fragility of an SSC is the conditional probability of its failure at a given hazard input level. The input could be earthquake motion, wind speed, or flood level. (From ASME/ANS PRA Standard).

Fragility Model: The fragility model generally used for a hazard specific PRA is a double lognormal model with three parameters: (1) the median capacity, (2) the logarithmic standard deviation of the aleatory (randomness) uncertainty in capacity, and (3), the logarithmic standard deviation of the epistemic (modeling and data) uncertainty in the median capacity.

Freedom of Information Act (FOIA): A Federal law that requires Federal agencies to provide, upon written request, access to records or information. Some material is exempt from FOIA, and FOIA does not apply to records that are maintained by State and local governments, or Federal contractors, grantees or private organizations or businesses.

Fuel assembly (fuel bundle, fuel element): A structured group of fuel rods (long, slender, metal tubes containing pellets of fissionable material), which provide fuel for nuclear reactors. Depending on the design, each reactor vessel may have dozens of fuel assemblies (also known as fuel bundles), each of which may contain 200 or more fuel rods.

Fuel reprocessing (recycling): The processing of reactor fuel to separate the unused fissionable material from waste material. Reprocessing extracts isotopes from spent nuclear fuel so they can be used again as reactor fuel. Commercial reprocessing is not practiced in the US, although it has been practiced in the past. However, the U.S. Department of Defense oversees reprocessing programs at DOE facilities such as in Hanford, WA, and Savannah River, SC.

Full-Time Equivalent (FTE): Human resources measurement equal to one staff person working full-time for 1 year.

Gaseous Diffusion: A uranium enrichment process used to prepare uranium for use in fabricating fuel for nuclear reactors by separating its isotopes (as gases) based on their slight difference in velocity. Lighter isotopes diffuse faster through a porous membrane or vessel than do heavier isotopes. This process involves filtering uranium hexafluoride (UF6) gas to separate uranium-234 and uranium-235 from uranium-238, in order to increase the percentage of uranium-235 from 1% to 3%. The only gaseous diffusion plant in operation in the United States is in Paducah, KY. A similar plant near Piketon, OH, was closed in March 2001. Both plants are leased by the United States Enrichment Corporation (USEC) from the DOE and regulated by the NRC since March 4, 1997.

Half-life (radiological): The time required for half the atoms of a particular radioisotope to decay into another isotope that has half the activity of the original radioisotope. A specific half-life is a characteristic property of each radioisotope. Measured half-lives range from millionths of a second to billions of years, depending on the stability of the nucleus. Radiological half-life is related to, but different from the biological half-life and the effective half-life.

High-Level Radioactive Waste (HLW): The highly radioactive materials produced as byproducts of fuel reprocessing or of the reactions that occur inside nuclear reactors. HLW includes: Irradiated spent nuclear fuel discharged from commercial nuclear power reactors. The highly radioactive liquid and solid materials resulting from the reprocessing of spent nuclear fuel, which contain fission products in concentration that includes some reprocessed HLW from defense activities, and a small quantity of HLW from the reprocessed commercial radioactive waste. Other highly radioactive materials, which the Commission may determine require permanent isolation.

Irradiation: Exposure to ionizing radiation. Irradiation may be intentional, such as in cancer treatments or in sterilizing medical instruments. Irradiation may also be accidental, such as being exposed to an unshielded source. Irradiation does not usually result in radioactive contamination, but damage can occur, depending on the dose received.

Isotope: Two or more forms (or atomic configurations) of a given element that have identical atomic numbers (the same number of protons in their nuclei) and the same, or very similar chemical properties but different atomic masses (different numbers of neutrons in their nuclei) and distinct physical properties. Thus, carbon-12, carbon-13, and carbon-14 are isotopes of the element carbon. The numbers denote the approximate atomic masses. Among their distinct physical properties, some isotopes (known as radioisotopes) are radioactive because their nuclei emit radiation as they strive toward a more stable nuclear configuration. For example, carbon-12 and carbon-13 are stable, but carbon-14 is unstable and radioactive.

Licensee: A company, organization, institution, or other entity to which the NRC has granted a general or specific license to construct or operate a nuclear facility, or to receive, possess, use, transfer, or dispose of source, byproduct, or special nuclear material.

Licensing basis: The collection of documents or technical criteria that provides the basis upon which the NRC issues a license to construct or operate a nuclear facility; or to conduct operations involving the emission of radiation; or to receive, possess, use, transfer, or dispose of source, byproduct, or special nuclear material.

Light-Water Reactor (LWR): A term used to describe reactors using ordinary water as a coolant, including boiling-water reactors (BWRs) and pressurized-water reactors (PWRs), the most common types used in the United States.

Low-Level Radioactive Waste (LLW): A general term for a wide range of items that have become contaminated with radioactive material or have become radioactive through exposure to neutron radiation. A variety of industries, hospitals and medical institutions, educational and research institutions, private or government laboratories, and nuclear fuel cycle facilities generate LLW as part of their day-to-day use of radioactive materials. Some examples include radioactively contaminated protective shoe covers and clothing; cleaning rags, mops, filters, and reactor water treatment residues; equipment and tools; medical tubes, swabs, and hypodermic syringes; and carcasses and tissues from laboratory animals. The radioactivity in these wastes can range from just above natural background levels to much higher levels, such as seen in parts from inside the reactor vessel in a nuclear power plant. Low-level waste is typically stored onsite by licensees. The storage is required either until it has decayed away and can be disposed of as ordinary trash, or until the accumulated amount becomes large enough to warrant shipment to a low-level waste disposal site.

Maximum dependable capacity (gross): The maximum amount of electricity that the main generating unit of a nuclear power reactor can reliably produce during the summer or winter (usually summer, but whichever represents the most restrictive seasonal conditions, with the least electrical output). The dependable capacity varies during the year because temperature variations in cooling water affect the unit's efficiency. Thus, this is the gross electrical output as measured (in Watts unless otherwise noted) at the output terminals of the turbine generator.

Maximum dependable capacity (net): The gross maximum dependable capacity of the main generating unit in a nuclear power reactor, minus the amount used to operate the station. Net maximum dependable capacity is measured in watts unless otherwise noted.

Megawatt (MW): A unit of power equivalent to one million watts.

Megawatt-hour (MWh): One million Watt-hours.

Moderator: In a nuclear fission process, scientists have known that the high speed neutrons (> 5 million electron volts — Mev) have to be slowed down to the speed of thermal neutrons (< 0.3 ev) for the chain reaction to continue effectively. What is required is a material that has the ability to slow down the high speed neutrons quickly, and has little tendency to absorb neutrons. Such a material is called a "moderator." Effective moderators are: ordinary water, heavy water, graphite, beryllium, and certain organic materials.

Monitoring of Radiation: Periodic, or continuous determination of the amount of ionizing radiation or radioactive contamination in a region. Radiation monitoring is a safety measure to protect the health and safety of the public and the environment through the use of bioassay, alpha scans, and other radiological survey methods to monitor air, surface water and ground water, soil and sediment, equipment surfaces, and personnel.

Natural Uranium: Uranium containing the relative concentrations of isotopes found in nature (0.7% uranium-235, 99.3% uranium-238, and a trace amount of uranium-234 by mass). In terms of radioactivity, however, natural uranium contains approximately 2.2% uranium-235, 48.6% uranium-238, and 49.2% uranium-234. Natural uranium can be used as fuel in nuclear reactors.

Net electric generation: The gross amount of electric energy produced by a generating station, minus the amount used to operate the station. Note: Electricity required for pumping at pumped-storage plants is regarded as electricity for station operation and is deducted from gross generation. Net electric generation is measured in watt-hours (Wh), except as otherwise noted.

Net Summer Capacity: The steady hourly output that generating equipment is expected to supply to system load, exclusive of auxiliary power, as demonstrated by measurements at the time of peak demand (summer). Net summer capacity is measured in watts unless otherwise noted.

Nonpower reactor (research and test reactor): A nuclear reactor that is used for research, training, or development purposes (which may include producing radioisotopes for medical and industrial uses) but has no role in producing electrical power. These reactors, which are also known as research and test reactors, contribute to almost every field of science, including physics, chemistry, biology, medicine, geology, archeology, and ecology.

NRC Operations Center: The primary center of communication and coordination among the NRC, its licensees, State and Tribal agencies, and other Federal agencies, regarding operating events involving nuclear reactors or materials. Located in Rockville, MD, the Operations Center is staffed 24 hours a day by employees trained to receive and evaluate event reports and coordinate incident response activities.

Nuclear Energy Agency (NEA): A specialized agency within the Organization for Economic Co-operation and Development (OECD), which was created to assist its Member countries in maintaining and further developing the scientific, technological, and legal bases for safe, environmentally friendly, and economical use of nuclear energy for peaceful purposes. The NEA's current membership consists of 28 countries in Europe, North America, and the Asia-Pacific region, which account for approximately 85% of the world's installed nuclear capacity.

Nuclear Energy Institute (NEI): A unified nuclear energy industry group that addresses and resolve nuclear regulatory issues and related technical matters that provide high levels of reliability and economic efficiency in nuclear power plant operations.

Nuclear fuel: Fissionable material that has been enriched to a composition that will support a self-sustaining fission chain reaction when used to fuel a nuclear reactor, thereby producing energy (usually in the form of heat or useful radiation) for use in other processes.

Nuclear poison (or neutron poison): In reactor physics, a substance (other than fissionable material) that has a large capacity for absorbing neutrons in the vicinity of the reactor core. This effect may be undesirable in some reactor applications because it may prevent or disrupt the fission chain reaction, thereby affecting normal operation. However, neutron-absorbing materials (commonly known as "poisons") are intentionally inserted into some types of reactors to decrease the reactivity of their initial fresh fuel load. Adding poisons, such as control rods or boron, is described as adding "negative reactivity" to the reactor.

Nuclear power plant: A thermal power plant, in which the energy (heat) released by the fissioning of nuclear fuel is used to boil water to produce steam. The steam spins the propeller like blades of a turbine that turns the shaft of a generator to produce electricity.

Nuclear/Radiological Incident Annex: An annex to the National Response Framework, which provides for a timely, coordinated response by Federal agencies to nuclear or radiological accidents or incidents within the United States. This annex covers radiological dispersal devices and improvised nuclear devices, as well as accidents involving commercial reactors or weapons production facilities, lost radioactive sources, transportation accidents involving radioactive material, and foreign accidents involving nuclear or radioactive material.

Nuclear reactor: The heart of a nuclear power plant or non-power reactor, in which nuclear fission may be initiated and controlled in a self-sustaining chain reaction to generate energy or produce useful radiation. Although there are many types of nuclear reactors, they all incorporate certain essential features, including the use of fissionable material as fuel, a moderator (such as water) to increase the likelihood of fission (unless reactor operation relies on fast neutrons), a reflector to conserve escaping neutrons, coolant provisions for heat removal, instruments for monitoring and controlling reactor operation and protective devices (such as control rods and shielding).

Nuclear waste: A subset of radioactive waste that includes unusable byproducts produced during the various stages of the nuclear fuel cycle, including extraction, conversion, and enrichment of uranium; fuel fabrication; and use of the fuel in nuclear reactors. Specifically, these stages produce a variety of nuclear waste materials, including uranium mill tailings, depleted uranium, and spent (depleted) fuel, all of which are regulated by the NRC. By contrast, "radioactive waste" is a broader term, which includes all wastes that contain radioactivity, regardless of how they are produced. It is not considered "nuclear waste" because it is not produced through the nuclear fuel cycle and is generally not regulated by the NRC.

Occupational dose: The internal and external dose of ionizing radiation received by workers in the course of employment in such areas as fuel cycle facilities, industrial radiography, nuclear medicine, and nuclear power plants. These workers are exposed to varying amounts of radiation, depending on their jobs and the sources with which they work. The NRC requires its licensees to limit occupational exposure to 50 mSv (5000 mrem) per year. Occupational dose does not include the dose received from natural background sources, doses received as a medical patient or participant in medical research programs or "second-hand doses" received through exposure to individuals treated with radioactive materials.

Outage: The period during which a generating unit, transmission line, or other facility is out of service. Outages may be forced or scheduled, and full or partial.

Pellet, fuel: A thimble-sized ceramic cylinder (approximately 3/8-inch in diameter and 5/8-inch in length), consisting of uranium (typically uranium oxide, UO2), which has been enriched to increase the concentration of uranium-235 (U-235) to fuel a nuclear reactor. Modern reactor cores in pressurized-water reactors (PWRs) and boiling-water reactors (BWRs) may contain up to 10 million pellets, stacked in the fuel rods that form fuel assemblies.

Performance-based regulation: A regulatory approach that focuses on desired, measurable outcomes, rather than prescriptive processes, techniques, or procedures. Performance-based regulation leads to defined results without specific direction regarding how those results are to be obtained. At the NRC, performance-based regulatory actions focus on identifying performance measures that ensure an adequate safety margin and offer incentives for licensees to improve safety without formal regulatory intervention by the agency.

Performance indicator: A quantitative measure of a particular attribute of licensee performance that shows how well a plant is performing when measured against established thresholds. Licensees submit their data quarterly; the NRC regularly conducts inspections to verify the submittals and then uses its own inspection data plus the licensees' submittals to assess each plant's performance.

Possession-only license: A license, issued by the NRC, that authorizes the licensee to possess specific nuclear material but does not authorize its use or the operation of a nuclear facility.

Power uprate: The process of increasing the maximum power level a commercial nuclear power plant may operate. This power level, regulated by the NRC, is included in the plant's operating license and technical specifications. A licensee may only change its maximum power output after the NRC approves an uprate application. The NRC's analyses must demonstrate that the plant could continue to operate safely with its proposed new configuration. When all requisite conditions are fulfilled, the NRC may grant the power uprate by amending the plant's operating license and technical specifications.

Pressurized-water reactor (PWR): A nuclear power reactor design in which very pure water is heated to a very high temperature by fission, kept under high pressure (to prevent it from boiling), and converted to steam by a steam generator (rather than by boiling, as in a BWR). The resulting steam is used to drive turbines, which activate generators to produce electrical power. A PWR essentially operates like a pressure cooker, where a lid is tightly placed over a pot of heated water, causing the pressure inside to increase as the temperature increases. As the steam cannot escape, it keeps the water from boiling at the usual 100°C (212°F). About two-thirds of the operating nuclear reactor power plants in the United States are PWRs (see Fig. 1.1 and Appendix 1A of this book).

Probabilistic risk assessment (PRA): A systematic method for assessing three questions that the NRC uses to define "risk." These questions consider (1) what can go wrong, (2) how likely it is, and (3) what its consequences might be. These questions allow the NRC to understand likely outcomes, sensitivities, areas of importance, system interactions, and areas of uncertainty, which the staff can use to identify risk-significant scenarios. The NRC uses PRA to determine a numeric estimate of risk to provide insights into the strengths and weaknesses of the design and operation of a nuclear power plant.

Rad (radiation absorbed dose): One of the two units used to measure the amount of radiation absorbed by an object or a person, known as the "absorbed dose," which reflects the amount of energy that radioactive sources deposit in materials through which they pass. The rad is the amount of energy (from any type of ionizing radiation) deposited in any medium (e.g., water, tissue, air). An absorbed dose of 1 rad means that 1 gram of material absorbed 100 ergs of energy (a small but measurable amount) as a result of exposure to radiation. The related international system unit is the Gray (Gy), where 1 Gy is equivalent to 100 rad.

Radiation, ionizing: A form of radiation, which includes alpha particles, beta particles, gamma rays and x-rays, neutrons, high-speed electrons, and high-speed protons. Compared to non-ionizing radiation, such as found in ultraviolet light or microwaves, ionizing radiation is considerably more energetic. When ionizing radiation passes through material such as air, water, or living tissue, it deposits enough energy to break molecular bonds and displace (or remove) electrons. This electron displacement may lead to changes in living cells. Given this ability, ionizing radiation has a number of beneficial uses, including treating cancer or sterilizing medical equipment. However, ionizing radiation is potentially harmful if not used correctly, and high doses may result in severe skin or tissue damage. It is for this reason that the NRC strictly regulates commercial and institutional uses of the various types of ionizing radiation.

Radiation, nuclear: Energy given off by matter in the form of tiny fast-moving particles (alpha particles, beta particles, and neutrons) or pulsating electromagnetic rays or waves (gamma rays) emitted from the nuclei of unstable radioactive atoms. All matter is composed of atoms, which are made up of various parts; the nucleus contains minute particles called protons and neutrons, and the atom's outer shell contains other particles called electrons. The nucleus carries a positive electrical charge, while the electrons carry a negative electrical charge. These forces work toward a strong, stable balance by getting rid of excess atomic energy (radioactivity). In that process, unstable radioactive nuclei may emit energy, and this spontaneous emission is called nuclear radiation. All types of nuclear radiation are also ionizing radiation, but the reverse is not necessarily true; for example, x-rays are a type of ionizing radiation, but they are not nuclear radiation because they do not originate from atomic nuclei. In addition, some elements are naturally radioactivez as their nuclei emit nuclear radiation as a result of radioactive decay, but others become radioactive by being irradiated in a reactor. Naturally occurring nuclear radiation is indistinguishable from induced radiation.

Radiation source: A radioactive material or byproduct that is specifically manufactured or obtained for the purpose of using the emitted radiation. Such sources are commonly used in teletherapy or industrial radiography; in various types of industrial gauges, irradiators, and gamma knives, and as power sources for batteries (such as those used in spacecraft). These sources usually consist of a known quantity of radioactive material, which is encased in a manmade capsule, sealed between layers of nonradioactive material, or firmly bonded to a nonradioactive substrate to prevent radiation leakage. Other radiation sources include devices such as accelerators and X-ray generators.

Radiation therapy (radiotherapy): The therapeutic use of ionizing radiation to treat disease in patients. Although most radiotherapy procedures are intended to kill cancerous tissue or reduce the size of a tumor, therapeutic doses may also be used to reduce pain or treat benign conditions. For example, intervascular brachytherapy uses radiation to treat clogged blood vessels. Other common radiotherapy procedures include gamma stereotactic radiosurgery (gamma knife), teletherapy, and iodine treatment to correct an overactive thyroid gland. These procedures use radiation sources, regulated by the NRC and its Agreement States, which may be applied either inside or outside the body. In either case, the goal of radiotherapy is to deliver the required therapeutic or pain relieving dose of radiation with high precision and for the required length of time, while preserving the surrounding healthy tissue.

Radiation warning symbol: An officially prescribed magenta or black trefoil on a yellow background, which must be displayed where certain quantities of radioactive materials are present or where certain doses of radiation could be received.

Radioactive contamination: Undesirable radioactive material (with a potentially harmful effect) that is either airborne or deposited in (or on the surface of) structures, objects, soil, water, or living organisms (people, animals, or plants) in a concentration that may harm people, equipment, or the environment.

Radioactive decay: The spontaneous transformation of one radioisotope into one or more different isotopes (known as "decay products" or "daughter products"), accompanied by a decrease in radioactivity compared to the parent material. This transformation takes place over a defined period of time (known as a "half-life"), as a result of electron capture; fission; or the emission of alpha particles, beta particles or photons (gamma radiation or x-rays) from the nucleus of an unstable atom. Each isotope in the sequence known as a "decay chain") decays to the next until it forms a stable, less energetic end product. In addition, radioactive decay may refer to gamma-ray and conversion electron emission, which only reduces the excitation energy of the nucleus.

Radioactivity: The property possessed by some elements (such as uranium) of spontaneously emitting energy in the form of radiation as a result of the decay (or disintegration) of an unstable atom. Radioactivity is also the term used to describe the rate at which radioactive material emits radiation. Radioactivity is measured in units of Becquerel or disintegrations per second.

Radiography: The use of sealed sources of ionizing radiation for nondestructive examination of the structure of materials. When the radiation penetrates the material, it produces a shadow image by blackening a sheet of photographic film that has been placed behind the material, and the differences in blackening suggest flaws and unevenness in the material.

Radioisotope (Radionuclide): An unstable isotope of an element that decays or disintegrates spontaneously, thereby emitting radiation. Approximately 5000 natural and artificial radioisotopes have been identified.

Radiopharmaceutical: A pharmaceutical drug that emits radiation and is used in diagnostic or therapeutic medical procedures. Radioisotopes that have short half-lives are generally preferred to minimize the radiation dose to the patient and the risk of prolonged exposure. In most cases, these short-lived radioisotopes decay to stable elements within minutes, hours, or days, allowing patients to be released from the hospital in a relatively short time.

Reactor core: The central portion of a nuclear reactor, which contains the fuel assemblies, water, and control mechanisms, as well as the supporting structure. The reactor core is where fission takes place.

Reactor Oversight Process (ROP): The process by which the NRC monitors and evaluates the performance of commercial nuclear power plants. Designed to focus on those plant activities that are most important to safety, the process uses inspection findings and performance indicators to assess each plant's safety performance.

Regulation: The governmental function of controlling or directing economic entities through the process of rulemaking and adjudication.

Regulatory Information Conference (RIC): An annual NRC conference that brings together NRC staff, regulated utilities, materials users, and other interested stakeholders to discuss nuclear safety topics and significant and timely regulatory activities through informal dialogue to ensure an open regulatory process.

REM (Roentgen equivalent man): One of the two standard units used to measure the dose equivalent (or effective dose), which combines the amount of energy (from any type of ionizing radiation) that is deposited in human tissue, along with the medical effects of the given type of radiation. For beta and gamma radiation, the dose equivalent is the same as the absorbed dose. By contrast, the dose equivalent is larger than the absorbed dose for alpha and neutron radiation, because these types of radiation are more damaging to the human body. Thus, the dose equivalent (in rems) is equal to the absorbed dose (in rads) multiplied by the quality factor of the type of radiation {Title 10 of the Code of Federal Regulations, Section 20.1004, ("Units of Radiation Dose")}. The related international system unit is the sievert (Sv), where 100 rem is equivalent to 1 Sv.

Renewable resources: Natural, but limited, energy resources that can be replenished, including biomass, hydro, geothermal, solar, and wind. These resources are virtually inexhaustible but limited in the amount of energy that is available per unit of time. In the future, renewable resources could also include the use of ocean thermal, wave, and tidal action technologies. Utility renewable resource applications include bulk electricity generation, onsite electricity generation, distributed electricity generation, non-grid connected generation, and demand-reduction (energy efficiency) technologies. The Information Digest (NUREG-1350) has included conventional hydroelectric and storage hydroelectric in a separate category from other resources.

Risk: The combined answer to three questions that consider (1) what can go wrong, (2) how likely it is, and (3) what its consequences might be. These three questions allow the NRC to understand likely outcomes, sensitivities, areas of importance, system interactions, and areas of uncertainty, which can be used to identify risk-significant scenarios.

Risk-based decision making: An approach to regulatory decision-making that considers only the results of a probabilistic risk assessment.

Risk-informed decision making: An approach to regulatory decision-making, in which insights from probabilistic risk assessment are considered with other engineering insights.

Risk-informed regulation: An approach to regulation taken by the NRC, which incorporates an assessment of safety significance or relative risk. This approach ensures that the regulatory burden imposed by an individual regulation or process is appropriate to its importance in protecting the health and safety of the public and the environment.

Risk-significant: "Risk-significant" can refer to a facility's system, structure, component, or accident sequence that exceeds a predetermined limit for contributing to the risk associated with the facility. The term also describes a level of risk exceeding a predetermined "significance" level.

Safeguards: The use of material control and accounting programs to verify that the special nuclear material is properly controlled and accounted for, as well as the physical protection (or physical security) equipment and security forces. As used by the International Atomic Energy Agency (IAEA), this term also means verifying that the peaceful use commitments made in binding nonproliferation agreements, both bilateral and multilateral, are honored.

Safeguards information (SGI): A special category of sensitive unclassified information that must be protected. Safeguards information concerns the physical protection of operating power reactors, spent fuel shipments, strategic special nuclear material, or other radioactive material.

Safety-related: In the regulatory arena, this term applies to systems, structures, components, procedures, and controls (of a facility or process) that are relied upon to remain functional during and following design-basis events. Their functionality ensures that key regulatory criteria, such as levels of radioactivity released, are met. Examples of safety-related functions include shutting down a nuclear reactor and maintaining it in a safe-shutdown condition.

Safety-significant: When used to qualify an object, such as a system, structure, component, or accident sequence, this term identifies that object as having an impact on safety, whether determined through risk analysis or other means that exceeds a predetermined significance criterion.

SAFSTOR: A method of decommissioning in which a nuclear facility is placed and maintained in a condition that allows the facility in a safe configuration and subsequently decontaminated (deferred decontamination) to levels that permit release for unrestricted use.

Scram: The sudden shutting down of a nuclear reactor, usually by rapid insertion of control rods, either automatically or manually by the reactor operator. Also known as a "reactor-trip," "scram" is actually an acronym for "safety control rod axe man," given to the worker assigned to insert the emergency rod on the first reactor (the Chicago Pile) in the United States.

Sensitive unclassified non-safeguards information (SUNSI): Information that is generally not publicly available and that encompasses a wide variety of categories, such as proprietary information, personal and private information, or information subject to attorney-client privilege.

Shutdown: A decrease in the rate of fission (and heat/energy production) in a reactor, usually by the insertion of control rods into the core.

Source material: Uranium or thorium, or any combination thereof, in any physical or chemical form, or ores that contain, by weight, one-twentieth of 1% (0.05%) or more of (1) uranium, (2) thorium, or (3) any combination thereof. Source material does not include special nuclear material.

Special nuclear material: Plutonium, uranium-233, or uranium enriched in the isotopes uranium-233 or uranium-235.

Spent fuel pool (SFP): SFP is an underwater storage and cooling facility for spent (depleted) fuel assemblies that have been removed from a reactor. The facility is also used for temporary storage of new fuel assemblies.

Subcriticality: The condition of a nuclear reactor system, in which nuclear fuel no longer sustains a fission chain reaction (that is, the reaction fails to initiate its own repetition, as it would in a reactor's normal operating condition). A reactor becomes subcritical when its fission events fail to release a sufficient number of neutrons to sustain an ongoing series of reactions, possibly, as a result of increased neutron leakage or poisons.

Teletherapy: Treatment in which the source of the therapeutic radiation is at a distance from the body. Because teletherapy is often used to treat malignant tumors deep within the body by bombarding them with a high-energy beam of gamma rays (from a radioisotope such as cobalt-60) projected from outside the body, it is often called "external beam radiotherapy."

Title 10 of the *Code of Federal Regulations* (10 CFR): Four volumes of the *Code of Federal Regulations* (CFR) address energy-related topics. Parts 1 to 199 contain the regulations (or rules) established by the NRC. These regulations govern the transportation and storage of nuclear materials; use of radioactive materials at nuclear power plants, research and test reactors, uranium recovery facilities, fuel cycle facilities, waste repositories, and other nuclear facilities; and use of nuclear materials for medical, industrial, and academic purposes.

Transient: A change in the reactor coolant system temperature, pressure, or both, attributed to a change in the reactor's power output. Transients can be caused by (1) adding or removing neutron poisons, (2) increasing or decreasing electrical load on the turbine generator, or (3) accident conditions.

Transuranic waste: Material contaminated with transuranic elements—artificially made, radioactive elements, such as neptunium, plutonium, americium, and others—that have atomic numbers higher than uranium in the periodic table of elements. Transuranic waste is primarily produced from recycling spent fuel or using plutonium to fabricate nuclear weapons.

Tritium: A radioactive isotope of hydrogen. Because it is chemically identical to natural hydrogen, tritium can easily be taken into the body by any ingestion path. It decays by emitting beta particles and has a half-life of about 12.5 years.

INDEX

Page numbers followed by f and t indicate figures and tables, respectively.

C

Q

R

S

T

U

V

W

Y

Z